Ramified Integrals, Singularities and Lacunas

Mathematics and Its Applications

Volume 315

Ramified Integrals, Singularities and Lacunas

by

V.A. Vassiliev
Research Institute for System Studies (NIISI), Moscow, Russia

SPRINGER-SCIENCE+BUSINESS MEDIA, B.V.

Library of Congress Cataloging-in-Publication Data

Vasil'ev, V. A., 1956-
Ramified integrals, singularities, and lacunas / by V.A. Vassiliev.
p. cm. -- (Mathematics and its applications ; v. 315)
Includes bibliographical references and index.
ISBN 978-94-010-4095-2 ISBN 978-94-011-0213-1 (eBook)
DOI 10.1007/978-94-011-0213-1
1. Integral geometry. 2. Integral transforms. I. Series: Mathematics and its applications (Kluwer Academic Publishers) ; v. 315.
QA672.V37 1995
516.3'62--dc20 94-35631

ISBN 978-94-010-4095-2

Printed on acid-free paper

TABLE OF CONTENTS

INTRODUCTION

Many special functions occuring in physics and partial differential equations can be represented by integral transformations: the fundamental solutions of many PDE's, Newton–Coulomb potentials, hypergeometric functions, Feynman integrals, initial data of (inverse) tomography problems, etc. The general picture of such transformations is as follows. There is an analytic fibre bundle $E \to T$, a differential form ω on E, whose restrictions on the fibres are closed, and a family of cycles in these fibres, parametrized by the points of T and depending continuously on these points. Then the integral of the form ω along these cycles is a function on the base.

The analytic properties of such functions depend on the monodromy action, i.e., on the natural action of the fundamental group of the base in the homology of the fibre: this action on the integration cycles defines the ramification of the analytic continuation of our function.

The study of this action (which is a purely topological problem) can answer questions about the analytic behaviour of the integral function, for instance, is this function single-valued or at least algebraic, what are the singular points of this function, and what is its asymptotics close to these points.

In this book, we study such analytic properties of three famous classes of functions:

the *volume functions*, which appear in the Archimedes–Newton problem on integrable bodies;

the *Newton–Coulomb potentials*, and

the *Green functions of hyperbolic equations* (studied, in particular, in the Hadamard–Petrovskii–Atiyah–Bott–Gårding lacuna theory).

We start with an introduction to the Picard–Lefschetz theory, which is the main topological tool in the calculation of the monodromy action. The necessary background, including the analysis of differential forms and homology theory, can be found in the excellent book [Pham 67], whose philosophy we widely use. This theory is a part of singularity theory (= mathematical catastrophe theory), so we include also the necessary facts of this theory.

In Chapter I we prove also several basic facts of the "stratified Picard–Lefschetz

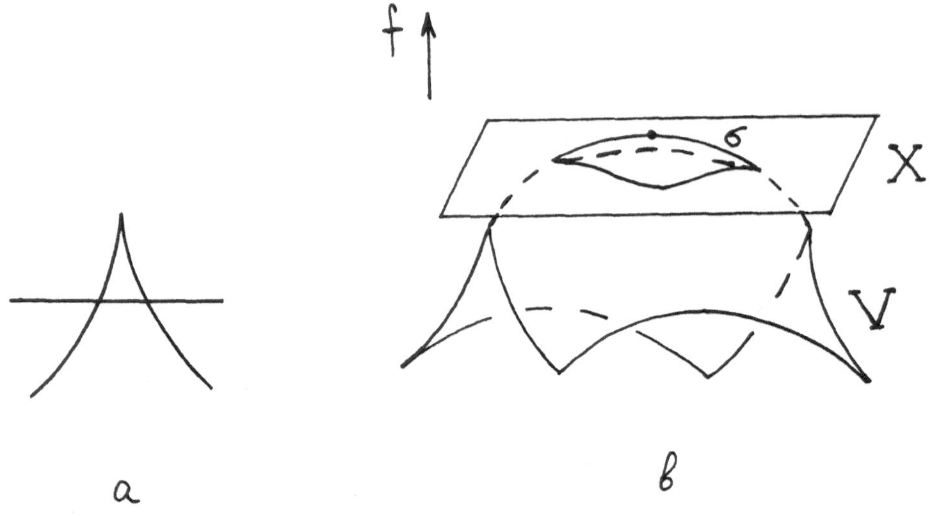

Fig. 1. Morse critical point of a function on a stratified variety and a transversal slice

theory", which is in the same relation with the standard one as the stratified Morse theory of [Goresky & MacPherson 86] is with the classical Morse theory.

Namely, let f be a holomorphic Morse function on a stratified analytic variety V, and x the critical point of its restriction to a stratum; see Figure 1. Then the monodromy operator in the homology of the spaces $V \cap f^{-1}(\cdot)$, defined by a small loop around the corresponding critical value, can be reduced to a similar operator for a transversal slice of our stratum; see Chapter I, § 11.

Historically the first integral representation was considered by Archimedes and Newton: this is the (two-valued) function on the space of hyperplanes in $\mathbf{R}^n$, which is determined by an arbitrary convex domain and is equal to the volume cut off from this domain by a hyperplane; see Figure 2. Archimedes' result implies, in particular, that if this domain is the standard disc in $\mathbf{R}^3$, then this function is algebraic. Newton's theorem says that if $n = 2$ and this body is bounded by any compact C^∞-smooth curve in $\mathbf{R}^2$, then this volume function cannot be algebraic. It is remarkable that Newton's proof already contains the concepts of monodromy theory: it is based on the fact that moving the cutting lines along some closed paths in the space of lines we get an infinite number of different values of our function.

In Chapter II we prove that Newton's result is true also in all even-dimensional spaces: the volume function defined by any convex compact hypersurface in $\mathbf{R}^{2n}$ cannot be algebraic. In the odd-dimensional case we give many topological obstruc-

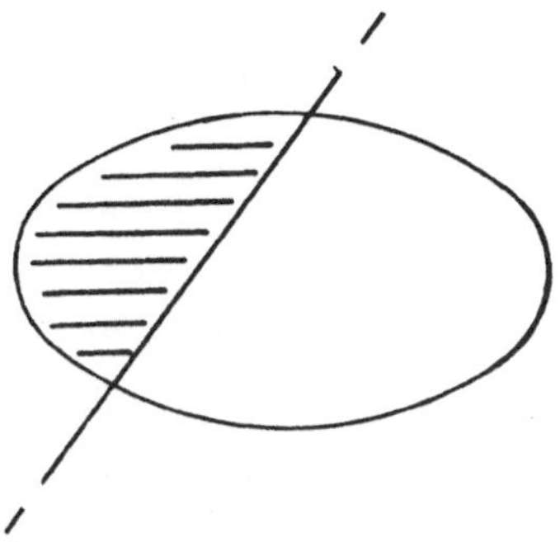

Fig. 2. Volume function on the space of hyperplanes

tions for this function to be algebraic: all of them are formulated in terms of the local geometry of the complexification of the bounding hypersurface. These obstructions lead to the following conjecture: in the odd-dimensional space Archimedes' example (the ellipsoids in $\mathbf{R}^{2n-1}$) is the only example in which the volume function can be algebraic. At least, this conjecture is true for all surfaces whose complexifications are nonsingular manifolds in $\mathbf{C}P^{2n-1}$.

In Chapter III we study the analytic properties of the Newton–Coulomb potential of hyperbolic algebraic surfaces in $\mathbf{R}^n$. Newton's theorem says that if this surface is a sphere, then the potential is constant inside it. Ivory [Ivory 1809] extended this result to the case of ellipsoids, and Arnold and Givental to the case of arbitrary hyperbolic hypersurfaces in $\mathbf{R}^n$; see [Arnold 83, 85, 87, 87'], [Givental 84], [Arnold & Vassiliev 89].

An algebraic hypersurface $A \subset \mathbf{R}^n$ of degree d is called *hyperbolic* with respect to a point O of its complement if any line through O in $\mathbf{R}^n$ intersects A in exactly d points (one of which can be infinitely distant). In this case A is hyperbolic also with respect to all other points of the same connected component of $\mathbf{R}^n \setminus A$. This component is called the *hyperbolicity domain* of A and is always convex. For instance, the ellipsoids and two-component hyperboloids are hyperbolic with respect to appropriate points.

The *standard charge* of a hyperbolic surface $A \subset \mathbf{R}^n$ distinguished by a polynomial equation $f = 0$ is equal to the limit of homogeneous charges between the surfaces $f = 0$ and $f = \varepsilon$ with density $1/\varepsilon$ and the signs equal to 1 or -1 depending on the parity of the number of the corresponding component of the surface, counting from the boundary of the hyperbolicity domain.

Theorem (see [Arnold 85]). *The standard charge of a smooth hyperbolic surface* *does not attract the points in the hyperbolicity domain. Moreover, the same holds for the product of the standard charge and a polynomial of degree* $\leq d-2$ *(where* d *is the degree of the surface).*

Also, after coming out into other domains, Newton's potential inherits some good algebraic properties. Let us number these domains starting with the hyperbolicity domain (which will be domain 0) and moving to more and more interior ones.

Theorem. *In the* k*-th domain of a smooth hyperbolic curve of degree* d *in* $\mathbf{R}^2$ *the attraction force defined by the standard charge coincides with a sum of two algebraic vector-functions, each of which is at most* C_d^k*-valued. Moreover, the same holds for the attraction force defined by the charge which is the product of the standard charge and any polynomial function of degree* $\leq d-2$ *(or of arbitrary degree if the hyperbolic curve is compact).*

Theorem. *The attraction force defined by the standard charge of a smooth hyperbolic hypersurface of degree 2 in* $\mathbf{R}^n$ *is algebraic in any component of its complement in* $\mathbf{R}^n$*.*

Theorem. *If* $d \geq 3, n \geq 3$*, and* $d+n \geq 8$*, then the attraction force defined by the standard charge of a generic hyperbolic hypersurface* A *of degree* d *in* $\mathbf{R}^n$ *cannot coincide with an algebraic function in some component of* $\mathbf{R}^n \setminus A$ *other than the hyperbolicity domain.*

Conjecture. *The requirement* $d+n \geq 8$ *in the last theorem is unnecessary.*

The proofs are based again on monodromy theory: the potential function has an integral representation and hence the number of values of its analytic continuation is estimated by the order of the corresponding monodromy group.

One more problem of the same kind comes from the theory of hyperbolic equations: the problem on sharp fronts and lacunas.

Recall that a PDO (with constant coefficients) is *hyperbolic* if it has a fundamental solution whose support belongs to a proper cone in the "positive half-space" (in which the Cauchy problem is posed). The most famous hyperbolic operator is the wave operator

$$\frac{\partial^2}{\partial t^2} - c^2 \sum \frac{\partial^2}{\partial x_i^2},$$

which describes the wave propagation with speed c. The *fundamental solution*, i.e., the elementary wave arising from an instant pointwise perturbation, has a singularity on the cone

$$c^2 \cdot t^2 = \sum x_i^2, \quad t \geq 0$$

in the (t,x)-space. The asymptotic behaviour of the wave (when the argument (t,x) tends to this cone) depends strongly on the parity of the number of variables. In four-dimensional space–time the signal is accepted only instantly, when it reaches the immovable observer. On the contrary, in the odd-dimensional case the signal will sound all the time after the first meeting. Both variants of the behaviour of waves have analogues for arbitrary hyperbolic equations: in the language of the general theory one says that in the even-dimensional case the interior component of the complement of the cone of singularities (=of the wave front) is a lacuna, while in the odd-dimensional case we have a diffusion from the side of this component. The exterior component is a lacuna for all dimensions (and all hyperbolic operators).

I.G. Petrovskii discovered that these phenomena have an explanation in terms of integral representations. Indeed, the fundamental solution (or at least all its sufficiently high partial derivatives) have an integral representation (Herglotz–Petrovskii–Leray formula); the fact that a domain in $\mathbf{R}^n$ is a lacuna is usually related to the fact (called the *Petrovskii condition*) that the corresponding integration cycle (the *Petrovskii cycle*) is homologous to zero; see [Petrovskii 45], [ABG 70].

Atiyah, Bott and Gårding [ABG 73] considered the local version of this problem. A local (close to a point x of the wave front) component of the complement of the front is called a (holomorphic) *local lacuna* (and, which is the same, the front is *sharp* from the side of this component) if the restriction of the fundamental solution to this component has a smooth analytic continuation onto a whole neighbourhood of the point x. In [ABG 73] a local version of the Petrovskii condition is formulated: some localization of the Petrovskii homology class responsible for the behaviour of the fundamental solution close to the point x must vanish. Atiyah, Bott and Gårding proved that this condition is sufficient to ensure the sharpness, and conjectured that it is also necessary. We prove that this conjecture is true for all points of wave fronts of almost all hyperbolic operators; on the other hand, there exist examples of very degenerate operators for which this conjecture fails.

The sharpness of the front close to its point x depends on the local geometry of the front. For instance, close to smooth points this condition can be expressed in terms of the signature of the second fundamental form of the front (the Davydova–Borovikov condition, see [Davydova 45], [Borovikov 59], [Leray 62], [ABG 70]).

The singular points of wave fronts are classified in accordance with the classification of the singularities of their generating functions; see [Hörmander 71], [Zakalyukin 76], [AVG 82]. The simplest and most important singularities are the so-called *simple* singularities (of types $A_k, k \geq 2$, $D_k, k \geq 4$, E_6, E_7, E_8). For instance, the generic operators acting in the spaces $\mathbf{R}^n$, $n \leq 7$, have only such singularities. The local lacunas close to the first of these singularities, of types A_2 and A_3, were investigated by L.Gårding [Gårding 77]. Namely, close to these points the front is diffeo-

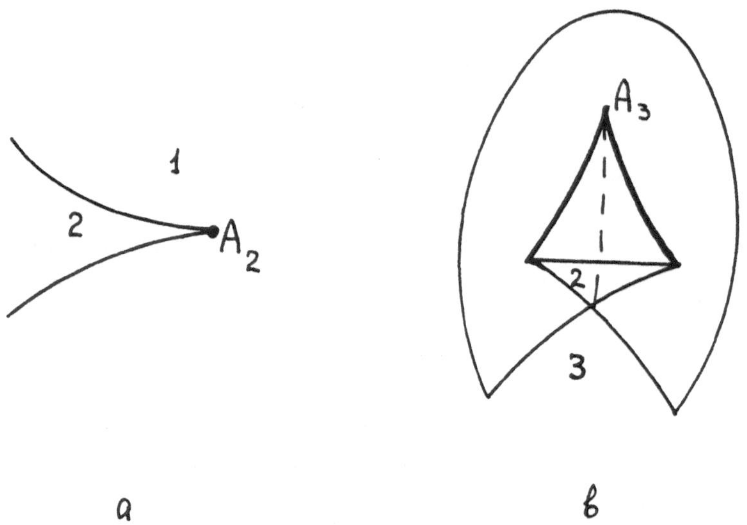

Fig. 3. Singularities A_2 and A_3 of wave fronts

morphic to the direct product of an Euclidean space and the semicubical parabola (respectively, the *swallowtail* shown in Figure 3b). The sharpness of a front close to these singularities depends on the parity of the dimension n of space-time where the operator acts, and on another index, denoted by i_+ and taking values "even" and "odd" (it is equal to the parity of the positive inertia index of the quadratic part of the standard generating function of the front at the singular point).

Close to points of type A_2 (see Figure 3a) only the component 2 can be a local lacuna: this happens only if n is odd and i_+ is even. Close to points of type A_3 there are only the following local lacunas: the component 2 if n is odd and i_+ is even, and the component 3 if i_+ is odd and n is arbitrary.

We solve the same problem close to all simple (and many nonsimple) singular points of wave fronts. The numbers of local lacunas close to these singularities, as well as the explicit description of these lacunas, are presented in Chapter V, § 2; see especially Tables 12, 13.

Here is a geometrical characterization of local lacunas, which follows from our calculations.

Theorem. *A local component of the complement of the wave front close to a simple singular point of a generic hyperbolic operator is a local lacuna if and only if*

a) the Davydova–Borovikov signature condition is satisfied close to all the non-singular points of the boundary of this component, and

b) this boundary does not contain points of type A_2 (semicubical edges) such that our component is the bigger one (i.e., situated as the component 1 *in* Figure 3a*) close to these points.*

Conjecture. *The same is also true for nonsimple singularities of generic operators.*

The necessity of these conditions for arbitrary singularities is obvious.

We present also a FORTRAN-algorithm, which finds the lacunas close to various singularities of fronts or proves the absence of them. This algorithm is purely combinatorial and is based again on the method of generating functions, see [Hörmander 71], [Zakalyukin 76], [AVG 82]. In the same way as the classification of fronts coincides with the classification of smooth functions, the local components of the complement of a front are in correspondence with the equivalence classes of different morsifications of the singularity of the generating function. Therefore, the problem of counting such components reduces to the investigation of different morsifications of real singularities.

This problem is posed for arbitrary real analytic singularities, and not only for the generating functions of fronts, because the localized Petrovskii classes can be defined in this general case.

The initial data of our algorithm are some discrete characteristics of some nonsingular morsification of the generating function (the number and Morse indices of its real critical points, the intersection matrix of all vanishing cycles, the homology class of the localization of the Petrovskii cycle, etc.). Then we make all the changes of this set of indices which could correspond to possible local surgeries of the morsification (such as the Morse birth-death surgeries and changes of the orders and signs of critical values on the real line). We say that we get a *formal lacuna* if after a chain of such transformations of indices the localized Petrovskii cycle becomes homologous to zero. Of course, in this case there remains the question of whether the formal lacuna corresponds to a real local lacuna, i.e., that our chain of formal transformations of indices can be realized by a series of real surgeries. But in the examples we have calculated this is always so.

Certainly, this algorithm can answer numerous other problems of the real singularity theory not related to the hyperbolic equations, for instance, does a given singularity have a morsification of some prescribed type, can two given morsifications lie in the same component of the complement of the discriminant variety, etc.

Problem. It seems very likely that the modern methods of the analysis of singular varieties (such as perverse sheaves, intersection homology, stratified Morse and Picard-Lefschetz theory, etc.) would lead to essential progress in the lacuna problem for degenerate hyperbolic equations. The first problem here is to define the localized Petrovskii class for nonisolated singularities of principal symbols and to study their local monodromy.

Notation. Throughout this book, $\tilde{H}_*$ denotes the homology group reduced modulo a point in the case of absolute homology and modulo the fundamental cycle in the case of the relative homology of a manifold (or an analytic variety) with boundary.

Theorems, definitions, etc. are numbered in each section, and the formulae in each chapter separately. The same Theorem 5 of Chapter 1, § 3 can be referred to as Theorem 5 (in the same section), Theorem 3.5 (in other sections of the same chapter), or Theorem I.3.5.

ACKNOWLEDGEMENTS

I thank V.I. Arnold who drew my attention to all the three problems of integral geometry investigated in this book; in particular, it was he who remarked that Newton's proof of the nonintegrability of plane curves is based essentially on the notion of monodromy and can probably be generalized by the modern methods of monodromy theory.

I thank A.M. Gabrielov, I.M. Gelfand, V.V. Goryunov, S.M. Gusein-Zade, W. Ebeling, Val.S. Kulikov and V.P. Palamodov for useful conversations, and also I.A. Andreeva, E.A. Kondratieva and A.L. Piatnitsky for their help in debugging the program and accomplishing the computer calculations. My special thanks are to I.M. Gelfand for the reasons mentioned in the second edition of [Vassiliev 92 & 94].

I am extremely grateful to my family for permanent multiform assistance and encouragement.

While writing this book I was partially supported by the AMS Former Soviet Union Aid Fund and by grant MQO 000 from the International Science Foundation.

ACKNOWLEDGEMENTS

Chapter I. PICARD–LEFSCHETZ–PHAM THEORY AND SINGULARITY THEORY

§ 1. Gauss–Manin connection in the homological bundles. Monodromy and variation operators

1.1. Homological and cohomological bundles.

Let $p : E \to T$ be a locally trivial fibre bundle with locally contractible base. The corresponding *homological bundle* is a fibre bundle with the same base, whose fibre over a point x is the group $H_*(p^{-1}(x))$. This bundle admits a natural local trivialization (i.e. a flat connection): the cycles in the fibres can be continuously transported in the neighbouring fibres, and the homology classes of the transported cycles do not depend on the choice of this transportation. The formal definition of this trivialization is as follows. (We denote the fibre $p^{-1}(x)$ by $\mathcal{F}_x$.) Consider an arbitrary contractible domain U in T. The bundle p over this domain is equivalent to the trivial one. Then, by the Künneth formula, for any point $x \in U$,

$$H_*(\mathcal{F}_x) = H_*(p^{-1}(U)).$$

Thus, all groups $H_*(\mathcal{F}_x)$ for all $x \in U$ are canonically identified with the group $H_*(p^{-1}(U))$ and hence also with each other. This identification defines the desired local trivialization of the homological bundle over U. It is easy to see that it does not depend on the choice of the trivialization of the original fibre bundle p over U.

In a similar way the *cohomological bundle* associated with the bundle p can be defined, as well as the homological and cohomological bundles associated with the locally trivial bundles of pairs $(E \supset K) \to T$.

Definition. The above-described natural flat connection in the homological (cohomological) bundle, associated with a locally trivial fibre bundle, is called the *Gauss–Manin connection.*

Of course, this connection respects the natural (dimensional) grading of these bundles.

1.2. The monodromy operator.

Let v be a distinguished point in T . Any loop in T with its origin and end at v defines some automorphism of the group $H_*(\mathcal{F}_v)$ (or $H^*(\mathcal{F}_v)$) : this automorphism is defined by the transportation of cycles (or cocycles) by the Gauss–Manin connection over this loop.

This automorphism depends only on the homotopy class of the loop, and hence defines a homomorphism of the fundamental group of the base into the group $\mathrm{Aut}H_*(\mathcal{F}_v)$ (or $\mathrm{Aut}H^*(\mathcal{F}_v)$). This homomorphism is called the *monodromy representation*, and its image is called the *monodromy group* associated with the bundle p.

Example. There are two nonequivalent fibre bundles over a circle whose fibres are circles: the torus and the Klein bottle. In the case of the torus (and, in general, in the case of any trivial bundle) the monodromy representation is trivial; in the case of the Klein bottle the monodromy representation in the group $H_0(S^1)$ is trivial, and in the group $H_1(S^1)$ the generator of the group $\pi_1(T)$ acts as multiplication by -1.

The monodromy operator in the homological bundle, corresponding to a loop u, is denoted by M_u.

1.3. The variation operator.

The simplest version of the variation operator is just the operator $M_u - \mathrm{Id}$, which indicates by how much the monodromy operator differs from the identical one. Its construction admits the following (very important) generalization.

Let K be a closed subspace in E , and $p : (E, K) \to T$ a locally trivial fibre bundle of pairs, such that its restriction to K is trivializable, and, moreover, K admits an open neighbourhood K_1 in E such that the fibre bundle of triplets $p : (E, K_1, K) \to T$ is again locally trivial and the trivialization of the bundle $p|_K$ can be extended to a trivialization of the bundle of pairs $p : (K_1, K) \to T$.

(For example, such a K_1 always exists if all the fibres of our bundle of pairs (E, K) are the pairs {a manifold; its boundary}: in this case K_1 is the space of the bundle of collars of these boundaries.)

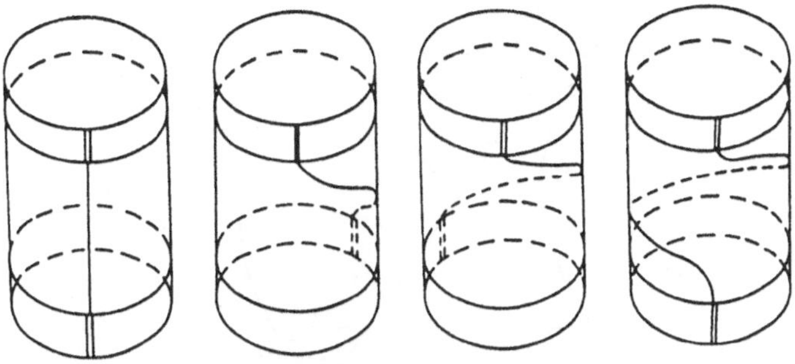

Fig. 4. Operator Var

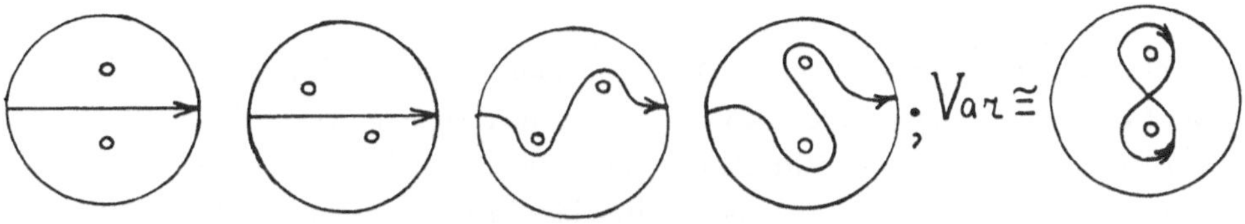

Fig. 5. Operator Var

Then for any loop $u \in \pi_1(T)$ the operator Var_u is well defined. It acts from the relative homology group $H_*(\mathcal{F}_v, \mathcal{F}_v \cap K)$ into the absolute homology group $H_*(\mathcal{F}_v \backslash K)$ and is defined as follows. (See Figures 4, 5. In these diagrams the fibres $\mathcal{F}_v$ are respectively the cylinder and the disc with two points removed; in both cases the base T is a circle and $K = \partial E$.)

First, suppose that T is the circle S^1 and u goes once round this circle. Let Γ be an element of the group $H_*(\mathcal{F}_v, \mathcal{F}_v \cap K)$, and γ an arbitrary relative cycle representing it. Using the local triviality of the bundle p, let us transport the cycle γ into all fibres over the loop u; let $\gamma_1 \subset \mathcal{F}_v$ be the result of this transportation over the whole circle S^1. Using the triviality of the bundle $p|_{K_1}$, we can choose this transportation in such a way that in the intersection with K_1 we have $\gamma_1 \equiv \gamma$. Then the chain $\gamma_1 - \gamma$ belongs to $F_v \setminus K$ and is a cycle there. The class of this cycle in the group $H_*(\mathcal{F}_v \setminus K)$ does not depend on the choice of the representative γ of the original class of the group $H_*(\mathcal{F}_v, \mathcal{F}_v \cap K)$ and of the local trivialization of the bundle p . Therefore, an operator $H_*(\mathcal{F}_v, \mathcal{F}_v \cap K) \to H_*(\mathcal{F}_v \setminus K)$ is well defined; it is denoted by Var_u .

In the case of an arbitrary base T the operator Var_u is constructed as follows. Any loop $u : (S^1, *) \to (T, v)$ induces a fibre bundle p' over S^1 from the bundle p, in particular, its fibre over the distinguished point $* \in S^1$ is identified with $\mathcal{F}_v$. For

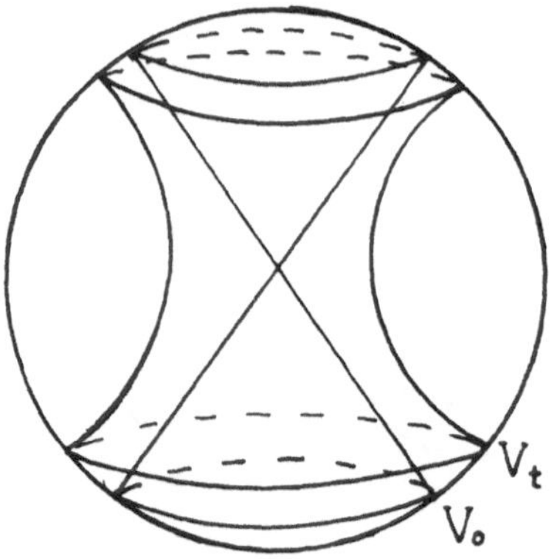

Fig. 6. Varieties V_t and V_0 of a complex Morse singularity

any relative cycle γ in $\mathcal{F}_v$ mod $\mathcal{F}_v \cap K_v$ consider its inverse image γ' in $(p')^{-1}(*)$, apply the previous construction to it, and map the resulting cycle $(\gamma_1' - \gamma') \subset p'^{-1}(*)$ into F_v by means of our identification. The cycle thus obtained belongs to $\mathcal{F}_v \setminus K$; the desired class $\mathrm{Var}_u(\gamma)$ is nothing but its class in the homology of this space.

§ 2. The Picard–Lefschetz formula. The Leray tube operation

2.1. Monodromy operator related to a Morse singularity.

Here is a very important example in which the variation operators appear.

Let $f : \mathbf{C}^n \to \mathbf{C}$ be a nondegenerate complex quadratic form, and $B \equiv B_\varepsilon$ a closed disc of radius ε with centre at 0 in $\mathbf{C}^n$. For any $t \in \mathbf{C}$ denote the set $f^{-1}(t) \cap B$ by V_t.

Proposition 1. *For arbitrary nonzero $t \in \mathbf{C}$ the set $f^{-1}(t)$ is smooth and diffeomorphic to the space TS^{n-1} of the tangent bundle of the sphere S^{n-1}. If $|t|$ is sufficiently small (in comparison with ε), $t \neq 0$, then the manifold $f^{-1}(t)$ intersects the boundary of B transversally, and its intersection with B is a smooth manifold with boundary and is diffeomorphic to the manifold T_1S^{n-1} of tangent vectors to the sphere S^{n-1} whose lengths do not exceed 1.*

In particular, the groups $\tilde{H}_i(V_t)$ and $\tilde{H}_i(V_t, \partial V_t)$ are isomorphic to $\mathbf{Z}$ for $i = n-1$ and are trivial for other i; see Figure 6. (Recall the notation $\tilde{H}_*$ for the reduced homology, see the Introduction.)

Denote by $\langle\cdot,\cdot\rangle$ the intersection index in the manifold V_t with respect to the complex orientation of this manifold.

Denote the generators of the groups $\tilde{H}_{n-1}(V_t)$ and $\tilde{H}_{n-1}(V_t, \partial V_t)$ by Δ and ∇ respectively, and choose their orientations in such a way that the intersection index $\langle \nabla, \Delta \rangle$ equals 1.

For example, let $f = x_1^2 + \cdots + x_n^2$, and let t be a small positive number. Then $V_t \cap \mathbf{R}^n$ is a sphere of radius $t^{1/2}$ in $\mathbf{R}^n$; the class of this sphere generates the group $\tilde{H}_{n-1}(V_t)$.

Since all nondegenerate quadratic forms in $\mathbf{C}^n$ are equivalent, the generator in the group $\tilde{H}_{n-1}(V_t)$ for an arbitrary form f of this kind can also be realized by a topological sphere. This sphere is called the *vanishing sphere* of the singularity f: this name is due to the fact that as t tends to 0 this sphere contracts to a point and its limiting position defines a zero class in the group $\tilde{H}_{n-1}(V_0)$. (This is not surprising, because the set V_0 is contractible.)

For another example, let $f = -x_1^2 - \cdots - x_{n-1}^2 + x_n^2$ and let t be a small positive number; then $f^{-1}(t) \cap \mathbf{R}^n$ is a two-component hyperboloid, and any connected component of the set $V_t \cap \mathbf{R}^n$ generates the group $\tilde{H}_{n-1}(V_t, \partial V_t)$.

Proposition 2. *The self-intersection index of the cycle Δ is equal to 0 if n is even, 2 if $n \equiv 1(mod\, 4)$, and -2 if $n \equiv 3(mod\, 4)$.*

Indeed, by Proposition 1 this index is equal to the self-intersection index of the zero section of the tangent bundle of an $(n-1)$-dimensional sphere. It is well-known that for an arbitrary compact manifold the self-intersection of the zero section of its tangent bundle, calculated with respect to the standard (symplectic) orientation of this bundle, is equal to the Euler characteristic of the manifold. It is easy to verify that for $n \equiv 1(\mathrm{mod}\ 4)$ the symplectic orientation of the tangent bundle of S^{n-1} coincides with the complex 3orientation of the manifold V_t, and for $n \equiv 3(\mathrm{mod}\ 4)$ these orientations are opposite. This implies Proposition 2.

Now let T be a disc in $\mathbf{C}^1$ of sufficiently small radius δ , and T' the punctured disc $T \setminus \{0\}$. The map $f : (f^{-1}(T') \cap B) \to T'$ is a locally trivial fibre bundle with fibre V_t; it is called the *Milnor fibration* of the singularity f. Moreover, its restriction to the subset $f^{-1}(T') \cap \partial B$ can be extended to a locally trivial (and hence globally trivializable) fibre bundle over the entire disc T. Thus, we are in the conditions of the previous section: for the base T we take the punctured disc T', and for the spaces E and K the spaces $f^{-1}(T') \cap B$ and $f^{-1}(T') \cap \partial B$ respectively.

Denote by u the loop $\{\delta e^{i\tau}\}$, $\tau \in [0, 2\pi]$, in T'.

Proposition 3. *The monodromy operator $M_u : \tilde{H}_{n-1}(V_t) \hookleftarrow$ maps the generator Δ of this group into itself if n is even and into $-\Delta$ if n is odd.*

Indeed, let $l_1, \dots, l_n$ be the basis vectors of a coordinate system in $\mathbf{C}^n$, in which f is equal to $x_1^2 + \cdots + x_n^2$. For the family of cycles $\Delta_\tau \in f^{-1}(\delta e^{i\tau}), \tau \in [0, 2\pi]$, which realize the monodromy, we can take the spheres of radius $\delta^{1/2}$ in the real n-dimensional planes in $\mathbf{C}^n$ spanned by the vectors $e^{i\tau/2}l_1, \dots, e^{i\tau/2}l_n$. These basis vectors define the orientations of the indicated planes; the spheres Δ_τ in them can be oriented in such a way that the frame {the exterior normal to the sphere; the tangent frame of the sphere, defining its desired orientation} defines the positive (just defined) orientation of the plane. Hence, the monodromy preserves the orientation of the sphere Δ if and only if the frame $(-l_1, \dots, -l_n)$ defines the same orientation of $\mathbf{R}^n$ as the frame $(l_1, \dots, l_n)$.

Proposition 4. *The operator*

$$\mathrm{Var}_u : \tilde{H}_{n-1}(V_t, \partial V_t) \to \tilde{H}_{n-1}(V_t) \tag{1}$$

maps the generator ∇ of the group $\tilde{H}_{n-1}(V_t, \partial V_t)$ into the generator

$$(-1)^{n(n+1)/2}\Delta \in \tilde{H}_{n-1}(V_t) \equiv \tilde{H}_{n-1}(V_t - \partial V_t).$$

In the case when n is odd this follows easily from Propositions 2 and 3. Indeed, by Proposition 3, in this case the operator $(M_u - \mathrm{Id})$ maps Δ into -2Δ; on the other hand, this operator can be regarded as the composition of the obvious map $\tilde{H}_{n-1}(V_t) \to \tilde{H}_{n-1}(V_t, \partial V_t)$ (sending Δ into $\langle \Delta, \Delta \rangle \nabla$) and the desired map (1).

For a proof of Proposition 4 in the general case see § 11.

This proposition is called the *Picard–Lefschetz formula.*

2.2. Equivalent statements.

There are some more fibrations related to the Milnor fibration of a Morse singularity which also define the monodromy and variation operators. All of them are derived from the fibre bundle over T', the fibre of which over a point t is the diagram of spaces

$$\begin{array}{ccc} B & \supset & V_t \\ \cup & & \cup \\ \partial B & \supset & \partial V_t \end{array} \tag{2}$$

The total space of this fibre bundle consists of four subspaces in the space of the trivial bundle $T' \times B \to T'$ such that for any $t \in T'$ the intersections of the fibre $t \times B \simeq B$ of this bundle with these subspaces are the sets $t \times B$, $t \times V_t$, $t \times \partial B$, $t \times \partial V_t$ respectively. It is easy to see that three of these subsets, the first, the third and the fourth, can be extended to fibre bundles over the whole T, and hence define a trivializable bundle of triplets.

Therefore in all the multitude of composite fibre bundles that appear from the diagram of bundles (2), the only nontrivial bundles are those in which V_t participates. Namely, they are the following bundles:

a) the above-considered bundles of manifolds V_t and of pairs $(V_t, \partial V_t)$,

b) the bundles of spaces $B \setminus V_t$ and of pairs $(B \setminus V_t, \partial B \setminus \partial V_t)$,

c) the bundles of pairs (B, V_t) and of pairs $(B, \partial B \cup V_t)$.

These bundles also define monodromy and variation operators, corresponding to the generator of the group $\pi_1(T')$. Namely, all these monodromy operators map the homology groups of all these spaces and pairs into themselves and are isomorphisms, and the variation operators related to the bundles b) and c) act as follows:

$$\tilde{H}_*(B \setminus V_\delta, \partial B \setminus \partial V_\delta) \to \tilde{H}_*(B \setminus V_\delta), \tag{3}$$

$$\tilde{H}_*(B, \partial B \cup V_\delta) \to \tilde{H}_*(B, V_\delta). \tag{4}$$

The operator (3) is defined by the general scheme from § 1, and the operator (4) by an elementary modification of it. In particular, the monodromy operators acting on the groups $\tilde{H}_*(B \setminus V_\delta, \partial B \setminus \partial V_\delta)$ and $\tilde{H}_*(B, \partial B \cup V_\delta)$ are equal to $\mathrm{Id} + j \circ \mathrm{Var}$, where j is the map from the exact sequence of the pair $(B \setminus V_\delta, \partial B \setminus \partial V_\delta)$ (respectively, of the triple $(B, \partial B \cup V_\delta, V_\delta)$); the monodromy operators acting on the groups $\tilde{H}_*(B \setminus V_\delta)$ and $\tilde{H}_*(B, V_\delta)$ are equal to $\mathrm{Id} + \mathrm{Var} \circ j$.

The action of all the new monodromy and variation operators can be reduced to similar operators considered in subsection 2.1. Namely, the following assertion holds.

Proposition 5. *For any i there exists the commutative diagram*

$$\begin{array}{ccccc} \tilde{H}_i(B, \partial B \cup V_\delta) & \to & \tilde{H}_{i-1}(V_\delta, \partial V_\delta) & \to & \tilde{H}_i(B \setminus V_\delta, \partial B \setminus \partial V_\delta) \\ \downarrow & & \downarrow & & \downarrow \\ \tilde{H}_i(B, V_\delta) & \to & \tilde{H}_{i-1}(V_\delta) & \to & \tilde{H}_i(B \setminus V_\delta), \end{array} \tag{5}$$

in which all vertical arrows are variation operators, and all horizontal arrows are isomorphisms and commute with the monodromy operators. In particular, all these groups are isomorphic to $\mathbf{Z}$ *if $i = n$ and are trivial for other i.*

In this diagram the left horizontal isomorphisms are boundary operators (the first of them is the composition of the boundary operator from the exact sequence of the triple $(B, \partial B \cup V_\delta, \partial B)$ and the excision isomorphism $\tilde{H}_*(\partial B \cup V_\delta, \partial B) \to \tilde{H}_*(V_\delta, \partial V_\delta)$).

Example 1. If $f = x_1^2 + \cdots + x_n^2$ and $t > 0$, then for the generator of the group $\tilde{H}_n(B, V_t)$ we can take the disc of radius $t^{1/2}$ in $\mathbf{R}^n$; this disc is called the *vanishing cell* corresponding to the vanishing cycle $\Delta \in \tilde{H}_{n-1}(V_t)$; see [Pham 67].

The right-hand isomorphisms in (5) are given by the *tube operators* (or the *Leray coboundary operators*) which we now define.

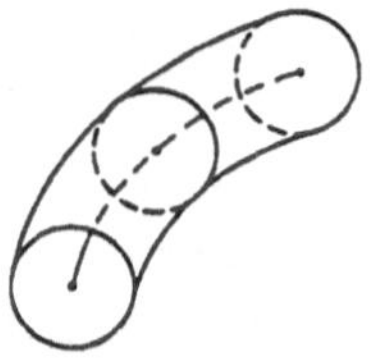

Fig. 7. Leray tube operation

2.3. Tube operator.

Let M be a smooth complex manifold, and V a complex submanifold of it of complex codimension 1. Then to any l-dimensional cycle $a \subset V$ there corresponds an $(l+1)$-dimensional cycle in $M \setminus V$, see Figure 7. Namely, regard a tubular neighbourhood of the manifold V in M as a fibre bundle over V whose fibre is a complex disc. These discs, and hence also their boundary circles, admit a natural orientation which is induced from the complex orientation of the normal bundle of V in M. The union of these circles over all points of a is a closed surface in $M \setminus V$; let us denote it by $t(a)$. The orientation of simplices of the chain a together with the orientation of the circles defines an orientation of the corresponding pieces of the surface $t(a)$: for an orienting frame of this surface we take a frame of which the first vector is tangent to the circle, and the last l vectors, being projected onto V, become a positively oriented tangent frame to a.

Hence, if the cycle a is oriented, then so is the cycle $t(a)$. The correspondence $a \to t(a)$ defines a homomorphism

$$\tilde{H}_l(V) \to \tilde{H}_{l+1}(M \setminus V), \qquad (6)$$

which is called the *tube operator* or *Leray coboundary homomorphism*.

This is exactly the operator providing the right-hand isomorphism in the lower line of (5); the right-hand isomorphism in the upper line is provided by a similar operation over the relative cycles (here we use the fact that V_δ is transversal to ∂B).

The word "coboundary" in the name of the operator is due to the following circumstance. The groups participating in the diagram (5) (and similar groups for arbitrary $t \in T'$) can be organized into one more diagram:

$$\begin{array}{ccccc}
\tilde{H}_n(B, V_t) & \otimes & \tilde{H}_n(B \setminus V_t, \partial B \setminus \partial V_t) & \to & \mathbf{Z} \\
\downarrow \partial & & \uparrow t & & \updownarrow \\
\tilde{H}_{n-1}(V_t) & \otimes & \tilde{H}_{n-1}(V_t, \partial V_t) & \to & \mathbf{Z} \\
\downarrow t & & \uparrow \partial & & \updownarrow \\
\tilde{H}_n(B \setminus V_t) & \otimes & \tilde{H}_n(B, \partial B \cup V_t) & \to & \mathbf{Z}.
\end{array} \qquad (7)$$

Here the left and middle vertical arrows are the horizontal isomorphisms from (5), given by the boundary or tube operators, the pairings are the intersection indices, and the right vertical arrows are multiplications by $(-1)^n$.

Proposition 6. *This diagram is commutative.*

This proposition in a more general situation will be proved in § 3.6.

Corollary 1. *Let $f = x_1^2 + \cdots + x_n^2$. Then for any small $t > 0$ (respectively, $t < 0$) the group $\tilde{H}_n(B \setminus V_t, \partial B \setminus \partial V_t)$ is generated by the class of the imaginary plane $i\mathbf{R}^n \cap B$ (respectively, of the real plane $\mathbf{R}^n \cap B$).*

Indeed, in the case $t > 0$ the intersection index of this cycle with the disc $\{x \in \mathbf{R}^n | x_1^2 + \cdots + x_n^2 \le t\}$ (which generates the group $\tilde{H}_n(B, V_t)$) is equal to 1; the proof for the case $t < 0$ is exactly the same.

§ 3. Local monodromy of isolated singularities of holomorphic functions

3.1. The classical monodromy operator. Milnor number.

The monodromy operator of a Morse singularity, considered in the previous section, is the simplest model example of a large family of similar operators, and simultaneously the main tool in their investigation.

Let $f : (\mathbf{C}^n, 0) \to (\mathbf{C}, 0)$ be a holomorphic function having an isolated singularity at $0 \in \mathbf{C}^n$ (i.e. $df(0) = 0$, but $df \neq 0$ in a punctured neighbourhood of 0). Again let ε be a sufficiently small positive number (so that the variety $f^{-1}(0)$ is transversal to the sphere $S_\varepsilon \equiv \partial B$ and to all smaller spheres centred at 0). Denote the set $f^{-1}(t) \cap B$ by V_t. If the number δ is sufficiently small with respect to ε, then for every t from the disc $T \equiv T_\delta = \{t : |t| \le \delta\}$ the set $f^{-1}(t)$ is transversal to the sphere S_ε. If additionally $t \neq 0$, then V_t is smooth inside B and hence is a smooth manifold with boundary $\partial V_t = V_t \cap \partial B$; if $t = 0$, then V_t has the unique singular point 0.

As in § 2, the monodromy and variation operators corresponding to the loop $\{\delta \cdot e^{it}\}, t \in [0, 2\pi]$, are well defined. The monodromy operators act on the groups $\tilde{H}_*(V_\delta)$ and $\tilde{H}_*(V_\delta, \partial V_\delta) \simeq \tilde{H}^*(V_\delta)$, while the variation operator maps $\tilde{H}_*(V_\delta, \partial V_\delta)$ into $\tilde{H}_*(V_\delta)$.

Theorem 1 (see [Milnor 68]). *If the singularity f is isolated, then the manifold V_t, $t \neq 0$, is homotopy equivalent to a wedge of finitely many spheres of dimension $n - 1$; in particular, the group $\tilde{H}_i(V_t)$ is trivial for $i \neq n - 1$ and is free Abelian for*

$i = n-1$. The variety V_0 is contractible and is homeomorphic to a cone over its boundary ∂V_0.

Corollary 1. *If the singularity f is isolated, and $0 < |t| < \delta$, then the group $\tilde{H}_i(V_\delta, \partial V_\delta)$ is trivial for $i \neq n-1$ and is free Abelian otherwise.*

This follows from the Poincaré duality theorem.

Corollary 2. *If the singularity f is isolated, then the group $\tilde{H}_i(\partial V_t)$ is trivial for all i other than $n-1, n-2$.*

This follows from Theorem 1, Corollary 1, and the exact sequence of the pair $(V_t, \partial V_t)$.

The rank of the group $\tilde{H}_*(V_t)$ is called the *Milnor number* of the singularity f and is denoted by $\mu(f)$, so that $\tilde{H}_{n-1}(V_t) \simeq \mathbf{Z}^{\mu(f)}$.

Here is another definition of the Milnor number. The gradient vector field that assigns to any point x the vector $(df/dx_1, \dots, df/dx_n)$ maps the sphere $S_\varepsilon \equiv \partial B$ into $\mathbf{C}^n$. Since our sungularity is isolated, the image of this map lies in the space $\mathbf{C}^n \setminus \{0\}$, which can be contracted onto the unit sphere $S_1 = S_1^{2n-1}$.

Proposition 1 (see [Milnor 68]). *The degree of the map*

$$\frac{grad f}{|grad f|} : S_\varepsilon \to S_1$$

is equal to the Milnor number of the singularity f.

The calculation of the monodromy and variation operators in the case of an arbitrary isolated singularity can be reduced to the special case considered in the previous section by "splitting" a complicated singularity into several Morse singularities.

Definition 1. A critical point of the function f is called *Morse* if the second derivative of f at this point is a nondegenerate quadratic form. A function is called *Morse* if all its critical points are Morse, and *strictly Morse* if additionally all its values at critical points are distinct.

Proposition 2 (Morse lemma). *In a neighbourhood of a Morse critical point, any holomorphic function in $\mathbf{C}^n$ can be expressed in the standard form*

$$z_1^2 + \cdots + z_n^2 + const \tag{8}$$

in appropriate local holomorphic coordinates.

Therefore, the topological behaviour of this function in a neighbourhood of a Morse point reduces to the one considered in the previous section.

By Proposition 2.1, the Milnor number of the Morse critical point is equal to 1. This is the unique example: the Milnor numbers of all non-Morse isolated critical points are strictly greater than 1 (and it is convenient to define the Milnor numbers of nonisolated singularities to be equal to infinity).

Lemma 1. *Any function $\mathbf{C}^n \to \mathbf{C}$ can be transposed by an arbitrarily small perturbation into a function of which all critical points are Morse and have different critical values. For such a perturbation we can take the function $f + l$, where l is an appropriate linear form with arbitrarily small coefficients.*

The proof follows easily from Sard's lemma, applied to the gradient map $\text{grad} f : \mathbf{C}^n \to \mathbf{C}^n$. Indeed, if $-l$ is not a critical value of this mapping, then the function $f + l$ is (nonstrictly) Morse. If a morsification of the form $f + l$ has occasionally coinciding critical values at different Morse critical points, then this coincidence can be destroyed by adding an arbitrarily small linear form which takes different values at all these critical points.

Definition 2. A small perturbation of the singularity f, which splits it into Morse singularities only, is called a *morsification* of f; if additionally all critical values of this morsification are distinct, then it is called a *strict morsification.*

Suppose that the morsification $\tilde{f}$ of the singularity f is very close to f, so that $|\text{grad}(\tilde{f} - f)| < |\text{grad} f|/2$ everywhere on S_ϵ. Then by the "argument principle" the index of the normed gradient map $\text{grad}\tilde{f}/|\text{grad}\tilde{f}| : S_\epsilon \to S_1$ is again equal to $\mu(f)$. On the other hand, it is easy to see that it is equal to the number of Morse points of the perturbed function $\tilde{f}$ in B, and we get the following statement.

Proposition 3. *The number of Morse points of any small morsification of the singularity f, lying in the disc B, is equal to the Milnor number of the singularity f; in particular, it does not depend on the choice of the morsification.*

Consider a strict morsification $\tilde{f}$ of the singularity f. If $\tilde{f}$ is sufficiently close to f, then all its $\mu(f)$ critical values lie strictly inside the disc $T \equiv T_\delta$ (and the corresponding critical points lie in B).

Denote the set of these critical points by $s(\tilde{f})$; see Figure 8. Consider the one-parameter family of functions $f_{(\tau)} \equiv (1 - \tau)f + \tau\tilde{f}$, $\tau \in [0, 1]$, connecting f and $\tilde{f}$. For any point z of the circle ∂T, the family of manifolds $(f_{(\tau)})^{-1}(z) \cap B$, $\tau \in [0, 1]$, realizes an isotopy between V_z and $\tilde{V}_z \equiv \tilde{f}^{-1}(z) \cap B$; in particular, it defines a canonical isomorphism between the homology groups (both absolute and relative) of these manifolds. The transportation of the value z along the circle ∂T defines the monodromy and variation operators, acting on the homologies of both manifolds V_δ and $\tilde{V}_\delta$; it is easy to see that these actions in the homologies of these two manifolds commute with the above-described isotopies of them. Thus, the calculation of this action for the manifold V_δ reduces to a similar calculation for $\tilde{V}_\delta$.

But the action of these operators on the homology of $\tilde{V}_\delta$ is a composition of elementary operators, corresponding to the loops going around the points of the set $s(\tilde{f})$; since all these points are Morse, these operators can be reduced to the Picard–Lefschetz formula described in the previous section. For the exact definition of these operators see the next subsection.

3.2. The monodromy group related to a morsification of a singularity.

Choose the point δ as the distinguished point in the disc T. The group $\pi_1(T \setminus s(\tilde{f}))$ is a free group with $\mu(f)$ generators. Any element of this group (i.e. a loop in $T \setminus s(\tilde{f})$ which starts and finishes at the point δ) defines monodromy and variation operators in the homology groups of the manifold $\tilde{V}_\delta$. In accordance with the general terminology, the monodromy representation related to the morsification $\tilde{f}$ is the corresponding representation of the group $\pi_1(T \setminus s(\tilde{f}))$ in the space $\tilde{H}_{n-1}(\tilde{V}_\delta)$. Thus, the classical monodromy operator, considered in the previous subsection, is a special case of a whole family of operators: it corresponds to the element $\{\partial T\}$ of the group $\pi_1(T \setminus s(\tilde{f}))$. Now we show how to calculate all the similar operators, corresponding to all elements of this group.

First of all, we choose a convenient system of generators in this group.

Definition 3 (see [Gabrielov 73], [Brieskorn 70]). A *distinguished system of paths*, related to a morsification $\tilde{f}$, is a system of μ smooth paths in T, connecting the point δ with all μ points of the set $s(\tilde{f})$ and numbered by the numbers $1, 2, \ldots, \mu$, such that

1) these paths do not self-intersect, and intersect each other only at their common point δ. In particular, their intersection points with the set $s(\tilde{f})$ are only their endpoints;

2) these paths go from the point δ with different tangents, and the angle between the i-th path and the vertical ray which goes up from the point δ increases as i increases. See Figure 8.

Definition 4. A *simple loop*, related to the i-th path of a distinguished system, is the loop in $T \setminus s(\tilde{f})$ which goes from the point δ along the i-th path almost to its endpoint, turns once around this endpoint counterclockwise along a small circle, and returns along the same path to the point δ. See Figure 9.

It is obvious that the set of simple loops related to the paths of a distinguished system generates the group $\pi_1(T \setminus s(\tilde{f}))$. Now we calculate the action of the monodromy and variation operators in the homology groups of $\tilde{V}_\delta$, defined by a simple loop.

Let us consider the family of varieties $\tilde{V}_\theta$, where θ goes along some path of the

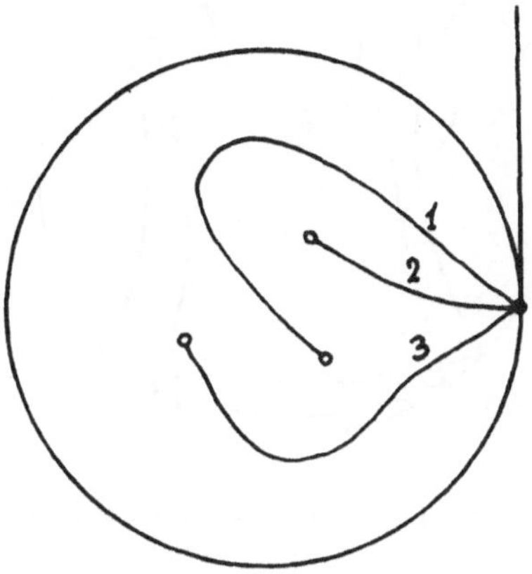

Fig. 8. Distinguished system of paths

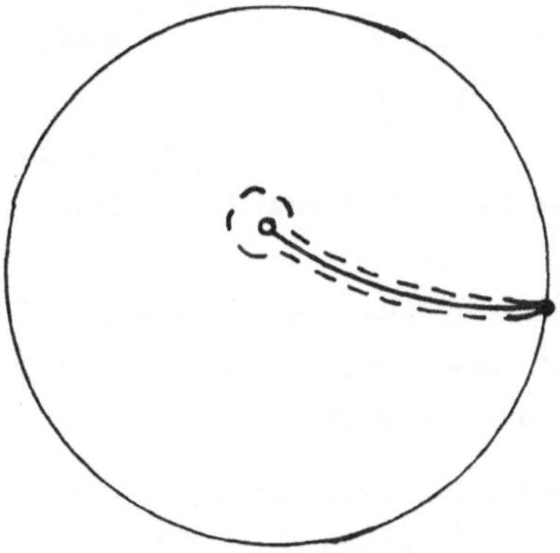

Fig. 9. Simple loop

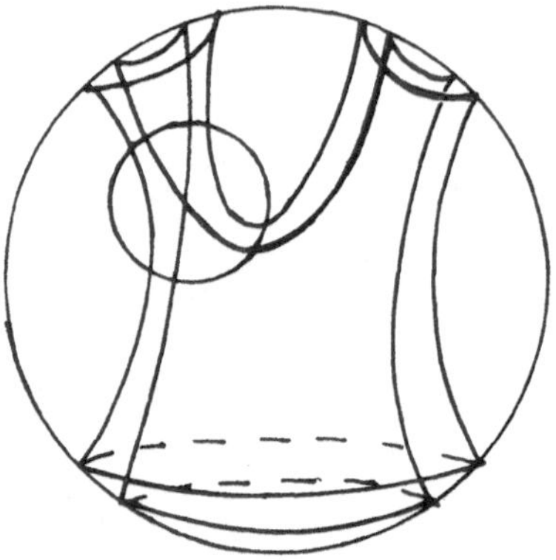

Fig. 10. Critical and noncritical level sets

distinguished system. For almost all points θ of this path but the last one, these varieties are isotopic to $\tilde{V}_\delta$, and only in the last instant (when θ becomes the critical value s_i) does something change.

This change happens in an arbitrarily small neighbourhood of the Morse critical point corresponding to the critical value s_i. By Proposition 2, in this neighbourhood our function $\tilde{f}$ can be reduced to the form $z_1^2+\cdots+z_n^2$ by a change of local coordinates and addition of a constant function. But everything that happens with such a function was studied explicitly in § 2. Namely, let β be a very small neighbourhood of the Morse critical point, and let the value θ be (very)^2 close to s_i, but not equal to it. Then the group $\tilde{H}_{n-1}(\tilde{V}_\theta \cap \beta)$ is isomorphic to $\mathbf{Z}$ and is generated by an embedded sphere, whose radius is of order $|\theta - s_i|^{1/2}$. (See Figure 10; note that in the disc β this picture coincides with Figure 6.) This sphere is called the *vanishing sphere* in $\tilde{V}_\theta$: the name is related to the fact that this sphere contracts to a point when θ tends to s_i.

Transporting this sphere by virtue of the isotopy of the manifolds $\tilde{V}_\cdot$ over our distinguished path, we get an embedded sphere in the manifold $\tilde{V}_\delta$. This sphere is called a *cycle, vanishing over our path.*

Lemma 2. *The vanishing sphere defines a nontrivial element in the group $\tilde{H}_{n-1}(\tilde{V}_\theta)$ (and hence also in the group $\tilde{H}_{n-1}(\tilde{V}_\delta)$).*

Indeed, it is sufficient to prove this statement for θ very close to s_i; moreover, we can suppose that $\theta - s_i$ is a positive real number. By the exact sequence of the pair $(B, \tilde{V}_\theta)$, $\tilde{H}_{n-1}(\tilde{V}_\theta) \simeq \tilde{H}_n(B, \tilde{V}_\theta)$, and it is sufficient to prove the nontriviality in the last group of the relative cycle, which spans our vanishing sphere in the disc B. To do this, consider the holomorphic volume form

$$dx = dx_1 \wedge \ldots \wedge dx_n. \tag{9}$$

It is easy to see that the integral of this form (as well as of any other holomorphic n-form in B) along any relative cycle in $(B, \tilde{V}_\theta)$ depends only on the homology class of this cycle, or, equivalently, on the class of its boundary in the group $\tilde{H}_{n-1}(\tilde{V}_\theta)$. Therefore it is sufficient to prove that the integral of this form along our spanning cycle is not equal to 0.

By the Morse lemma, close to the considered critical point our function $\tilde{f}$ can be reduced to the form $z_1^2 + \cdots + z_n^2 + s_i$, and for the spanning cycle we can take the disc of radius $(\theta - s_i)^{1/2}$ in $\mathbf{R}^n$ (in coordinates z_i). Then the desired integral is equal to the integral along this disc of the diferential form $J(z)dz_1 \wedge \ldots \wedge dz_n$, where $J(z)$ is the Jacobian determinant of our transformation of coordinates. It is easy to see that this integral is equal to

$$c \cdot J(0)(\theta - z_i)^{n/2} + O(\theta - s_i)^{1+n/2},$$

where c is the volume of the unit disc. Since $J(0) \neq 0$, we get the lemma.

In fact, a much stronger fact is valid.

Theorem 2 (see [Brieskorn 70]). *For any distinguished system of paths in D_δ, the collection of cycles vanishing along all the paths of this system forms a basis in the group $\tilde{H}_{n-1}(\tilde{V}_\delta)$.*

Definition 5. The basis in $\tilde{H}_{n-1}(\tilde{V}_\delta)$, consisting of cycles vanishing along some distinguished system of paths is called a *distinguished basis.*

Now we are going to study how the transportation along a simple loop acts on the homology of $\tilde{V}_\delta$. Again, this problem can be reduced to the study of the local monodromy and variation close to the corresponding critical point. Let the point $\theta \in \mathbf{C}^1$ of the i-th distinguished path be very close to its endpoint, i.e., to the critical value s_i, let β be the above-considered disc around the corresponding critical point, in which $\tilde{f}$ can be reduced to the canonical form, and let Δ be an arbitrary element of the group $\tilde{H}_{n-1}(\tilde{V}_\theta)$. Let us calculate the result of the transportation of the element Δ along a very small circle around the point s_i.

For any value of θ' that is also very close to s_i (including $\theta' = s_i$) let us divide the variety $V_{\theta'}$ into two parts: $V_{\theta'} \cap \beta$ and $V_{\theta'} \setminus \beta$. The second of these varieties does not change topologically when θ' moves close to s_i, therefore we can trivialize the fibre bundle determined by the function $\tilde{f}$ on the set

$$\tilde{f}^{-1}(\{\text{a very small neighbourhood of } s_i\}) \cap (B \setminus \beta).$$

In particular, if the element Δ can be realized by a cycle placed completely outside β, then the monodromy around the critical value s_i takes it into itself. In the general case, Δ can be represented by a cycle whose intersection with β coincides with the (taken with some multiplicity $\lambda(\Delta)$) cycle ∇_i, which generates the group $\tilde{H}_{n-1}(\tilde{V}_\theta \cap \beta, \tilde{V}_\theta \cap \partial\beta)$. (Recall that this group is isomorphic to $\mathbf{Z}$; see § 2.)

In formal terms, the class in $\tilde{H}_{n-1}(\tilde{V}_\theta \cap \beta, \tilde{V}_\theta \cap \partial\beta)$ of the intersection of Δ with β can be defined as the image of the class Δ under the factorization homomorphism

$$\tilde{H}_{n-1}(\tilde{V}_\theta) \to \tilde{H}_{n-1}(\tilde{V}_\theta, \tilde{V}_\theta \setminus \beta) \equiv \tilde{H}_{n-1}(\tilde{V}_\theta \cap \beta, \tilde{V}_\theta \cap \partial\beta).$$

The number $\lambda(\Delta)$ here is equal to the intersection index $\langle \Delta, \Delta_i \rangle$ in $\tilde{V}_\theta$ of the cycle Δ and the cycle Δ_i, which vanishes along the rest of our path at the critical point with the critical value s_i.

The transportation of the cycle Δ over the circle which goes around s_i can be realized in such a way that the part of Δ lying outside β (and on the boundary of β) is transported in accordance with the chosen trivialization (and hence returns to its original position), and in the intersection with β the behaviour of this cycle is defined by the action of the operator Var (related to the considered Morse critical point) on the element $\langle \Delta, \Delta_i \rangle \cdot \nabla_i$. This action is expressed by the Picard–Lefschetz formula (see § 2). Writing this formula (and transporting the result along the distinguished path from the fibre $\tilde{V}_\theta$ to the fibre $\tilde{V}_\delta$) we get the following version of the Picard–Lefschetz formula.

Theorem 3. *The monodromy along a simple loop, corresponding to the i-th path of a distinguished system of paths, transforms any cycle $\Delta \in \tilde{H}_{n-1}(\tilde{V}_\delta)$ into the cycle*

$$\Delta + (-1)^{n(n+1)/2} \langle \Delta, \Delta_i \rangle \cdot \Delta_i, \tag{10}$$

where Δ_i is the cycle that vanishes along our path.

Corollary. *The monodromy representation in the group $\tilde{H}_{n-1}(\tilde{V}_\delta)$ preserves the bilinear form on this group defined by the intersection indices.*

This follows immediately from formula (10).

In the case of odd n, when the intersection form is symmetric, this formula has the following interpretation: the monodromy operator defined by the simple loop acts on the space (or lattice) $\tilde{H}_{n-1}(\tilde{V}_\delta)$ as the reflection in the plane orthogonal (in the sense of this form) to the cycle Δ_i.

Theorem 3 has also a relative variant, which can be proved in exactly the same way:

Theorem 3′. *The variation operator along a simple loop, corresponding to the i-th path of an distinguished system of paths, transforms a relative cycle $\nabla \in \tilde{H}_{n-1}(\tilde{V}_\delta, \partial\tilde{V}_\delta)$ into the cycle*

$$(-1)^{n(n+1)/2}\langle \nabla, \Delta_i \rangle \cdot \Delta_i, \tag{11}$$

where Δ_i is the cycle that vanishes along our path.

Corollary (see [Lamotke 75], [Gusein-Zade 77]). *The variation operator of an isolated singularity f, defined by the loop ∂T, is an isomorphism between the groups $\tilde{H}_{n-1}(V_\delta, \partial V_\delta)$ and $\tilde{H}_{n-1}(V_\delta)$.*

Indeed, let us choose a basis in $\tilde{H}_{n-1}(V_\delta, \partial V_\delta)$ which is Poincaré dual to the distinguished basis of vanishing cycles in $\tilde{H}_{n-1}(V_\delta)$. The loop ∂T is the composition of the simple loops corresponding to all paths of the distinguished system, therefore it follows immediately from (11) that the matrix of the operator Var in these chosen bases will be triangular with numbers $(-1)^{n(n+1)/2}$ on the diagonal.

Since the simple loops generate the group $\pi_1(T \setminus s(\tilde{f}))$, we have learned (in principle) how to calculate the monodromy and variation operators along arbitrary elements of this group: we only need to know the intersection indices of vanishing cycles in $\tilde{V}_\delta$.

3.3. Some general facts on the intersection matrices of the vanishing cycles.

The intersection indices of the vanishing cycles Δ_i defined by a distinguished system of paths form a $\mu \times \mu$ matrix: the *intersection matrix of vanishing cycles*, with the number $\langle \Delta_i, \Delta_j \rangle$ in the cell (i, j). We list here some natural properties of these matrices.

Symmetry. If n is odd, then the intersection matrix of vanishing cycles is symmetric, and if n is even, then it is skew-symmetric.

Stabilization. Let $f : (\mathbf{C}^n, 0) \to (\mathbf{C}, 0)$ be a function-germ with an isolated singularity at the origin; then its q-th stabilization $f^{(q)} : (\mathbf{C}^{n+q}, 0) \to (\mathbf{C}, 0)$ is defined by the formula

$$f^{(q)}(z_1, \ldots, z_n, u_1, \ldots, u_q) = f(z_1, \ldots, z_n) + u_1^2 + \cdots + u_q^2.$$

If $\tilde{f}$ is a morsification of the singularity f, then the function

$$\tilde{f}^{(q)}(z_1, \ldots, z_n, u_1, \ldots, u_q) \equiv \tilde{f}(z_1, \ldots, z_n) + u_1^2 + \cdots + u_q^2$$

is a morsification of $f^{(q)}$ and has the same critical values. Set $\tilde{V}_\delta^q \equiv \tilde{f}^{(q)-1}(\delta) \cap B \subset \mathbf{C}^{n+q}$.

Let $\Delta_1, \ldots, \Delta_\mu$ be a distinguished basis in the group $\tilde{H}_{n-1}(\tilde{V}_\delta)$, and $\Delta_1^q, \ldots, \Delta_\mu^q$ the distinguished basis in $\tilde{H}_{n-1+q}(\tilde{V}_\delta^q)$ of which the elements vanish along the same system of paths as the cycles Δ_i.

A system of paths determines the vanishing cycles in the homology groups only up to the signs (which depend on the choice of orientations of the corresponding vanishing spheres).

Theorem 4 (see [Gabrielov 73]). *For an appropriate choice of orientations of the cycles Δ_i, Δ_i^q, the intersection matrices $\langle \Delta_i, \Delta_j \rangle$ and $\langle \Delta_i^q, \Delta_j^q \rangle$ are related by the following equations:*

$$\langle \Delta_i^q, \Delta_j^q \rangle = [\operatorname{sign}(i-j)]^q (-1)^{qn+q(q-1)/2} \langle \Delta_i, \Delta_j \rangle \tag{12}$$

for $i \neq j$; the case $i = j$ is determined by Proposition 2 of § 2.

It follows easily from this theorem that the matrix $\langle \Delta_i^q, \Delta_j^q \rangle$ is periodic in q with period 4, and the increase of q by 2 changes this matrix to its negative.

The Dynkin diagram is a convenient way to describe the intersection matrices of the singularities.

Suppose first that $n \equiv 3 (\mathrm{mod}\ 4)$, and $\Delta_1, \ldots, \Delta_\mu$ is a distinguished basis in the group $\tilde{H}_{n-1}(\tilde{V}_\delta)$ of the singularity $f : (\mathbf{C}^n, 0) \to \mathbf{C}, 0)$. Then the *Dynkin diagram* of this basis is a graph with μ numbered vertices. The vertices with the numbers i and j are connected by k continuous segments if $\langle \Delta_i, \Delta_j \rangle = k > 0$, and by $-k$ dashed segments if $\langle \Delta_i, \Delta_j \rangle = k < 0$; if $\langle \Delta_i, \Delta_j \rangle = 0$, then one draws no segments.

For $n \not\equiv 3 (\mathrm{mod}\ 4)$, the Dynkin diagram of a singularity f of n variables is defined as the Dynkin diagram of an arbitrary stabilization

$$f^{(q)}(z_1, \ldots, z_n, u_1, \ldots, u_q) = f(z_1, \ldots, z_n) + u_1^2 + \cdots + u_q^2$$

of it such that $n + q \equiv 3 (\mathrm{mod}\ 4)$.

Proposition 4 (see [Gabrielov 74]). *The Dynkin diagram of any isolated singularity defined by an arbitrary distinguished basis is a connected graph.*

The explicit calculation of Dynkin diagrams is a crucial point in determining the monodromy operators; some methods of this calculation will be described in § 4.

3.4. Deformations of singularities. The discriminant.

The monodromy representation of the group $\pi_1(T \backslash s(\tilde{f}))$ has a natural generalization: the monodromy representation of the fundamental group of the complement of the discriminant variety of an arbitrary deformation of the singularity f. Here are the definitions of all these objects.

Definition 6. A *deformation* of a singularity $f : (\mathbf{C}^n, 0) \to (\mathbf{C}, 0)$ is any holomorphic function $F : (\mathbf{C}^n \times \mathbf{C}^l, 0) \to (\mathbf{C}, 0)$ such that $F(\cdot, 0) \equiv f$.

It is natural to regard such a deformation as the family of functions $f_\lambda \equiv F(\cdot, \lambda)$, $\lambda \in \mathbf{C}^l$; in particular, $f = f_0$. The parameter space $\mathbf{C}^l$ is called the *base* of the deformation F. Again let B be the small disc centred at the origin in $\mathbf{C}^n$, considered above. If the original singularity f is isolated, and $\mu(f)$ is its Milnor number, then for any λ sufficiently close to 0 the sum of the Milnor numbers of the corresponding function f_λ at the critical points belonging to B is equal to $\mu(f)$: this follows from the elementary topological considerations about the index of the gradient map. In particular, the number of these points does not exceed $\mu(f)$.

To any point $\lambda \in \mathbf{C}^l$ there corresponds the variety $V_\lambda = f_\lambda^{-1}(0) \cap B$. This variety is smooth if and only if f_λ has no critical points with critical value 0 in B.

Definition 7. The *discriminant* of the deformation F is the set of points $\lambda \in \mathbf{C}^l$ such that the function $f_\lambda = F(\cdot, \lambda)$ has a critical point in B with critical value 0. The discriminant of the deformation F is denoted by $\Sigma(F)$.

Example. Let $n = 1, f \equiv x^k$, and suppose that the deformation F has the form

$$F(x, \lambda_1, \ldots, \lambda_k) = x^k + \lambda_1 x^{k-1} + \cdots + \lambda_{k-1} x + \lambda_k. \tag{13}$$

Then $\Sigma(F)$ is the set of those λ for which the polynomial (13) has a multiple root, i.e. $\Sigma(F)$ is the usual discriminant variety of the general polynomial of degree k.

It is easy to see that this discriminant variety is diffeomorphic to the direct product of a complex line and the intersection of it with the base of the subdeformation of the deformation (13) distinguished by the condition $\lambda_1 = 0$. In the cases $k = 3$ and $k = 4$ the real analogues of the discriminant varieties in these reduced deformations are respectively the semicubical parabola $\{(\lambda_2, \lambda_3) | 4\lambda_2^3 + 27\lambda_3^2 = 0\}$ and the swallowtail, i.e. the surface shown in Figure 3b.

Let D_δ be a disc with centre at the origin in the parameter space $\mathbf{C}^l$, which is so small that for any $\lambda \in D_\delta$ the sum of the Milnor numbers of the critical points of f_λ lying inside B is exactly equal to $\mu(f)$ and the variety V_λ is transversal to ∂B. Let us fix a distinguished point $v \in D_\delta \setminus \Sigma(F)$. Then, as above, for any loop $u \in \pi_1(D_\delta \setminus \Sigma(F))$ the monodromy operator

$$M_u : \tilde{H}_{n-1}(V_v) \to \tilde{H}_{n-1}(V_v)$$

and the variation operator

$$\mathrm{Var}_u : \tilde{H}_{n-1}(V_v, \partial V_v) \to \tilde{H}_{n-1}(V_v)$$

are well defined. In particular, the monodromy representation

$$\pi_1(D_\delta \setminus \Sigma(F)) \to \mathrm{Aut}\tilde{H}_{n-1}(V_v)$$

appears.

Example 1. Let $l = 1$ and suppose that the deformation F has the form $F(x, \lambda) = f(x) - \lambda$. Then $\Sigma(F) = \{0\}$, $\pi_1(D_\delta \setminus \Sigma(F)) = \mathbf{Z}$, and the action of the generator of the latter group is exactly the classical monodromy operator considered in subsection 3.1.

Example 2. Let $l = 2$, let $\tilde{f}$ be a strict morsification of f which is very close to f, and $F(x; \lambda_1, \lambda_2) \equiv (1 - 2\lambda_1/\delta)f(x) + (2\lambda_1/\delta)(\tilde{f}(x) - f(x)) - \lambda_2$. The coordinate λ_2 on the line $\{(\lambda_1, \lambda_2) : \lambda_1 = \delta/2\} \subset \mathbf{C}^2$ identifies this line with the set of values of the function $\tilde{f}$. The intersection points of this line with $\Sigma(F)$ are exactly the points of the set $s(\tilde{f})$; see Figure 8. The action of the monodromy (and variation) along the loops lying in this complex line outside $\Sigma(F)$ can be naturally identified with the similar operators considered in subsection 3.2 above. Thus, the monodromy representation of the group $\pi_1(D_\delta \setminus \Sigma(F))$, which we consider now, is a generalization of the representation of $\pi_1(D_\delta \setminus \Sigma(F))$ considered in subsections 3.1 and 3.2. A little later we shall see that the calculation of this representation can be reduced to the calculations from subsection 3.2.

3.5. Versal deformations.

The set of all deformations of an isolated singularity f is very large, but it contains some "maximal" deformations such that all other deformations can be reduced to them: these are the so-called versal (or semi-universal) deformations. Let us give the exact definitions.

Definition 8. A deformation $F'(x, \lambda)$ of the function f is *equivalent* to the deformation $F(x, \lambda)$ if it can be represented in the form $F'(x, \lambda) \equiv F(U(x, \lambda), \lambda)$, where $U : (\mathbf{C}^n \times \mathbf{C}^l, 0) \to (\mathbf{C}^n, 0)$ is a smooth family (parametrized by λ) of diffeomorphisms $\mathbf{C}^n \to \mathbf{C}^n$ defined near 0, which is a deformation of the identical diffeomorphism (i.e. $U(\cdot, 0) \equiv \mathrm{Id}$).

Definition 9. A deformation $\Phi(x, \kappa)$, $\kappa \in \mathbf{C}^k$, of the same singularity f is *induced* from the deformation F by a smooth transformation of parameters $\theta : (\mathbf{C}^k, 0) \to (\mathbf{C}^l, 0)$ if $\Phi(x, \kappa) \equiv F(x, \theta(\kappa))$.

Definition 10. A deformation $F(x, \lambda)$ of the function-germ $f(x)$ is called *versal* if any other deformation $\Phi(x, \kappa)$ is equivalent to a deformation induced from F (that is, it can be represented in the form

$$\Phi(x, \kappa) = F(U(x, \kappa), \theta(\kappa)) \tag{14}$$

for some U and θ.)

Theorem 5 (see [Tyurina 69], [MATHER], [AVG 82]). *Any isolated singularity f with the Milnor number $\mu(f) < \infty$ admits l-parameter versal deformations for any $l \geq \mu(f)$ and does not admit versal deformations depending on fewer than $\mu(f)$ parameters.*

There is an easy way of constructing versal deformations of a given isolated singularity. To describe this method we need a new definition of the Milnor number.

Definition 11. The *gradient ideal* of a singularity $f : (\mathbf{C}^n, 0) \to (\mathbf{C}, 0)$ is the ideal in the algebra $\mathbf{C}[[x_1, \dots, x_n]]$ of formal power series generated by all the partial derivatives of the function f. The *local algebra* Q_f of the singularity f is the quotient algebra of the algebra of formal power series by the gradient ideal:

$$Q_f = \mathbf{C}[[x_1, \dots, x_n]]/\{\partial f/\partial x_1, \dots, \partial f/\partial x_n\}.$$

Theorem 6 (see [Milnor 68], [Palamodov 67]). *The dimension of the local algebra Q_f of a singularity f is equal to its Milnor number $\mu(f)$ if this singularity is isolated, and is infinite if it is nonisolated.*

Theorem 7 (see [MATHER], [AVG 82]). *For any isolated singularity f the following deformation of f is versal:*

$$f + \sum_{i=1}^{\mu(f)} \lambda_i \varphi_i, \tag{15}$$

where λ_i are the parameters of the deformation, and φ_i are the functions whose classes in the algebra Q_f generate a basis of this algebra.

It is easy to see that a basis in the local algebra can always be realized by the classes of some monomials; in this case among these monomials there will surely be the monomial 1.

Definition 12. A deformation of an isolated singularity f is called *miniversal* if it is versal and the number of its parameters is equal to $\mu(f)$.

Proposition 5. *If F and Φ are versal deformations of the singularity f, F is miniversal, and θ is the map of the base $\mathbf{C}^k$ of Φ into the base $\mathbf{C}^l$ of F inducing a deformation equivalent to Φ from F, then θ is a submersion close to the point $0 \in \mathbf{C}^k$.*

This assertion follows from the equivalence of the notions of versal and infinitesimally versal deformations; see [AVG 82].

The map θ, inducing the deformation F' from F, maps $\Sigma(F')$ into $\Sigma(F)$. Hence it defines a homomorphism of the fundamental groups of complements of discriminants, and induces the monodromy representation of the group $\pi_1(D_\delta \setminus \sigma(F'))$ from the similar group for F. In particular, Theorem 5 implies that the calculation of the monodromy representation of the fundamental group of the complement of the discriminant of an arbitrary deformation of f can be reduced to the calculation of a similar representation for an arbitrary versal deformation of it.

The latter calculation can be reduced to a problem considered in subsection 3.2. Indeed, let F be a versal deformation of f, which contains together with any function f_λ also all functions of the form $f_\lambda - c$, $c \in \mathbf{C}^1$. Let the distinguished point v of the set $D_\delta \setminus \Sigma(F)$ be such that the corresponding function f_v is a strict morsification of f. Let L be the line in $\mathbf{C}^l$ which consists of all functions of the form $f_v - c$; c can be regarded as a coordinate on L. Then the set $L \cap \Sigma(F)$ consists of exactly $\mu(f)$ points: the values of the coordinate c at these points are the critical values of the function f_v. In particular, $\pi_1(L \cap D_\delta \setminus \Sigma(F))$ is a free group with $\mu(f)$ generators, and its monodromy representation is described by the Picard–Lefschetz formula.

Proposition 6. *The identical embedding $L \setminus \Sigma(F) \to D_\delta \setminus \Sigma(F)$ induces an epimorphism of the fundamental groups.*

This is a local variant of the theorem of Zariski about the realization of the fundamental group of the complement of a complex analytic variety; see [Zariski 37], [Varchenko 72], [Hamm & Lê 73], [Goresky & MacPherson 86].

Corollary. *The monodromy representation*

$$\pi_1(D_\delta \setminus \Sigma(F)) \to Aut(H_*(V_v))$$

preserves the intersection form.

This follows immediately from Proposition 6 and the Corollary of Theorem 3.

This proposition together with the existence of versal deformations reduces the monodromy representation of the fundamental group of the complement of the discriminant of an arbitrary deformation to the similar representation of the free group $\pi_1(T \setminus s(\tilde{f}))$ considered in subsection 3.2 above. Nevertheless it is often useful not to forget the fundamental group of the complement of the discriminant: for instance, using the information about this group we can guess that some element of the group $\pi_1(T \setminus s(\tilde{f}))$ after the identical embedding $L \to D_\delta$ becomes a contractible loop in the complement of the discriminant, and this reduces the calculations. See for example [Hefez & Lazzeri 74].

Theorem 8 (see [Gabrielov 74]). *The discriminant variety of any versal deformation of an isolated singularity is irreducible, and its normalization is smooth and contractible.*

Namely, for such a normalization we can always take the (germ at the origin of) the subset in $\mathbf{C}^n \times \mathbf{C}^l$ consisting of pairs (x, λ) such that the function f_λ has a singular point with zero critical value at x.

Proposition 4 is a direct corollary of this theorem. Moreover, the following improvement of Proposition 4 also follows immediately from this theorem.

Corollary (see [Gusein-Zade 77]). *The monodromy group of an isolated singularity acts transitively on the set of vanishing cycles in the homology of the local nonsingular level variety V_λ, that is, for any two vanishing cycles Δ and Δ' there exists an element of the monodromy group of the singularity that maps Δ into $\pm\Delta'$. If n is odd, $\pm\Delta'$ here can be replaced by Δ'.*

3.6. The equivalent formulations.

The monodromy representation of the group $\pi_1(D_\delta \setminus \Sigma(F))$ was defined earlier for a fibre bundle whose fibre over a point λ is the variety V_λ. As in § 2.2, together with this fibre bundle we can consider a few more bundles, whose fibres over λ are respectively the set $B \setminus V_\lambda$, the pair (B, V_λ), the pair $(B, V_\lambda \cup \partial B)$, and the pair $(B \setminus V_\lambda, \partial B \setminus V_\lambda)$.

Proposition 7. *Propositions 5 and 6 from § 2 remain valid if in the first of them we change the lower index δ to v (where v is the distinguished point in $D_\delta \setminus \Sigma(F)$ and vertical arrows are the operators* Var *along arbitrary fixed loop in $D_\delta \setminus \Sigma(F)$), and in the diagram (7) we change t to λ (where λ is an arbitrary point of the set $D_\delta \setminus \Sigma(F)$) so that it becomes the following diagram (7′):*

$$\begin{array}{ccccc}
\tilde{H}_n(B, V_\lambda) & \otimes & \tilde{H}_n(B \setminus V_\lambda, \partial B \setminus \partial V_\lambda) & \to & \mathbf{Z} \\
\downarrow \partial & & \uparrow t & & \updownarrow \\
\tilde{H}_{n-1}(V_\lambda) & \otimes & \tilde{H}_{n-1}(V_\lambda, \partial V_\lambda) & \to & \mathbf{Z} \\
\downarrow t & & \uparrow \partial & & \updownarrow \\
\tilde{H}_n(B \setminus V_\lambda) & \otimes & \tilde{H}_n(B, \partial B \cup V_\lambda) & \to & \mathbf{Z}.
\end{array} \tag{7'}$$

The assertion about Proposition 5 follows immediately from the definitions of the operator Var and the Leray tube operator; let us prove the assertion about Proposition 6.

For the generators of the group $\tilde{H}_n(B, V_\lambda)$ we can take the classes of cones over the vanishing cycles: if Δ_α is an oriented sphere in V_λ which vanishes over a critical value α after the transportation over some path in $\mathbf{C}^1$, then the corresponding cone ("Lefschetz thimble") δ_α is the chain swept by this cycle during this transportation and oriented in such a way that the chosen orientation of the sphere coincides with the orientation of it as for the boundary of δ_α.

To prove the assertion about the upper two rows of the diagram (7′), it is sufficient to show that for any such cone δ_α and any relative cycle $\nu \subset V_\lambda$ there holds the identity $\langle \delta_\alpha, t(\nu)\rangle = (-1)^n \langle \Delta_\alpha, \nu\rangle$. We can suppose that Δ_α and ν are transversal in V_λ. The function f_λ allows us to identify a tubular neighbourhood of V_λ in B with the direct product of V_λ and a small disc centred at the point 0 of the space $\mathbf{C}^1$ of values of this function. Let w be the coordinate in this space. Without loss of generality we can assume that the path of the distinguished system coincides close to 0 with the axis $\{\mathrm{Re}\, w = 0, \mathrm{Im}\, w > 0\}$, and δ_α and $t(\nu)$ coincide with the direct products of Δ_α and this axis and of ν and the small circle $\{\varepsilon e^{is}\}$; by the definitions of the cycle δ_α and the tube operation, the axis must be oriented towards the decrease of Im w, and the circle towards the increase of the parameter s. To each intersection point of the cycles Δ_α and ν there corresponds a unique intersection point of δ_α and $t(\nu)$, which lies in the manifold $f_\lambda^{-1}(\varepsilon \cdot i) \cap B$. The obvious counting of their orientations at this point proves the assertion. A similar assertion about the lower two rows of the diagram (7′) is proved in exactly the same way; here the generators of the group $\tilde{H}_n(B, \partial B \cup V_\lambda)$ can be swept by the generators of the group $\tilde{H}_{n-1}(V_\lambda, \partial V_\lambda)$ in transporting them over any path connecting the value 0 with the "infinity" in the set of noncritical values of f_λ.

3.7. On the deformations of real singularities.

All definitions from subsection 3.4 can be carried out word-for-word in the case of the singularities of real variables: it is only necessary to change $\mathbf{C}$ everywhere to $\mathbf{R}$. Instead of the holomorphic functions and deformations here it is possible to consider similar C^∞-smooth objects.

If $f : (\mathbf{R}^n, 0) \to (\mathbf{R}, 0)$ is a real singularity, then for any nonnegative integers p, q the (p,q)-stabilization $f^{(p,q)}$ of f is defined: this is the function of $n+p+q$ variables

$$f^{(p,q)}(x_1, \ldots, x_n, u_1, \ldots, u_{p+q}) =$$

$$= f(x_1, \ldots, x_n) - u_1^2 - \cdots - u_p^2 + u_{p+1}^2 + \cdots + u_{p+q}^2.$$

Also, if $F = F(x, \lambda)$ is a deformation of f, then its (p,q)-stabilization $F^{(p,q)}$ is a deformation of $f^{(p,q)}$ defined by the formula

$$F^{(p,q)}(x, u, \lambda) \equiv F(x, \lambda) - u_1^2 - \cdots - u_p^2 + u_{p+1}^2 + \cdots + u_{p+q}^2.$$

§ 4. Intersection form and complex conjugation in the vanishing homology of real singularities in two variables

4.1. The method of Gusein–Zade and A'Campo of calculating the Dynkin diagrams of real singularities.

Definition 1. A holomorphic function $f : \mathbf{C}^n \to \mathbf{C}$ is called *real* if it takes real values on the subspace $\mathbf{R}^n \subset \mathbf{C}^n$.

S.M. Gusein–Zade and, somewhat later, N. A'Campo developed an effective method of calculating the intersection matrices of vanishing cycles for the real singularities of two variables (and of multidimensional singularities stably equivalent to them), see [Gusein-Zade 74] and [A'Campo 75]. Here we describe this method.

Definition 2. A holomorphic function $f : (\mathbf{C}^2, \mathbf{R}^2, 0) \to (\mathbf{C}, \mathbf{R}, 0)$ with an isolated singularity at 0 is *completely real* if all irreducible local components of the complex curve $f^{-1}(0)$ are invariant under complex conjugation. A (nonstrict) morsification $\tilde{f}$ of f is called a *sabirization* if it also takes only real values on $\mathbf{R}^2$, all its $\mu(f)$ critical points lie in $\mathbf{R}^2$, and the critical values of $\tilde{f}$ at all saddlepoints (respectively, minima or maxima) are equal to 0 (respectively, are negative or positive).

Theorem 1 (see [Gusein-Zade 74′], [A'Campo 75]). *Any isolated completely real singularity* $f : (\mathbf{C}^2, \mathbf{R}^2, 0) \to (\mathbf{C}, \mathbf{R}, 0)$ *admits a sabirization.*

Suppose that f is a real isolated singularity and $\tilde{f}'$ a sabirization of it. Let $\tilde{f}$ be a strict morsification of f that is very close to $\tilde{f}'$, so that all critical values of $\tilde{f}$ at the saddlepoints are again more (respectively, less) than the values at all minima (respectively, maxima), the critical points of $\tilde{f}_1$ are in obvious correspondence with the critical points of $\tilde{f}'$ from which they were obtained by this small perturbation, and the flow grad $\tilde{f}$ is topologically equivalent to that for grad $\tilde{f}'$, in particular, has the same graph of separatrices connecting different critical points.

Let us choose a distinguished system of paths connecting the noncritical value δ with the critical values of $\tilde{f}$ in such a way that they completely (except only for the endpoints) lie in the domain where Im > 0; see Figure 11. The elements of the coresponding distinguished basis in the one-dimensional homology of the manifold $\tilde{V}_\delta \equiv \tilde{f}^{-1}(\delta) \cap B$ are in obvious correspondence with some basis elements in the similar homology group $H_1(\tilde{V}'_\delta)$, where $\tilde{V}'_\delta \equiv \tilde{f}'^{-1}(\delta) \cap B$: this correspondence is realized by the obvious isotopy of these close manifolds. Thus some basis of vanishing cycles in the latter homology group is also well defined. The intersection indices of these basis vanishing cycles in the manifolds $\tilde{V}_\delta$, $\tilde{V}'_\delta$ can be calculated in

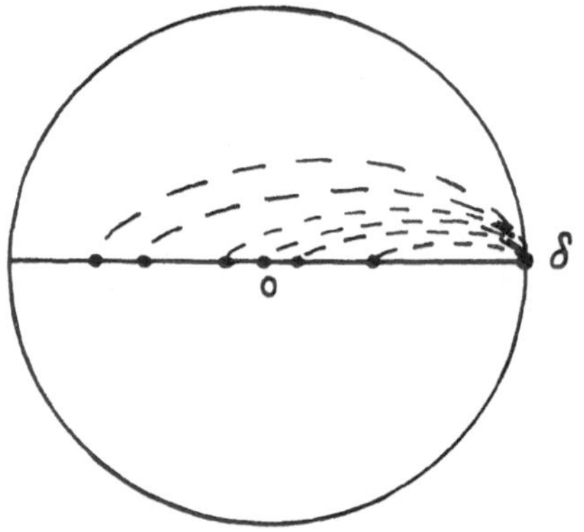

Fig. 11. Distinguished system of paths for a real morsification with real critical values

the following way in terms of the topological picture of the gradient flow of either of the functions $\tilde{f}$ or $\tilde{f}'$.

From any saddlepoint of $\tilde{f}$ two separatrices of the gradient vector-field go up, and two more separatrices go down.

Theorem 2 (see [Gusein–Zade 74], [A'Campo 75]). *The intersection index of two vanishing cycles in $\tilde{V}_\delta$, defined by the paths of our distinguished system and vanishing at two minima (respectively, at two maxima, two saddlepoints), is equal to 0.*

For some canonical choice of orientations of the vanishing cycles, the intersection index of two cycles, the first of which vanishes at a saddlepoint and the second at a minimum (respectively, maximum), is equal to the number of separatrices (respectively, minus the number of separatrices) going from this saddlepoint to this minimum (maximum), while the intersection index of two cycles, the first of which vanishes at a minimum point and the second at a maximum point, is equal to the number of connected components (in the space of orbits of the flow grad $\tilde{f}$*) of such orbits that go from this minimum point to this maximum point.*

Example. Let $f = x^3 + y^4$. Let us choose a sabirization $\tilde{f}$ of f in the form $x^3 - 3\alpha^4 x + y^4 - 4\alpha^3 y^3 + 2\alpha^6$. The topological picture of its zero level set is shown in Figure 12. Hence, by the previous theorem (and the stabilization theorem from § 3.3) the Dynkin diagram of the singularity $\tilde{f}$ (in the described distinguished basis) is as shown in Figure 13.

Proof of Theorem 2. Let ν be a negative number that is greater than all critical values of $\tilde{f}$ at the minimum points, but less than the values at the saddlepoints.

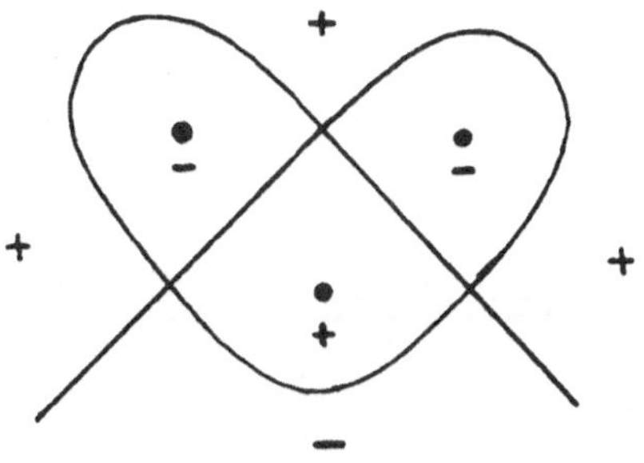

Fig. 12. A sabirization of $f = x^3 + y^4$

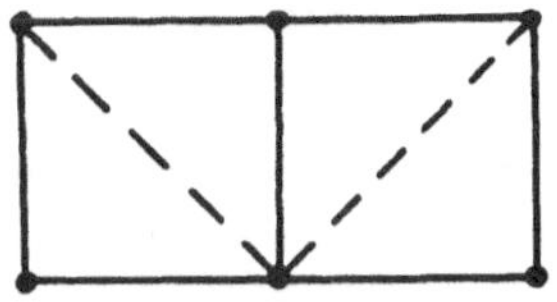

Fig. 13. Dynkin diagram for the morsification of Figure 12

In the same way as we have defined the vanishing cycles in the homology of the manifold $\tilde{V}_\delta$, we can define the vanishing cycles in the homology of $\tilde{V}_\nu \equiv \tilde{f}^{-1}(\nu) \cap B$; here we suppose again that the distinguished system of paths that determines these cycles lies in the half-plane Im > 0; see Figure 14.

This system of paths can be obtained from the one shown in Figure 11 by a continuous homotopy, hence the intersection indices of the cycles defined by them in $\tilde{V}_\nu$ coincide with the intersection indices of the corresponding (i.e. vanishing at the same critical points) cycles in $\tilde{V}_\delta$, defined by the paths from Figure 11.

The real trajectories of the field grad $\tilde{f}$ that go up from a minimum point cut a circle in the manifold $\tilde{V}_\nu$. It is easy to see that the homology classes of these circles (taken with appropriate orientations) coincide with the classes of the cycles vanishing at the corresponding minimum points. In particular, the intersection indices of these vanishing cycles are equal to 0 : indeed, these circles do not intersect in $\tilde{V}_\nu$.

Let us fix once and for all an orientation in $\mathbf{R}^2$, and let us orient these circles in such a way that at any point of them the frame {grad $\tilde{f}$; positively oriented tangent vector } is positively oriented in $\mathbf{R}^2$.

The cycles in $\tilde{V}_\nu$ that vanish at the saddlepoints can be realized in the following way. Close to the saddlepoint, $\tilde{f}$ can be written in the appropriate coordinates in the form $x^2 - y^2 + \alpha$; these coordinates can be chosen in such a way that the frame $\{\partial/\partial x, \partial/\partial y\}$ is positively oriented. Let the number ν be less than α but very

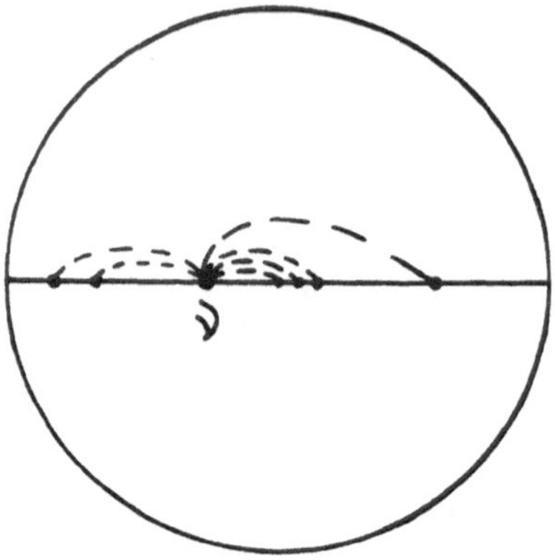

Fig. 14

close to α. Then for a cycle in $\tilde{V}_\nu$, vanishing at our saddlepoint, we can take the parametrized circle

$$x(t) = (\alpha - \nu)^{1/2} i \cdot \sin t, \quad y(t) = (\alpha - \nu)^{1/2} \cos t, \tag{16}$$

$t \in [0, 2\pi]$. These circles are contained in small neighbourhoods of the corresponding saddlepoints, in particular, their intersection indices are equal to 0.

Any of them intersects the plane $\mathbf{R}^2$ in exactly two points, corresponding to the parameter values $t = 0$ and $t = \pi$. These points intersect exactly the components of the set $\tilde{V}_\nu \cap \mathbf{R}^2$ into which there enter the lower separatrices of the field grad $\tilde{f}$ issuing from our saddlepoint. This (after a trivial counting of the intersection indices) proves the part of Theorem 2 concerning the intersection indices of cycles vanishing at the minima and saddlepoints.

The part of the theorem concerning the intersection indices of cycles that vanish at the maxima and the saddlepoints can be proved in exactly the same way. Namely, we choose the number η slightly more than all the critical values at the saddlepoints (but less than the values at all maxima). The further calculations completely repeat the previous ones. Here we realize the cycles in $\tilde{V}_\eta$ vanishing at the saddlepoints by circles of the form

$$x(t) = (\eta - \alpha)^{1/2} \cos t, \; y(t) = (\eta - \alpha)^{1/2} i \cdot \sin t. \tag{17}$$

For these calculations to be concordant with the previous ones, we need to prove that our cycles in $\tilde{V}_\nu$ and $\tilde{V}_\eta$ are concordant, i.e. that the identification of the groups $H_1(\tilde{V}_\nu)$ and $H_1(\tilde{V}_\eta)$ given by the Gauss–Manin connection over a small arc, connecting the points ν and η in the half-plane $\{\text{Im} > 0\}$, takes the class of the circle

(16) into the class of the circle (17). This can easily be proved, using the fact that the transportation of our cycles can be realized in arbitrarily small neighbourhoods of our saddlepoints: in particular, in the neighbourhoods in which $\tilde{f}$ admits the local representation $x^2 - y^2 + \alpha$.

Now we only need to prove the part of the theorem concerning the intersections of cycles vanishing at the minima with cycles vanishing at the maxima. This part follows immediately from the previous ones by means of the following new concept.

4.2. Complex conjugation in the vanishing homology of a real singularity.

Let f be a real singularity, $f : (\mathbf{C}^n, \mathbf{R}^n, 0) \to (\mathbf{C}, \mathbf{R}, 0)$, and $\tilde{f}$ its real morsification. Let ν be a noncritical real value of the function $\tilde{f}$; then the complex conjugation maps the manifold $\tilde{V}_\nu$ into itself and, in particular, defines an automorphism of the group $\tilde{H}_{n-1}(\tilde{V}_\nu)$. Obviously, this automorphism σ is an involution, and hence the space $\tilde{H}_{n-1}(\tilde{V}_\nu; Q) \equiv \tilde{H}_{n-1}(\tilde{V}_\nu) \otimes Q$ splits into the direct sum of two subspaces, H_+ and H_-, on which σ acts trivially (respectively, as multiplication by -1).

Proposition 1. *If n is even, then the subspaces H_+ and H_- are isotropic with respect to the (skew) intersection form in the homology of $\tilde{V}_\nu$ (i.e. the restriction of the intersection form on any of these subspaces is trivial). If n is odd, then the spaces H_+ and H_- are orthogonal with respect to the (symmetric) intersection form.*

This follows immediately from the definitions and from the fact that in the case of odd n the complex conjugation preserves the complex orientation of the manifold $\tilde{V}_\nu$ (with respect to which the intersection index is calculated), and in the case of even n it changes this orientation.

Again let $\tilde{f}$ be a morsification of f that is very close to a sabirization of it, $n = 2$, and ν a noncritical real value that is a little less than the (zero) critical values of $\tilde{f}$ at all saddlepoints, but more than the values at the minima. See Figure 14. It is easy to see that the cycles in the manifold $\tilde{V}_\nu$ that vanish at the minimum points along the paths shown in this picture belong to the space H_+, and the cycles vanishing at the saddlepoints belong to H_-. The cycles vanishing at the maxima do not belong to either H_+ or H_-, but for any such cycle γ the cycle $\gamma + \sigma(\gamma)$ belongs to H_+. In particular, if Δ is a cycle vanishing at a minimum, then by Proposition 1 $\langle \Delta, \gamma + \sigma(\gamma) \rangle = 0$. But the cycle $\sigma(\gamma)$ can be realized as the cycle vanishing at the same maximum point along the path that is complex conjugate to the path along which there vanishes the cycle γ. Deforming this path in an appropriate way in the complement to the set of critical values of $\tilde{f}$, we see that the cycle $\sigma(\gamma)$ can be obtained from the cycle γ by the Gauss–Manin connection along a loop with beginning and end at the point ν, which goes firstly in the half-plane $\{\text{Im} > 0\}$ along the real axis, then intersects the axis at a point separating the set of critical values of $\tilde{f}$ at all the saddlepoints and the set of values at the maxima, and finally returns to ν in the half-plane $\{\text{Im} < 0\}$. By the Picard–Lefschetz formula and the statements of Theorem 2 proved above,

$$\sigma(\gamma) = \gamma + \sum_i \langle \gamma, \Delta_i \rangle \Delta_i ,$$

where the summation is taken over all saddlepoints of $\tilde{f}$, and Δ_i are the cycles vanishing at these saddlepoints. Substituting this formula in the identity $\langle \Delta, \gamma + \sigma(\gamma) \rangle = 0$, we find that the intersection index $\langle \Delta, \gamma \rangle$ is equal to half the number of triplets consisting of

1) a saddlepoint of $\tilde{f}$,

2) an upper separatrix of grad $\tilde{f}$, going from this saddlepoint to the maximum point at which the cycle γ vanishes, and

3) a lower separatrix, going from the same saddlepoint to the minimum point at which the cycle Δ vanishes.

But half of the number of such triplets is just the number of connected components of the set of trajectories of the gradient field, going from our minimum point to our maximum point: indeed, the domain in $\mathbf{R}^2$ swept out by such a connected set of trajectories is bounded exactly by two pairs of separatrices as in the definition of a triple above. Theorem 2 is completely proved.

4.3. The relative cycles defined by the real points of distant positive and negative levels.

The same consideration allows us to calculate the intersection indices of the vanishing cycles with two important relative cycles in $\tilde{V}_\nu$. Namely, let $\tilde{f}$ be the same morsification of f as above, and suppose that the real noncritical value ν (respectively, η) is less (respectively, greater) than all critical values of $\tilde{f}$. Denote by $\Pi(\nu)$ the element of the group $\tilde{H}_{n-1}(\tilde{V}_\nu, \partial \tilde{V}_\nu)$ determined by the curve $\tilde{V}_\nu \cap \mathbf{R}^2$ oriented in such a way that the frame {grad $\tilde{f}$, (the tangent orienting vector of this curve)} defines the positive orientation in $\mathbf{R}^2$. In a similar way the element $\Pi(\eta) \in \tilde{H}_{n-1}(\tilde{V}_\eta, d\tilde{V}_\eta)$ is defined. Choose a distinguished system of paths connecting the point ν (respectively, η) with the critical values in such a way that they lie completely (except for the endpoints) in the half-plane $\{\text{Im} > 0\}$.

Theorem 3. *The intersection index of the cycle $\Pi(\nu)$ with the cycles vanishing along this system of paths at a critical point is equal to 0 if this point is a minimum; if this is a saddlepoint, then this index is equal to minus the number of lower* *separatrices of the field* grad $\tilde{f}$ *going from this saddlepoint and not finishing at any minimum point; if this is a maximum point, then this index is equal to the number of connected components of the set of trajectories of the field* grad $\tilde{f}$ *going down from this maximum and not entering any minimum point.*

The intersection indices of $\Pi(\eta)$ with the vanishing cycles can be described in exactly the same way, apart from changing all minima to maxima, lower separatrices to upper, and vice versa.

The proof repeats that of Theorem 2.

§ 5. Classification of real and complex singularities of functions

5.1. Simple singularities.

Denote by $\mathcal{M}^2$ the space of germs of C^∞ functions $(\mathbf{R}^n, 0) \to (\mathbf{R}, 0)$ having a singularity at the point 0 (i.e. $df(0) = 0$). Denote by D_0 the group of germs of smooth diffeomorphisms $(\mathbf{R}^n, 0) \to (\mathbf{R}^n, 0)$. The group D_0 acts in an obvious way on the space $\mathcal{M}^2$. The most standard classification of singularities is the classification up to this action. Two singularities of functions of n variables, belonging to the same orbit of this action, are called *D_0-equivalent.*

Two singularities of function germs $(\mathbf{R}^n, a) \to \mathbf{R}$, $(\mathbf{R}^n, b) \to \mathbf{R}$ are *D-equivalent* if they are taken into each other by a germ of the diffeomorphism $(\mathbf{R}^n, a) \leftrightarrow (\mathbf{R}^n, b)$.

Here is a few more rough equivalence relation that allows us to identify the singularities of different numbers of variables.

Definition 1. Two singularities $f : (\mathbf{R}^n, 0) \to (\mathbf{R}, 0)$ and $g : (\mathbf{R}^m, 0) \to (\mathbf{R}, 0)$ are called *stably D_0-equivalent* if they become D_0-equivalent after adding some non-degenerate quadratic forms of additional variables.

For example, the functions $x^3 + y^2$, $2x^3 - y^2$ of two variables x, y are stably equivalent to each other and to the singularity x^3 of one variable, but are not stably equivalent to the function x^3 considered as a function of two variables.

It is easy to see that stably equivalent singularities have the same Milnor numbers.

The most important singularities are the so-called *simple singularities.* These singularities can be reduced by the action of the group D_0 to one of the normal forms given in the Table 1. The symbols A_k, D_k, E_k in the left column of the table are the standard notation for the corresponding classes of simple singularities. The

Table 1. Normal forms of real simple singularities

Notation	Normal form	corank
A_1	Q	0
$A_{2k};\quad k \geq 1$	$x_1^{2k+1} + Q$	1
$\pm A_{2k-1};\quad k \geq 2$	$\pm(x_1^{2k} + Q)$	1
$D_{2k}^-;\quad k \geq 2$	$x_1^2 x_2 - x_2^{2k-1} + Q$	2
$D_{2k}^+;\quad k \geq 2$	$x_1^2 x_2 + x_2^{2k-1} + Q$	2
$\pm D_{2k+1};\quad k \geq 2$	$\pm(x_1^2 x_2 - x_2^{2k} + Q)$	2
$\pm E_6$	$\pm(x_1^3 + x_2^4 + Q)$	2
E_7	$x_1^3 + x_1 x_2^3 + Q$	2
E_8	$x_1^3 + x_2^5 + Q$	2

lower indices in this notation are the (common) Milnor numbers of the singularities of these classes. The number given in the third column of the table is the *corank* of the singularity, i.e. the deficiency of the quadratic form defined by the 2-jet of the singularity; this number can be defined also as the minimal number of variables on which there can depend a singularity stably equivalent to the singularities of this class. In all the normal forms of Table 1, Q denotes a nondegenerate quadratic form on the additional variables (which do not appear in the normal form before Q); for this quadratic form we always can take

$$\pm x_{c+1}^2 \pm \cdots \pm x_n^2 ,$$

where the signs $+$ or $-$ can be chosen arbitrarily, and c is the corank of f.

Proposition 1. *The codimension of the set of all nonsimple singularities in the space $\mathcal{M}^2$ is equal to 6 if $n \geq 3$, is equal to 7 if $n = 2$, and is infinite if $n = 1$.*

If we consider the similar classification of complex singularities, then several classes of Table 1 merge into the same class; see Table 2. In this table, Q always denotes the sum of the squares of additional variables, $Q = +x_{c+1}^2 + \cdots + x_n^2$.

It follows from Theorem 7 of § 3 that the versal deformations of the simple singularities (real or complex) can be chosen in the form shown in the Table 3. (In this table f_0 everywhere denotes the original simple singularity, written in the corresponding normal form of Table 1 or 2.)

The simple singularities are distinguished among all other singularities by several nice properties; here are some of them.

Definition 2. The *modality* of a singularity $f \in \mathcal{M}^2$ is the dimension of the space of orbits of the group D_0 in a sufficiently small neighbourhood of f in $\mathcal{M}^2$ (or,

Table 2. Normal forms of complex simple singularities

Notation	Normal form	corank
A_1	Q	0
A_k	$x_1^{k+1} + Q\,;\quad k \geq 2$	1
D_k	$x_1^2 x_2 + x_2^{k-1} + Q\,;\quad k \geq 4$	2
E_6	$x_1^3 + x_2^4 + Q$	2
E_7	$x_1^3 + x_1 x_2^3 + Q$	2
E_8	$x_1^3 + x_2^5 + Q$	2

Table 3. Versal deformations of simple singularities

A_k	$f_0 + \lambda_1 + \lambda_2 x_1 + \cdots + \lambda_k x_1^{k-1}$
D_k	$f_0 + \lambda_1 + \lambda_2 x_1 + \lambda_3 x_2 + \lambda_4 x_2^2 + \cdots + \lambda_k x^{k-2}$
E_6	$f_0 + \lambda_1 + \lambda_2 x_1 + \lambda_3 x_2 + \lambda_4 x_1 x_2 + \lambda_5 x_2^2 + \lambda_6 x_1 x_2^2$
E_7	$f_0 + \lambda_1 + \lambda_2 x_1 + \lambda_3 x_2 + \lambda_4 x_1 x_2 + \lambda_5 x_2^2 + \lambda_6 x_2^3 + \lambda_7 x_2^4$
E_8	$f_0 + \lambda_1 + \lambda_2 x_1 + \lambda_3 x_2 + \lambda_4 x_1 x_2 + \lambda_5 x_2^2 + \lambda_6 x_1 x_2^2 + \lambda_7 x_2^3 + \lambda_8 x_1 x_2^3$

equivalently, the smallest number m such that in some neighbourhood of $\mathcal{M}^2$ there is no $(m+1)$-parameter family of pairwise D_0-nonequivalent singularities).

Obviously, the modality is the same for any two D_0-equivalent singularities.

Theorem 1 (see [Arnold 72]). *The singularities of modality* 0 *of functions* $\mathbf{R}^n \to \mathbf{R}$ *(respectively,* $\mathbf{C}^n \to \mathbf{C}$*) are exactly the simple singularities given in Table 1 (respectively, in Table 2). All isolated singularities whose Milnor numbers are less than 8 are simple.*

Proposition 2. *For odd n, the quadratic form defined by the intersection indices in the vanishing homology group of an isolated singularity of n variables is definite if and only if this singularity is simple. (For a simple singularity, this form is positive definite if* $n \equiv 1 (mod\ 4)$ *and negative definite if* $n \equiv 3 (mod\ 4)$*.)*

Proposition 3. *In the appropriate distinguished basis, the Dynkin diagram of a simple singularity of the class* A_k, D_k *or* E_k *coincides with the classical Dynkin diagram of the simple Lie algebra having the same notation (see* [Bourbaki 68]*); this justifies the notation of these classes.*

Proposition 4 (see [A'Campo 75′]). *If the Dynkin diagram defined by some distinguished basis of an isolated singularity has no cycles, then this singularity is simple.*

Proposition 5. *For odd n, the monodromy group of an isolated singularity f of n variables is a finite subgroup of Aut* $\tilde{H}_{n-1}(\tilde{V}_\nu)$ *(see § 3) if and only if the singularity f is simple. For simple f, this subgroup is isomorphic to the Weyl group of the simple Lie algebra with the same notation* A_k, D_k *or* E_k *as the class of our singularity; see Table 2.*

Remark 1. When we work with simple singularities and their canonical deformations given by the formulae of Tables 1, 2 and 3, we can forget the requirement (which appears in the definitions of the singularity, of the Milnor fibre, and of the versal deformation) that everything is happening close to the origin. This is related to the following properties of these singularities and deformations.

Definition 3. A function $f : (\mathbf{R}^n, 0) \to (\mathbf{R}, 0)$ or $(\mathbf{C}^n, 0) \to (\mathbf{C}, 0)$ is called *quasihomogeneous* of degree d with weights $\alpha_1, \ldots, \alpha_n$ (where all α_i are positive numbers) if for any $t \in \mathbf{R}^1$ or $\mathbf{C}^1$ it satisfies the condition

$$t^d f(x_1, \ldots, x_n) = f(t^{\alpha_1} x_1, \ldots, t^{\alpha_n} x_n). \tag{18}$$

Here is an equivalent definition. Consider in the space $\mathbf{Z}^n_+$ the set of exponents of all nonzero monomials of the function f. This function is quasihomogeneous of degree d with weights α_i if and only if all these exponents $K = (k_1, \ldots, k_n)$ belong to the affine hyperplane in $\mathbf{Z}^n$ given by the condition

$$\alpha_1 k_1 + \cdots + \alpha_n k_n = d.$$

The geometrical properties of a quasihomogeneous function in the whole space $\mathbf{C}^n$ or $\mathbf{R}^n$ can be reduced to properties in an arbitrarily small neighbourhood of the origin in this space: indeed, they can be reduced to one another by the one-parameter group of quasihomogeneous dilations of the argument and target spaces. (The action of this group corresponding to the value t of the parameter is defined by the formula

$$T_t : (x_1, \ldots, x_n;\ y) \to (t_1^\alpha x_1, \ldots, t_n^\alpha x_n;\ t^d y), \tag{19}$$

where y is the coordinate in the line of values.)

Obvoiusly, all normal forms of Tables 1 and 2 are quasihomogeneous. Moreover, the normal forms of Table 3 for the deformations, considered as functions of x and λ, are also quasihomogeneous, therefore in the case of simple singularities the parameters of (these distinguished) deformations can be assumed to be arbitrary (and not only "sufficiently small").

5.2. Unimodal singularities. Proper modality.

The next (in decreasing importance) collection of singularities consists of the *unimodal* singularities, i.e. singularities of modality 1. This collection (in the real case)

Table 4. Real parabolic singularities

Notation	Normal form	Restrictions
P_8^1	$x_1^3+\alpha x_1x_2x_3+x_2x_3^2+x_2^2x_3+Q$	$\alpha>-3$
P_8^2	$x_1^3+\alpha x_1x_2x_3+x_2x_3^2+x_2^2x_3+Q$	$\alpha<-3$
$\pm X_9$	$\pm(x_1^4+\alpha x_1^2x_2^2+x_2^4+Q)$	$\alpha>-2$
X_9^1	$x_1x_2(x_1^2+\alpha x_1x_2+x_2^2)+Q$	$\alpha^2<4$
X_9^2	$x_1x_2(x_1+x_2)(x_1+\alpha x_2)+Q$	$\alpha\in(0,1)$
J_{10}^3	$x_1(x_1-x_2^2)(x_1-\alpha x_2^2)+Q$	$\alpha\in(0,1)$
J_{10}^1	$x_1(x_1^2+\alpha x_1x_2^2+x_2^4)+Q$	$\alpha^2<4$

consists of the classes of singularities whose (one-parameter) normal forms are listed in the Tables 4 – 7 (see [AVG 82]).

Definition 4. A singularity f is called *elliptic* (respectively, *parabolic, hyperbolic*) if the intersection form in the vanishing homology group of any singularity stably equivalent to f and depending on $n\equiv 3(mod\ 4)$ variables is negative definite (respectively, semidefinite, has positive inertia index equal to 1).

By Proposition 2, the elliptic singularities are exactly the simple singularities.

Proposition 6 (see [AVG 82]). *All parabolic and hyperbolic singularities are unimodal. All parabolic real singularities are listed in Table 4, and all hyperbolic are listed in Tables 5, 6. All unimodal singularities that are neither parabolic nor hyperbolic belong to 14 exceptional families listed in Table 7.*

(In this Table 7, the singularities E_{12} through W_{13} are of corank 2, and singularities Q_{10} through U_{12} are of corank 3.)

In this table the index 1 or 2 over P_8 denotes the number of connected components of the curve in $\mathbf{R}P^2$, defined by the corresponding cubic form, the index 2 or 1 over X_9 denotes the number of pairs of lines in $\mathbf{R}^2$ on which the corresponding quartic form vanishes, and 1 or 3 over J_{10} denotes the number of real plane curves on which the corresponding function of two variables is equal to 0.

The versal deformations of the parabolic singularities (of Table 4) can be chosen in the following form:

Definition 5. The μ = const *stratum* of the singularity f is the connected component of the set of singularities in $\mathcal{M}^2$ that contains f and has the same Milnor number as f. The *proper modality* of the singularity f is the dimension of the set of parameters λ of its miniversal deformation such that the function f_λ has a critical point with critical value 0 and with Milnor number equal to $\mu(f)$.

Table 5. Hyperbolic unimodal singularities of corank 2

Notation	Normal form	Restrictions
$\pm J_{2k-1}$	$\pm(x_1^3 \pm x_1^2x_2^2 + \alpha x_2^{2k-5} + Q)$	$\alpha > 0, k \geq 6$
$\pm J_{2k}^3$	$\pm(x_1^3 \pm x_1^2x_2^2 + \alpha x_2^{2k-4} + Q)$	$\alpha < 0, k \geq 6$
$\pm J_{2k}^1$	$\pm(x_1^3 \pm x_1^2x_2^2 + \alpha x_2^{2k-4} + Q)$	$\alpha > 0, k \geq 6$
$\pm Y_{2k,2l}^{+,+}$	$\pm(\alpha x_1^2x_2^2 + x_1^{2k} + x_2^{2l} + Q)$	$\alpha > 0, k \geq 2 < l$
$\pm Y_{2k,2l}^{+,-}$	$\pm(\alpha x_1^2x_2^2 + x_1^{2k} - x_2^{2l} + Q)$	$\alpha > 0, k \geq 2 \leq l, k+l \geq 5$
$\pm Y_{2k,2l}^{-,-}$	$\pm(\alpha x_1^2x_2^2 - x_1^{2k} - x_2^{2l} + Q)$	$\alpha > 0, k \geq 2 < l$
$\pm Y_{2k,2l+1}^{+}$	$\pm(\alpha x_1^2x_2^2 + x_1^{2k} + x_2^{2l+1} + Q)$	$\alpha > 0, k \geq 2 \leq l$
$\pm Y_{2k,2l+1}^{-}$	$\pm(\alpha x_1^2x_2^2 - x_1^{2k} + x_2^{2l+1} + Q)$	$\alpha > 0, k \geq 2 \leq l$
$\pm Y_{2k+1,2l+1}$	$\pm(\alpha x_1^2x_2^2 + x_1^{2k+1} + x_2^{2l+1} + Q)$	$\alpha > 0, k \geq 2 \leq l$
$\pm \tilde{Y}_{k,\mathbf{R}}$	$\pm[(x_1^2 + x_2^2)^2 + \alpha x_1^k + Q]$	$\alpha \neq 0, k \geq 5$

Table 6. Hyperbolic unimodal singularities of corank 3

Notation	Normal form	Restrictions
$P_{8+k} \equiv T_{3,3,3+k}$	$x_1^3 + x_1^2x_3 + x_2^2x_3 + \alpha x_3^{k+3} + Q$	$\alpha \neq 0, k > 0$
$R_{l,m} \equiv T_{3,l,m}$	$x_1(x_1^2 + x_2x_3) \pm x_2^l \pm \alpha x_3^m + Q$	$\alpha \neq 0, m \geq l > 4$
$\tilde{R}_m \equiv \tilde{T}_{3,m,m}$	$x_1(\pm x_1^2 + x_2^2 + x_3^2) + \alpha x_2^m + Q$	$\alpha \neq 0, m > 4$
$T_{p,q,r}$	$\alpha x_1x_2x_3 \pm x_1^p \pm x_2^q \pm x_3^r + Q$	$\alpha \neq 0,$ $p^{-1} + q^{-1} + r^{-1} < 1$
$\tilde{T}_{p,m} \equiv \tilde{T}_{p,m,m}$	$x_1(x_2^2 + x_3^2) \pm x^p + \alpha x_2^m + Q$	$\alpha \neq 0,$ $p^{-1} + 2m^{-1} < 1$

Table 7. Exceptional unimodal singularities

E_{12}	$x_1^3 + x_2^7 + \alpha x_1x_2^5 + Q$	W_{13}	$\pm x_1^4 + x_1x_2^4 + \alpha x_2^6 + Q$
E_{13}	$x_1^3 + x_1x_2^5 + \alpha x_2^8 + Q$	Q_{10}	$x_1^3 + x_2^2x_3 \pm x_3^4 + \alpha x_1x_3^3 + Q$
E_{14}	$x_1^3 \pm x_2^8 + \alpha x_1x_2^6 + Q$	Q_{11}	$x_1^3 + x_2^2x_3 \pm x_1x_3^3 + \alpha x_3^5 + Q$
Z_{11}	$x_1^3x_2 + x_2^5 + \alpha x_1x_2^4 + Q$	Q_{12}	$x_1^3 + x_2^2x_3 \pm x_3^5 + \alpha x_1x_3^4 + Q$
Z_{12}	$x_1^3x_2 + x_1x_2^4 + \alpha x_1^2x_2^3 + Q$	S_{11}	$x_3(x_1^2 + x_2x_3) \pm x_2^4 + \alpha x_2^3x_3 + Q$
Z_{13}	$x_1^3x_2 \pm x_2^6 + \alpha x_1x_2^5 + Q$	S_{12}	$x_3(x_1^2 + x_2x_3) + x_1x_3^3 + \alpha x_2^5 + Q$
W_{12}	$\pm x_1^4 + x_2^5 + \alpha x_1^2x_2^3 + Q$	U_{12}	$x_1(x_1^2 \pm x_2^2) \pm x_3^4 + \alpha x_1x_2x_3^2 + Q$

Table 8. Versal deformations of parabolic singularities

P_8	$f_0 + \lambda_0 + \lambda_1 x_1 + \lambda_2 x_2 + \lambda_3 x_3 + \lambda_4 x_1 x_2 + \lambda_5 x_1 x_3 + \lambda_6 x_2 x_3 + \lambda_7 x_1 x_2 x_3$
X_9	$f_0 + \lambda_0 + \lambda_1 x_1 + \lambda_2 x_2 + \lambda_3 x_1^2 + \lambda_4 x_1 x_2 + \lambda_5 x_2^2 + \lambda_6 x_1^2 x_2 + \lambda_7 x_1 x_2^2 +$ $+\lambda_8 x_1^2 x_2^2$
J_{10}	$f_0 + \lambda_0 + \lambda_1 x_1 + \lambda_2 x_2 + \lambda_3 x_1 x_2 + \lambda_4 x_2^2 + \lambda_5 x_1 x_2^2 + \lambda_6 x_2^3 + \lambda_7 x_1 x_2^3 +$ $+\lambda_8 x_2^4 + \lambda_9 x_1^2 x_2^2$

It follows easily from the definitions that the proper modality is equal to $\mu(f)-1$ minus the codimension of the $\mu =$ const stratum of f in $\mathcal{M}^2$.

Theorem 2 (see [Gabrielov 74]). *The modality of a complex isolated singularity is equal to its proper modality.*

A similar statement has not been proved (and is probably wrong) in the case of real singularities (see [Vassiliev & Serganova 91]) and becomes wrong if we slightly change the equivalence relation of the singularities, say, if we classify them up to $\mathcal{K}$-equivalence; see [AVG 82] and Definition 13.3 below.

Example. The parameter spaces of the deformations of parabolic singularities indicated in Table 8 intersect the corresponding $\mu =$ const stratum along a one-dimensional set which is distinguished by the equation $\lambda_0 = \cdots = \lambda_6$ in the case P_8, $\lambda_0 = \cdots = \lambda_7$ in the case X_9, and $\lambda_0 = \cdots = \lambda_8$ in the case J_{10}.

Theorem 3 (see [Looijenga 77]). *The discriminant variety of the deformation from Table 8 of any standard parabolic singularity is topologically trivial along the intersection of the parameter space of this deformation with the* $\mu =$ const *stratum.*

§ 6. Lyashko–Looijenga covering and its generalizations

6.1. Complex Lyashko–Looijenga covering.

Let $f : (\mathbf{C}^n, 0) \to (\mathbf{C}, 0)$ be a holomorphic function with an isolated singularity at 0, and $F : (\mathbf{C}^n \times \mathbf{C}^l, 0) \to (\mathbf{C}, 0)$ its deformation. Any function f_λ (where $\lambda \in \mathbf{C}^l$ is sufficiently close to 0) has $\mu(f)$ critical values at critical points close to the origin (if we count these critical values with multiplicities equal to the sums of the Milnor numbers over all critical points of f_λ having the given value).

Denote by X_μ the set of unordered collections of μ (in general, not distinct) points in $\mathbf{C}^1$. This set admits a natural structure of a smooth complex manifold:

indeed, the *Vieta map* identifies it with the space of all complex polynomials of the form

$$z^{\mu} + a_1 z^{\mu-1} + \cdots + a_{\mu}. \tag{20}$$

Hence, the map $v : (\mathbf{C}^l, 0) \to (X_{\mu}, (0, \ldots, 0))$ is well defined, which maps any λ to the set of critical values of the function f_{λ} (taken with the same multiplicities).

Proposition 1. *The map v is smooth.*

Proof. We can assume that the deformation F is versal: indeed, the map v commutes with the maps of the bases of deformations which induce these deformations from one another.

By Hartogs' lemma, it is sufficient to prove the smoothness only close to points λ such that f_{λ} has μ Morse critical points or $\mu - 2$ Morse points and one point of type A_2: indeed, the functions having other singularities form a set of codimension 2 in $\mathbf{C}^l$.

It follows easily from the definition of versality that a versal deformation of f is also a versal deformation for any critical point close to 0 of the function f_{λ} if λ is close to 0. This reduces our assertion to the study of only two singularities: of types A_1 (Morse) and A_2. These two cases are trivial and the proposition is proved.

Now suppose that f_{λ} has exactly $\mu(f)$ different (and hence Morse) critical points close to 0. Numbering these points in any way, we get a map $\bar{v}$ of a small neighbourhood U of the point λ in $\mathbf{C}^{\mu}$: this map assigns to the point λ' the (naturally numbered) set of critical values of the function $f_{\lambda'}$; these critical values will be denoted by $\nu_1(\lambda'), \ldots, \nu_{\mu}(\lambda')$.

Proposition 2. *If the deformation F is versal, then this map*

$$\bar{v} : U \to \mathbf{C}^{\mu} \tag{21}$$

is a submersion in the domain where it is defined.

Proof. Without loss of generality we can assume that the versal deformation has the form (15). It is easy to check that the partial derivative $\partial \nu_j / \partial \lambda_i$, calculated at the point λ', is equal to the value of the function φ_i at the j-th critical point of the function $F_{\lambda'}$. If our deformation is sufficiently large (say, if among the functions φ_i there are all monomials of degree $\leq \mu$ in $\mathbf{C}^n$), then the desired assertion follows from this fact and from the interpolation theorem.

But any deformation is equivalent to one induced from any miniversal deformation by an appropriate map of parameters; if our large deformation is versal, then this inducing map is a map onto, and commutes with the maps (21) defined in some domains in bases of both deformations. This proves our assertion for an arbitrary miniversal deformation and hence, by Proposition 5 of § 3, also for any versal deformation.

Definition 1. The *bifurcation diagram of functions* of the deformation F is the set of all $\lambda \in \mathbf{C}^l$ such that the function f_λ has fewer than $\mu(f)$ different critical values at the critical points close to the origin in $\mathbf{C}^n$.

The bifurcation diagram of functions of the deformation F is denoted by $\Delta(F)$. It consists of two components: the *caustic*, i.e. the set of λ such that f_λ has a non-Morse critical point, and the *Maxwell set*, i.e. the closure of the set of λ such that the values of f_λ at two different critical points coincide.

Notation. $B(\mathbf{C}^1, \mu)$ is the space of all subsets of cardinality μ in $\mathbf{C}^1$; this set is an open subset in X_μ.

Let $F : (\mathbf{C}^n \times \mathbf{C}^\mu, 0) \to (\mathbf{C}, 0)$ be a miniversal deformation of the singularity f. Proposition 2 implies that the restriction of the map $v : \mathbf{C}^\mu \to X_\mu$ to $\mathbf{C}^\mu \setminus \Delta(F)$ is a submersion, and its image belongs to $B(\mathbf{C}^1, \mu)$.

Theorem 1 (see [Looijenga 74]). *Suppose that the singularity f is simple, and F is its miniversal deformation written in the normal form of Table 3. Then the map v is proper, its restriction to the space $\mathbf{C}^\mu \setminus \Delta(F)$ is a finite covering over $B(\mathbf{C}^1, \mu)$, and the number of its preimages over any point of X_μ is finite and majorized uniformly by that over any point of $B(\mathbf{C}^1, \mu)$. In particular, for any $\lambda \in \mathbf{C}^\mu$ any smooth path in X_μ beginning at the point $v(\lambda)$ can be lifted to a path in $\mathbf{C}^\mu$ beginning at λ.*

Note that this lifted path can be nonsmooth if the lower path meets the discriminant (i.e. not belonging to $B(\mathbf{C}^1, \mu)$) points in X_μ.

Corollary 1. *For any simple singularity the space $\mathbf{C}^\mu \setminus \Delta(F)$ is a space of type $K(\pi, 1)$ for some group π.*

This follows immediately from Theorem 1, from the exact homotopy sequence of a covering, and from the fact that the space $B(\mathbf{C}^1, \mu)$ is a space of type $K(\pi, 1)$ (for the braid group with μ strings).

Corollary 2. *For any choice of the distinguished basis in the homology group of a nonsingular fibre of a Morse perturbation f_λ of a simple singularity, the intersection index of any two different basis vanishing cycles can be equal only to 0, 1 or -1.*

Indeed, let Δ_i and Δ_j be two cycles that vanish over the critical values α_i, α_j. Consider the path in X_μ beginning at $v(\lambda)$, along which all other critical values stay the same, and the values α_i, α_j move along the corresponding paths of the distingushed system towards the common start point of these paths. By the previous theorem, the corresponding path in X_μ can be lifted to a path in $\mathbf{C}^\mu$; the last point of this path is a function $f_{\lambda'}$ having exactly one critical value of multiplicity 2. If this critical value is reached at two different critical points, then the intersection index of the corresponding vanishing cycles is equal to 0. Otherwise we get one critical point with Milnor number equal to 2. By the Classification Theorem 1 of § 5, all such critical points are of class A_2. The intersection index of two vanishing cycles of any morsification of any function of this class is always equal to 1 or -1, and the corollary is proved.

The same statement can be proved by more direct methods, but the considerations used in the previous proof will be useful later.

A statement slightly weaker than that of Theorem 1 holds also for parabolic singularities. However, in this case the space of the covering is the complement of the bifurcation set not in a neighbourhood of the origin in the base of the miniversal deformation of a certain parabolic singularity, but in the whole space of functions of the form indicated in Table 8.

Theorem 2 (see [Jaworski 86]). *For any class of complex parabolic singularities P_8, X_9 or J_{10}, the map v, defined on the complement of the bifurcation diagram of functions in the space of all functions indicated in the corresponding line of Table 8, is a covering over $B(\mathbf{C}^1, \mu)$, where μ is equal to 8, 9 or 10, respectively.*

6.2. The real covering.

Suppose that a singularity f and its deformation F are *real*, i.e. $f(\mathbf{R}^n) \subset \mathbf{R}$ and $F(\mathbf{R}^n \times \mathbf{R}^l) \subset \mathbf{R}$. Then the restriction of the above-defined map v to $\mathbf{R}^l$ maps this space into the space of all unordered collections of μ points in $\mathbf{C}^1$ that are invariant under the action of complex conjugation. This space is naturally isomorphic to the space of *real* polynomials of the form (20), and hence admits a smooth structure; denote this space by $\mathrm{Re}X_\mu$. Again, if f is simple and is given by appropriate normal form of Table 1, and its deformation F is miniversal and is given by the corresponding normal form from Table 3, then the map $\mathrm{Re}\, v : \mathbf{R}^\mu \to \mathrm{Re}X_\mu$ is proper, and it is a covering over any connected component of the space $\mathrm{Re}X_\mu \cap B(\mathbf{C}^1, \mu)$.

§ 7. Complements of discriminants of real simple singularities (after E. Looijenga)

Let $f : (\mathbf{C}^n, \mathbf{R}^n, 0) \to (\mathbf{C}, \mathbf{R}, 0)$ be a real singularity, and $F : (\mathbf{C}^n \times \mathbf{C}^l, \mathbf{R}^n \times \mathbf{R}^l, 0) \to (\mathbf{C}, \mathbf{R}, 0)$ a real deformation of it.

Definition 1. The *real discriminant* of the deformation F is the set of all values of the parameter $\lambda \in \mathbf{R}^l$ such that the function $f_\lambda = F(\cdot, \lambda)$ has critical value 0 at some *real* critical point close to the origin in $\mathbf{R}^n$. This set is denoted by $\Sigma_{\mathbf{R}}(F)$.

If the singularity f (and the deformation F) is sufficiently complicated, then this set does not coincide with the set $\Sigma(F) \cap \mathbf{R}^l$: indeed, a real function can have imaginary critical points.

We present here (following Looijenga) the algebraic counting of connected components of the complements of real discriminants of real simple singularities.

Proposition 1. *Let F be a real versal deformation of a real singularity; then the codimension in $\mathbf{R}^l$ of the difference $(\Sigma(F) \cap \mathbf{R}^l) \setminus \Sigma_{\mathbf{R}}(F)$ is not less than 2 (or this difference is empty). In particular, the path-components of the complements of $\Sigma_{\mathbf{R}}(F)$ and of $\Sigma(F)$ in a small neighbourhood of the origin in $\mathbf{R}^l$ are in obvious one-to-one correspondence.*

This statement follows easily from the fact that for any point λ of this difference the function f_λ has two different critical points with critical value 0; see [Looijenga 78].

Theorem 1 (see [Looijenga 78]). *All components of the complement of the real discriminant of a real versal deformation of a real simple singularity are contractible.*

An enumeration of these components can be given in terms of the monodromy group and complex conjugation in the vanishing homology of these singularities.

The discriminants of a real deformation and of its stabilization (see § 3.7) are naturally identified with each other, therefore we can suppose without loss of generality that the dimension n is odd (say, equal to 3). In this case the intersection form $\langle \cdot, \cdot \rangle$ in the homology of a nonsingular fibre is symmetric and nondegenerate, and the (preserving it) monodromy group of the corresponding complex singularity is a finite subgroup of the orthogonal group $O_\mu = O(\tilde{H}_{n-1}(V_\lambda, \mathbf{Z}); \langle \cdot, \cdot \rangle)$; this subgroup is denoted by W_λ. For any nondiscriminant *real* λ this orthogonal group contains one more remarkable element, the action of complex conjugation; it is denoted by σ_λ. The pair consisting of the group W_λ and the involution σ_λ, considered up to W_λ-conjugations in the orthogonal group, depends only on the component containing λ of the complement of the discriminant, and, as we shall see later, determines this component.

Denote by N_λ the normalizer of the subgroup W_λ in O_μ. The groups W_λ (and hence also N_λ) for different $\lambda \in \mathbf{C}^l \setminus \Sigma(F)$ are isomorphic to each other. This isomorphism for two points λ, λ' can be specified by a path in $\mathbf{C}^l \setminus \Sigma(F)$ connecting λ and λ'. The induced isomorphisms of the quotient groups $N_\lambda/W_\lambda \sim N'_\lambda/W'_\lambda$ do not depend on the choice of the path, hence these quotient groups for all λ can be identified in a canonical way. Denote the resulting object by N/W.

Proposition 2 (see [Looijenga 78]). *1. The quotient group N/W is canonically isomorphic to the group of automorphisms of the standard Dynkin diagram of our* *simple singularity (i.e., it is trivial for singularities A_1, E_7, E_8, and isomorphic to the group $S(3)$ for D_4 and to $\mathbf{Z}_2$ in other cases).*

2. For any $\lambda \in \mathbf{R}^l \setminus \Sigma(F)$ the element σ_λ belongs to the group N_λ, and its class in the quotient group N/W is the same for all $\lambda \in \mathbf{R}^l \setminus \Sigma(F)$. In particular, the class $\{W\sigma\} \in N/W$ is well defined.

With any real point λ of the complement of the discriminant we associate the class containing σ of equivalent (up to W_λ-conjugation) involutions of the space $\tilde{H}_{n-1}(V_\lambda, \mathbf{Z})$ that belong to the coset $W\sigma \subset N$. For all λ from the same component of the complement of the discriminant these classes are naturally identified with each other.

Theorem 2 (see [Looijenga 78]). *For any real simple singularity of three variables and for any real versal deformation of it there is a one-to-one correspondence between the components of the complement of the discriminant in the space of real parameters of this deformation and the set of W-conjugacy classes of involutions of $\tilde{H}_{n-1}(V_\lambda, \mathbf{Z})$ that belong to the coset σW; this correspondence assigns to any component the action of complex conjugation in the homology of the manifold V_λ for any λ from this component.*

This correspondence is, in general, not the same for nonequivalent but stably equivalent singularities (i.e., it is not preserved by the natural identification of the nondiscriminant points in bases of versal deformations of these singularities; see § 3.7). Moreover, even the cosets $\{W\sigma\} \in N/W$ for such stably equivalent singularities may be different. On the other hand, multiplication of the function f (and the deformation) by -1, which also maps discriminant into discriminant, commutes with this correspondence.

Proposition 3 (see [Looijenga 78]). *The element σ belongs to the subgroup W only for the simple real singularities of three variables that can be reduced by a diffeomorphism and maybe by additional multiplication by -1 to one of the following normal forms:*

$x^{2m+1} + y^2 + z^2$ *for* A_{2m},
$x^{2m+2} \pm (y^2 + z^2)$ *for* A_{2m+1},
$x^{2m-1} - xy^2 + z^2$ *for* D_{2m},
$x^{2m} + xy^2 + z^2$ *for* D_{2m+1},
$x^4 + y^3 + z^2$ *for* E_6,
and all versions of E_7 and E_8.

In § 6 of Chapter V we present the morsifications f_λ of all these singularities such that the corresponding involution σ is the identity operator.

§ 8. Stratifications. Semialgebraic, semianalytic and subanalytic sets

Every analytic variety can be split into a locally finite family of smooth manifolds of different dimensions: first we take the set of all its nonsingular points, then consider the remaining set of singular points and take the set of its own nonsingular points, and so on. Such splittings of topological spaces are called *stratifications.* The stratifications are a convenient way of making the notions of "general position" and "transversality" precise: in particular, these notions can be generalized to the case of maps and dispositions of arbitrary analytic (and semianalytic) spaces.

In this section, we follow the books [Pham 67] and [Goresky & MacPherson 86].

8.1. Whitney stratifications.

Let M be a smooth real manifold, and S a closed subset of M.

Definition 1. A *primary stratification* of S is a sequence of closed sets $S \equiv S^{m_1} \supset S^{m_2} \supset \dots$, of dimensions $m_1 > m_2 > \dots$ respectively, such that

a) for every l, the set $S^{m_l} \setminus S^{m_{l+1}}$ is a smooth manifold (its path-connected components are called *strata*);

b) the boundary $\bar{A} \setminus A$ of every stratum A is the union of finitely many strata whose dimensions are strictly less than that of A;

c) the division of S into strata is locally finite.

$B < A$ (where A and B are two strata) is the notation for the fact that B is contained in the boundary of A; in this case one also says that B *adjoins* A.

The fact that a set has a primary stratification does not provide much useful information, because the strata of a primary stratification may adjoin each other in a quite exotic way; see for example [Pham 67], [Trotman 77], [Trotman 83]. It is therefore convenient to consider stratifications that satisfy certain conditions of regularity of adjoining, the so-called Whitney conditions.

Let S be a closed subset of Euclidean space supplied with a primary stratification, and $B < A$ two strata of this stratification; let x be a point of B.

Definition 2. B *regularly adjoins* A at the point x if for any pair of sequences of points $\{b_i\} \subset B$, $\{a_i\} \subset A$, converging to x and such that

i) the lines l_i spanning the vectors $b_i - a_i$ converge to a certain line l, and

ii) the tangent spaces $T_{a_i}A$ converge to a certain subspace τ,

the following conditions are satisfied:

a) $T_xB \subset \tau$,

b) $l \subset \tau$.

These two conditions are called *Whitney conditions.*

Now let S be a closed subset of an arbitrary smooth manifold M, and $B < A$ two strata of a primary stratification of S.

Definition 3. *B regularly adjoins A* at a point $x \in B$ if for some smooth proper embedding of M into Euclidean space the images of A and B are adjoined regularly at the image of x (it is easy to verify that this condition does not depend on the choice of embedding).

Our stratification is called a *Whitney stratification* if for any pair $B < A$ of its adjoining strata, B adjoins A regularly at any point of it.

Proposition 1 (see [Mather 70]). *The Whitney condition b) implies condition a).*

Here are several nice properties of Whitney stratifications.

Theorem 1. *The C^1-triviality of a pair of strata $B < A$ implies the regularity of their adjacency at all points of B.*

(C^1-triviality means that close to any point of B the pair (the closure of A; B) is ambient C^1-diffeomorphic to the product of a Euclidean space and a pair $(\bar{C}, *)$, where $\bar{C}$ is the closure of a manifold C not containing the point $*$. If in this definition we replace C^1-diffeomorphism by a homeomorphism, then we get the condition of *topological triviality*, which occurs in the next theorem.)

Theorem 2 (see [Thom 69]). *The regularity of the adjacency $B < A$ implies the topological triviality of the pair $(\bar{A}, B)$ along B.*

Theorem 3. *If the adjacency $B < A$ is regular at the point $x \in B$, and the map $\phi : N \to M$ of a smooth manifold N is transversal to B at a point $y \in \phi^{-1}(x)$, then this map is transversal to the stratum A at all points of N sufficiently close to y.*

Theorem 4 (see [Cheniot 72]). *A transversal intersection of two Whitney stratified sets is again a Whitney stratified set, whose strata are the intersections of strata of the two original spaces.*

Remark 1. *If S is a Whitney stratified subset of a smooth manifold M, then the partition of M into the strata of this stratification and the components of the complement of S is a Whitney stratification of M.*

8.2. Thom's isotopy theorem.

Let Z be a stratified subset of a smooth manifold M.

Definition 4. A smooth map $p : M \to T$ is called a *proper stratified submersion* (with respect to Z) if

a) the restriction $p|_Z$ is proper;

b) for any stratum of Z, the restriction of p to this stratum is a submersion.

Theorem 5 (Thom's first isotopy lemma; see [Thom 69], [Mather 70]). *Suppose that $p : M \to \mathbf{R}^t$ is a proper stratified submersion. Then there is a stratification-preserving homeomorphism*

$$Z \to \mathbf{R}^t \times (p^{-1}(0) \cap Z)$$

that is smooth on any stratum and commutes the map p with the obvious projection of the right-hand part onto $\mathbf{R}^t$. In particular, there exist stratification-preserving homeomorphisms between all fibres of the restriction $p|_Z$.

8.3. Semialgebraic, semianalytic and subanalytic sets.

Definition 5 (see for example [Goresky & MacPherson 86]). A subset S of a smooth (real or complex) analytic manifold M is called a *semianalytic subset* if for some covering of S by open sets $U \subset M$, every intersection set $S \cap U$ is the union of several path-components of sets of the form $g_i^{-1}(0) \setminus h_i^{-1}(0)$, where the functions g_i, h_i run over a finite collection of analytic functions on U. A subset S of a smooth (real or complex) analytic manifold M is called a *subanalytic subset* if for some covering of S by open sets $U \subset M$, every intersection set $S \cap U$ is the union of several path-components of sets of the form $f(G) \setminus f(H)$, where G and H belong to a finite collection $\mathcal{G}$ of subanalytic subsets of a certain smooth analytic manifold M', and $f : M' \to M$ is an analytic map whose restriction to the union of the sets from $\mathcal{G}$ is proper.

Tautologically, all semianalytic sets are subanalytic.

The semialgebraic subsets of algebraic manifolds are defined in exactly the same way as the semianalytic subsets of analytic manifolds; they are also "subalgebraic sets" for the following reason.

Theorem 6 (Tarski–Seidenberg lemma). *The image of a semialgebraic set under any algebraic map is semialgebraic.*

Remark. In the first essential work on subanalytic sets, [Gabrielov 68], they were called "$\mathcal{P}$-sets". Nevertheless, we use here the name established later.

Theorem 7 (see [Gabrielov 68], [Goresky & MacPherson 86]). *Any closed subanalytic (in particular, any closed semianalytic or semialgebraic) set admits a Whitney stratification.*

Theorem 8 (see [Gabrielov 68], [Lojasiewicz 72]). *The family of all subanalytic (respectively, semianalytic, semialgebraic) subsets of a fixed manifold is closed with* *respect to all the usual set-theoretic operations (i.e. intersection, union, and taking the complement), and also with respect to taking the closure. Any connected component of a subanalytic (semianalytic, semialgebraic) set is subanalytic (semianalytic, semialgebraic). The set of connected components of a subanalytic subset is locally finite in the ambient manifold.*

By Theorems 7 and 8, the dimension of a subanalytic set is well-defined: it is the highest dimension of strata of its closure. Denote by sing S the set of all points of S close to which S is not a smooth manifold of dimension dim S.

Theorem 9. *The set* sing S *of a semianalytic (subanalytic, semialgebraic) set belongs to the same class.*

Definition 6. Any locally finite partition of a semianalytic (subanalytic, semialgebraic) set into semianalytic (subanalytic, semialgebraic) subsets is called a *semianalytic (subanalytic, semialgebraic) partition* of it. These subsets are called the *parts* of the partition. A stratification is said to be semianalytic (subanalytic, semialgebraic) if it is a semianalytic (subanalytic, semialgebraic) partition.

Notation. For each pair of adjacent strata $B < A$ of an arbitrary primary stratification, irr (A, B) is the set of all points of B close to which this adjacency is not regular. irr (B) is the union of sets irr (A, B) over all strata A such that $B < A$.

Theorem 10 (see [Wall 75], [Goresky & MacPherson 86]). *For any two strata* $A < B$ *of an arbitrary primary semianalytic (respectively, subanalytic) stratification, the set* irr (A, B) *is semianalytic (respectively, subanalytic) and* dim irr$(A, B) <$ dim B.

Theorem 11. *For any semianalytic (respectively, subanalytic) partition of a closed semianalytic (subanalytic) set, there is a semianalytic (subanalytic) Whitney stratification subordinate to this partition. This stratification can be obtained from the initial partition by some combination of standard set-theoretical operations, partitions into connected components, taking the closure, and the operations* sing $(\cdot)$ *and* irr $(\cdot)$.

This theorem is a direct corollary of Theorems 8, 9 and 10. Theorem 7 follows immediately from it.

Theorem 12 (see [Remmert 56], [Remmert 57]). *The image of any complex analytic set under a proper analytic map is an analytic set.*

A similar assertion about real analytic sets is false.

Theorem 13 (see [Milnor 68]). *For every analytic subset* $M \subset \mathbf{R}^m$*, every point* $a \in M$*, and every sufficiently small disc* $B \subset \mathbf{R}^m$ *with centre at* a*, the set* $M \cap B$ *is* *contractible and, moreover, there exists a homeomorphism of this set onto the cone over its boundary $M \cap B \to C(M \cap \partial B)$ which is identical on $M \cap \partial B$ and maps the point a into the vertex of the cone.*

Theorem 14 (see [Lojasiewicz 65], [Lojasiewicz 72]). *Any semianalytic or semialgebraic set admits a triangulation subordinate to some Whitney stratification of it. In particular, any compact seminalytic or semialgebraic subset of a smooth manifold is an absolute neighbourhood retract.*

§ 9. Pham's formulae

In §§ 2 and 3 we have studied the action of the monodromy and variation operators, generated by the loops which go in the parameter spaces around the simplest (Morse) degenerations of complex hypersurfaces. This action is determined by the Picard–Lefschetz formula. Pham (see [Pham 65], [Pham 67]) obtained similar formulae for a whole series of degenerations of hypersurfaces; the Picard–Lefschetz case is one special case of Pham's series.

Namely, consider a compact complex manifold M^n and two collections $S_1, \dots, S_q$ and $X_1, \dots, X_r$ of hypersurfaces in M^n; suppose that these hypersurfaces depend smoothly on a parameter τ which runs over a complex manifold $V : S_i = S_i(\tau), X_i = X_i(\tau)$. (That is, hypersurfaces $\mathcal{S}_i$, $\mathcal{X}_j$ in $M^n \times V$ are defined such that $S_i(\tau) = \mathcal{S}_i \cap (M^n \times \{\tau\})$, $X_j(\tau) = \mathcal{X}_j \cap (M^n \times \{\tau\})$.)

For every τ, consider the group

$$h(\tau) \equiv \tilde{H}_*(M^n \setminus (S_1(\tau) \cup \dots \cup S_q(\tau)); X_1(\tau) \cup \dots \cup X_r(\tau)). \tag{22}$$

Suppose that for generic τ all these hypersurfaces are smooth and the union of them has only normal crossings. Denote the set of parameters τ corresponding to this generic situation by reg V, and its complement in V by $\Sigma(V)$. Consider the fibre bundle over reg V, the fibre of which is the set $S_1(\tau) \cup \dots \cup S_q(\tau) \cup X_1(\tau) \cup \dots \cup X_r(\tau)$ stratified by all intersections of these hypersurfaces. By Thom's isotopy theorem, applied to this fibre bundle, all groups $h(\tau)$ corresponding to such generic τ are isomorphic to each other. The group $\pi_1(\text{reg } V)$ acts naturally on the space $h(\tau_0)$ (where τ_0 is the distinguished point in reg V), and (following Pham) we study this action. (In §§ 2, 3 we considered special cases of this problem corresponding to $q + r = 1$.)

Pham ([Pham 65]) calculated the monodromy of $h(\tau)$ defined by small loops around the singular values of τ corresponding to the following $n + 1$ standard degenerations of the collection of surfaces S_i, X_i.

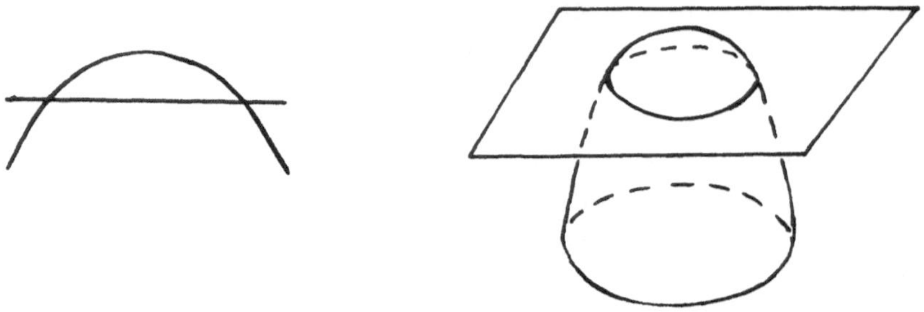

Fig. 15. Standard Pham degenerations of type P_2

P_1: only one hypersurface S_i or X_i participates in this degeneration, and for nongeneric τ this hypersurface has a Morse singular point (i.e. this hypersurface can be defined in some local coordinates by the equation $x_1^2 + \cdots + x_n^2 = 0$).

P_2: exactly two hypersurfaces participate in the degeneration; for all close values of τ these hypersurfaces are smooth, and for the singular values they have a simple tangency. See Figure 15, where a close nondiscriminant disposition of two such (real) surfaces in the cases $n = 2$ and $n = 3$ is shown.

P_3: three hypersurfaces are essential, which are always smooth and pairwise transversal, but for nongeneric values of τ they have one point of simplest abnormal crossing. See Figure 16.

. . . .

P_{n+1}: $n+1$ hypersurfaces, any n of which are in general position for all τ close to the considered nongeneric values, and for critical values all of them meet at the same point. See Figure 17.

The study of these degenerations and of the corresponding monodromy action can be reduced to the following $n+1$ model examples. In all these cases the parameter space is one-dimensional, so that τ is a complex number, $V = \mathbf{C}^1$.

In the case P_1 the only degenerating hypersurface is given (in some local coordinates in $M^n \times V$) by the formula

$$x_1^2 + \cdots + x_n^2 + \tau = 0.$$

In general, for any $i = 1, \ldots, n+1$ the model example for the degeneration type P_i is as follows: the first $i-1$ hypersurfaces participating in this degeneration do not depend on τ and are given by the formulae

$$x_1 = 0, \ldots, x_{i-1} = 0, \tag{23}$$

and the i-th hypersurface is distinguished by the equation

$$x_1 + \cdots + x_{i-1} + x_i^2 + \cdots + x_n^2 + \tau = 0\,; \tag{24}$$

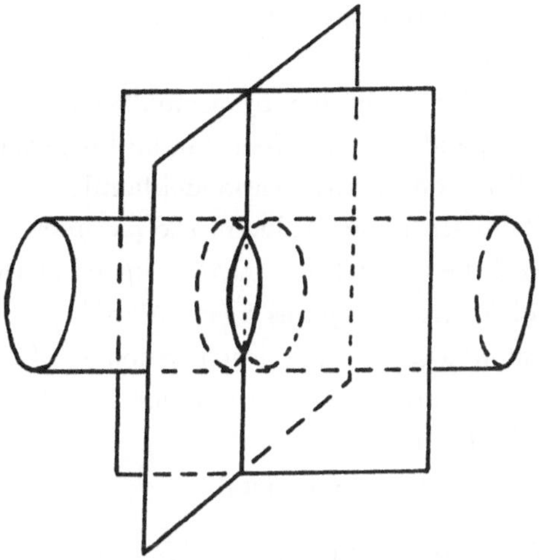

Fig. 16. Standard degeneration of type P_3

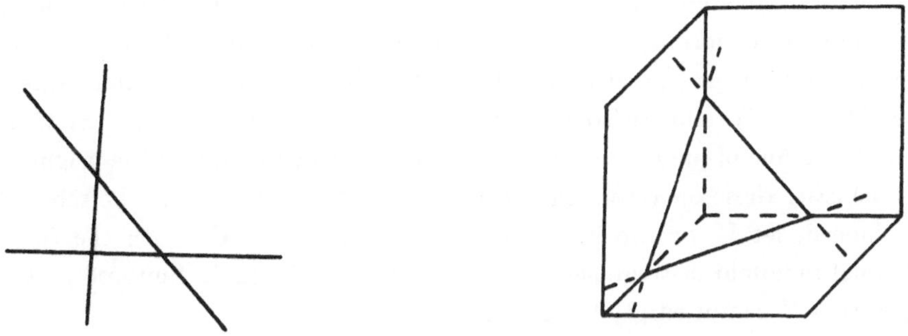

Fig. 17. Degenerations of type P_{n+1} for $n = 2$ and 3

see Figures 15 - 17 for the cases $n = 2, 3$, and $i \leq 4$.

These model one-parameter families are universal for the degenerations P_i, i.e., for any i, any collection of i hypersurfaces defining a singularity of type P_i can be reduced to canonical form (corresponding to the value $\tau = 0$ of the model example), and the one-parameter family from this example is a versal deformation (in a certain category) of this singularity.

The monodromy operator generated by a small loop which goes around the component of $\Sigma(V)$ corresponding to the i-th standard degeneration can be reduced to the similar (localized) action for the i-th model family.

Namely, let τ' be the discriminant value of the parameter τ, around which our small loop goes, and let B be a small disc in M^n centred at the degeneration point of the corresponding collection of varieties $S_i(\tau'), X_j(\tau')$.

Set $i = q + r$, so that we consider the degeneration P_i of Pham's series. Denote the union $S_1(\tau) \cup \cdots \cup S_q(\tau)$ by $S(\tau)$ and the union $X_1(\tau) \cup \cdots \cup X_r(\tau)$ by $X(\tau)$. Let τ_0 be the basepoint of the loop.

Then the monodromy and variation operators

$$M : \tilde{H}_*(B \setminus S(\tau_0), X(\tau_0)) \to \tilde{H}_*(B \setminus S(\tau_0), X(\tau_0)), \tag{25}$$

$$Var : \tilde{H}_*(B \setminus S(\tau_0), X(\tau_0) \cup \partial B) \to \tilde{H}_*(B \setminus S(\tau_0), X(\tau_0)) \tag{26}$$

corresponding to this loop are well defined in exactly the same way as in § 1. In particular, the monodromy action of this loop on the group (22) is equal to the sum of the identity operator and the composition of three operators: the obvious map $h(\tau_0) \to \tilde{H}_*(B \setminus S(\tau_0), X(\tau_0) \cup \partial B)$ (the reduction modulo the complement of B), the operator (26), and the inclusion map $\tilde{H}_*(B \setminus S(\tau_0), X(\tau_0)) \to h(\tau_0)$.

It is sufficient to study the action of this variation operator for the corresponding model family; let us formulate the corresponding problem explicitly. From now on, we suppose that our degenerating family $\{S, X\}$ is one of the model one-parameter families (23) - (24), and we consider the operators (25), (26) defined by a small circle C in the line of parameters τ, going counterclockwise around the origin.

In this case, these operators can be reduced to the usual Picard–Lefschetz formula. Indeed, let V_τ be the intersection of the disc $B \subset \mathbf{C}^n$ with the $(n-i)$-dimensional manifold distinguished by all equations (23), (24). Suppose first that $i \leq n$, so that V_τ is not empty for generic τ.

Proposition 1. *If $|\tau|$ is sufficiently small (but positive), then V_τ is diffeomorphic to the space of tangent vectors of length ≤ 1 to the sphere S^{n-i}. In particular, both groups $\tilde{H}_k(V_\tau)$, $\tilde{H}_k(V_\tau, \partial V_\tau)$ are isomorphic to $\mathbf{Z}$ if $k = n-i$ and are trivial otherwise.*

Indeed, this is just a part of Proposition 2.1.

Our small circle in the line of parameters τ also defines the natural monodromy and variation operators

$$M : \tilde{H}_*(V_\tau) \to \tilde{H}_*(V_\tau), \tag{27}$$

$$Var : \tilde{H}_*(V_\tau, \partial V_\tau) \to \tilde{H}_*(V_\tau). \tag{28}$$

These operators are described by the usual Picard–Lefschetz formula for a Morse singularity in $\mathbf{C}^{n-i+1}$: this follows from the explicit form of the manifolds V_τ.

Theorem 1 (see [Pham 65]). *There is a natural commutative diagram*

$$\begin{array}{lcl} \tilde{H}_*(V_\tau, \partial V_\tau) & \to & \tilde{H}_{i+*}(B \setminus S(\tau_0), X(\tau_0) \cup \partial B) \\ \downarrow & & \downarrow \\ \tilde{H}_*(V_\tau) & \to & \tilde{H}_{i+*}(B \setminus S(\tau_0), X(\tau_0)), \end{array} \tag{29}$$

all arrows in which are isomorphisms, and the vertical arrows are the variation operators from the formulae (28) and (26).

Indeed, the upper horizontal map is the composition of maps

$$\tilde{H}_l(V_\tau, \partial V_\tau) \equiv \tilde{H}_l(X_1 \cap \cdots \cap X_r \cap S_1 \cap \cdots \cap S_q, \partial B) \to$$

$$\tilde{H}_{l+1}(X_1 \cap \cdots \cap X_r \cap S_1 \cap \cdots \cap S_{q-1} \setminus S_q, \partial B) \to$$

$$\tilde{H}_{l+2}(X_1 \cap \cdots \cap X_r \cap S_1 \cap \cdots \cap S_{q-2} \setminus (S_{q-1} \cup S_q), \partial B) \to \dots$$

$$\tilde{H}_{l+q}(X_1 \cap \cdots \cap X_r \setminus (S_1 \cup \cdots \cup S_q), \partial B) \to$$

$$\tilde{H}_{l+q+1}(X_1 \cap \cdots \cap X_{r-1} \setminus (S_1 \cup \cdots \cup S_q), X_r \cup \partial B) \to \dots$$

$$\tilde{H}_{l+q+r}(B \setminus (S_1 \cup \cdots \cup S_q), X_1 \cup \cdots \cup X_r \cup \partial B) \equiv \tilde{H}_{l+q+r}(B \setminus S, X \cup \partial B),$$

where the first q homomorphisms are the tube operators, and the last r homomorphisms are inverse to the boundary operators in the exact sequences of corresponding triples; all these operators are isomorphisms.

The lower isomorphism in the diagram is given by the similar composition of operators, in which reduction modulo ∂B is always omitted.

In the special case, when $i = 1$, this theorem coincides with Proposition 5 of § 2.

Finally, let $i = n + 1$. In this case V_τ is empty, the variation operator (26) is equal to zero, and the monodromy operator (25) is the identity.

§ 10. Monodromy of hyperplane sections

In the next two sections we consider more carefully a special case of the problem stated at the beginning of § 9, which is especially important for integral geometry and the theory of hyperbolic equations. Let A and S be two fixed analytic subvarieties in $\mathbf{C}P^n$, $\dim S = n-1$, and let $P_C \equiv (\mathbf{C}P^n)^*$ be the space of all complex affine hyperplanes in $\mathbf{C}P^n$. For every $X \in P_C$, consider the relative homology group

$$\tilde{H}_*(\mathbf{C}P^n - S, A \cup X). \tag{30}$$

The main problem of these two sections is the study of the ramification of groups (30) when X runs through P_C. (In many important examples, S is just the improper hyperplane in $\mathbf{C}P^n$, so that $\mathbf{C}P^n - S = \mathbf{C}^n$; see Chapter II.)

First, we specify (following [Pham 67]) the ramification set of the corresponding monodromy.

10.1. The dual variety.

Consider an analytic subvariety $V \subset \mathbf{C}P^n$. Its *dual variety* in the space P_C is defined as follows. We fix some analytical Whitney stratification of V. For any stratum σ of this stratification we define the variety $\operatorname{tang}(\sigma) \subset P_C$ as the closure of the set of all hyperplanes which are nontransversal to σ at its nonsingular points, and define the set $\operatorname{tang}(V,$ the stratification) as the union of all sets $\operatorname{tang}(\sigma)$. Finally, we define the set $\operatorname{tang}(V)$ as the intersection of all sets $\operatorname{tang}(V, \{\text{a stratification}\})$ over all Whitney stratifications of V.

Definition 1. A plane X is *transversal* to V if it is a point of $P_C \setminus \operatorname{tang}(V)$.

Proposition 1. *The set tang(V, a stratification), defined by any analytic Whitney stratification of V, is a closed subanalytic set and admits a Whitney stratification, and its real codimension in P_C is at least 2.*

Indeed, consider the set of pairs $(a, X) \in \mathbf{C}P^n \times P_C$ such that $a \in V$ and X is tangent to the corresponding stratum of V at the point a. The closure of this set is a semianalytic subset of dimension at most $n-1$ in the compact manifold $\mathbf{C}P^n \times P_C$. The set $\operatorname{tang}(V,$ the stratification) is its projection into P_C, in particular, it admits a Whitney stratification by Theorem 8.7.

Let V be the union of two varieties, $V = A \cup S$, where S is a hypersurface. By Thom's isotopy theorem, for all $X \in P_C \setminus \operatorname{tang}(A \cup S)$ the corresponding groups (30) are isomorphic to each other.

As usual, the study of the monodromy action of the group $\pi_1(P_C \setminus \operatorname{tang}(V))$ on the spaces (30) can be localized in the following way. This group is generated by simple loops going around the nonsingular pieces of the set tang (V). Any point of such a nonsingular piece corresponds to a plane $\bar{X} \subset CP^n$ which has only one nontransversality point with our stratification. We fix a small disc B in $\mathbf{C}P^n$ around this nontransversality point a. Then the small loop around this nonsingular piece of tang(V) defines in the standard way the variation operator

$$\tilde{H}_*(B \setminus S, A \cup X \cup \partial B) \to \tilde{H}_*(B \setminus S, A \cup X), \tag{31}$$

and the global monodromy operator acting on the group (30) is equal to the identity operator plus the composition of three operators as in § 2.2.

As in § 9, in many cases an appropriate sequence of boundary and coboundary operators reduces this operator (31) to a similar operator

$$\tilde{H}_{*-2}(A \cap X \cap B, \partial B) \to \tilde{H}_{*-2}(A \cap X \cap B) \quad \text{if } a \in A \setminus S, \tag{31'}$$

$$\tilde{H}_{*-2}(S \cap X \cap B, \partial B) \to \tilde{H}_{*-2}(S \cap X \cap B) \quad \text{if } a \in S \setminus A, \tag{31''}$$

$$\tilde{H}_{*-3}(A \cap S \cap X \cap B, \partial B) \to \tilde{H}_{*-3}(A \cap S \cap X \cap B) \quad \text{if } a \in A \cap S. \tag{31'''}$$

We shall specify several such reduction theorems in subsection 10.4 below.

In subsection 10.5 we study the ramification of groups (30) with generic X close to simplest points of components tang(σ) $\subset$ tang($A \cup S$) that correspond to strata σ consisting of only one point. In § 11 we show how to reduce the monodromy around the tangencies with strata σ of positive dimension k to the previous problem, concerning certain subvarieties in a space of reduced dimension $n - k$, namely, in the transversal slice of A by a generic plane of that dimension.

10.2. Complex link of a singular point of a variety.

Let $A \subset \mathbf{C}P^n$ be a germ of a complex m-dimensional variety at the point a. Denote by $P_C(a)$ the set of all complex hyperplanes through a; obviously $P_C(a) \simeq \mathbf{C}P^{n-1}$.

Definition 2. A plane $Y \in P_C(a)$ is *generic* with respect to A if for some open disc $B \subset \mathbf{C}P^n$, centred at a, and some analytic Whitney stratification of $A \cap B$, there is a neighbourhood of the point Y in P_C that has no common points with all sets tang(σ), where σ are the strata of our stratification not coinciding with the point a.

Lemma 1. *For any analytic variety A and a point $a \in A$, the set of points $Y \in P_C(a)$ generic with respect to A is a connected open dense subset of $P_C(a)$.*

This follows immediately from Proposition 1.

Suppose that some stratum σ of an analytic Whitney stratification of the variety A consists of one point a. Let $Y \in P_C(a)$ be a generic plane with respect to A, and

B a sufficiently small closed disc in $\mathbf{C}P^n$ around a (so that Y is transversal to the variety $A \cap B$ outside the point a). Then any plane $X \in P_C$ that is parallel (with respect to some affine chart containing B) and sufficiently close to Y (but not equal to it) is transversal to the variety $A \cap B$. Moreover, by Thom's isotopy theorem, for all such X the pairs $(A \cap X \cap B, A \cap X \cap \partial B)$ are homeomorphic to each other.

Definition 3. Any of the spaces $A \cap X \cap B$ just constructed (and the topological type of them) is called the *complex link* of the point $a \in A$.

If the plane Y is fixed, then these spaces $A \cap X \cap B$ (and the planes X themselves) are parametrized by the points of a small punctured disc in a complex line transversal to Y at the point a; let ξ be a coordinate on this line such that Y corresponds to the value $\xi = 0$. The spaces $A \cap X \cap B$ form a locally trivial fibre bundle over this punctured disc. A small circle C in this disc, going counterclockwise around a, defines by the standard scheme of § 1 the operators of variation and monodromy:

$$\mathrm{Var}_C : \tilde{H}_*(A \cap X \cap B, \partial B) \to \tilde{H}_*(A \cap X \cap B), \tag{32}$$

$$M_C : \tilde{H}_*(A \cap X \cap B) \to \tilde{H}_*(A \cap X \cap B). \tag{33}$$

Remark. Suppose that the point σ is a "false" stratum, i.e. A also admits a Whitney stratification such that a is not a stratum of it. Then both operators (32), (33) are trivial (i.e. $M_C = \mathrm{Id}, \mathrm{Var}_C = 0$). Indeed, in this case the generic plane Y and all neighbouring parallel planes X_ξ are transversal to our stratum everywhere close to a, and our locally trivial fibre bundle over the punctured disc can be extended to that over the entire disc, in particular, it is trivializable. Therefore, only in the case when a is "more complicated" than all neighbouring points can the operators (32), (33) be nontrivial.

10.3. Complex link of a stratum. Lê number.

The previous notion of the complex link of a point a can be generalized to the strata of positive dimension. Indeed, for any such stratum σ of dimension k we take an arbitrary complex plane $\Xi \subset \mathbf{C}^n$ of complementary dimension $n - k$ transversal to σ at an arbitrary point $a \in \sigma$, and define the complex link of the stratum as that of the point $a \equiv \sigma \cap \Xi$ in the space Ξ.

Since the previous definitions are local, we can also define the complex links of a point and a stratum of a stratified subvariety in any smooth complex manifold by using any local coordinate system in this manifold: the homotopy types of complex links (regarded as stratified varieties) thus defined do not depend on the choice of the coordinates. The complex link of a stratum σ is denoted by $\mathcal{L}(\sigma)$.

It follows immediately from Thom's isotopy theorem that if σ is a stratum of a Whitney stratification, then the topological type of the pair {the complex link $\mathcal{L}(\sigma)$; its boundary $\partial\mathcal{L}(\sigma) \equiv \mathcal{L}(\sigma) \cap \partial B(a)$} does not depend on the choice of Ξ, $B(a)$ and Y as above. Moreover, in this case the intersections with strata of A define on this link a Whitney stratification, the topological type of which also does not depend on these choices.

Theorem 1 (see [Lê 75]). *Suppose that the l-dimensional variety A is a complete intersection (i.e. it can be distinguished by $n-l$ complex analytic equations; see § 13 below). Then the complex link of any stratum of A has the homotopy type of a wedge of several spheres of dimension $n-l$.*

In particular, this is always true in the case $l = n-1$: indeed, any complex analytic hypersurface is locally a complete intersection.

Definition 4. The number of spheres in the wedge from the prevous theorem is called the *Lê number* of the stratum σ.

10.4. Reduction of the dimension in the operator Var.

In this subsection we show how the variation operator (31) reduces to similar operators (31′), (31‴) acting on the homology groups of reduced dimensions. Let a, A, S, B, Y and X be as above: A and S are two complex analytic subvarieties in $\mathbf{C}P^n$, $\dim S = n-1$, the hyperplane Y is nontransversal to the stratified variety $A \cup S$ at a unique point a and belongs to only one component $\operatorname{tang}(\sigma)$ of the set $\operatorname{tang}(A \cup S)$ (where σ is the stratum containing a), B is a small disc around a, and X is a very close generic hyperplane parallel to Y in some local affine coordinates.

First, suppose that $a \in A \setminus S$, and also that the disc B does not intersect S.

Notation. Denote the set $X \cap B$ by X', and $A \cap B$ by A'.

Theorem 2. *There exist canonical homomorphisms* red′, red, *defining the vertical arrrows in the diagram*

$$\begin{array}{ccc} \tilde{H}_*(B, A' \cup X' \cup \partial B) & \to & \tilde{H}_*(B, A' \cup X') \\ \downarrow \text{red}' & & \downarrow \text{red} \\ \tilde{H}_{*-2}(A' \cap X', \partial B) & \to & \tilde{H}_{*-2}(A' \cap X') \end{array} \qquad (34)$$

(where the horizontal arrows are the variation operators (31), (31′)), defined by arbitrary loop $l \in \pi_1(D - \operatorname{tang}(P_C))$ such that for any l this diagram becomes commutative; here the right-hand homomorphism red *is an isomorphism.*

Indeed, red′ is the composition of maps

$$\tilde{H}_*(B, A' \cup X' \cup \partial B) \to \tilde{H}_{*-1}(A' \cup X', \partial B) \to \tilde{H}_{*-1}(A' \cup X', X' \cup \partial B) \to$$

$$\tilde{H}_{*-1}(A', (A' \cap X') \cup \partial B) \to \tilde{H}_{*-2}(A' \cap X', \partial B), \tag{35}$$

where the first, second and fourth mappings are taken from the appropriate exact sequences of triples, and the third is excision.

The homomorphism red is the composition of parallel isomorphisms

$$\tilde{H}_*(B, A' \cup X') \to \tilde{H}_{*-1}(A' \cup X') \to \tilde{H}_{*-1}(A', A' \cap X') \to \tilde{H}_{*-2}(A' \cap X'). \tag{36}$$

Proposition 2. *The first three mappings (35) and all the mappings (36) are isomorphisms.*

This assertion follows from the triviality of the groups $\tilde{H}_*(B, \partial B)$, $\tilde{H}_*(X', \partial B)$, $\tilde{H}_*(B)$ and $\tilde{H}_*(X')$, which form the third terms in the corresponding exact sequences.

Theorem 2 follows immediately from the construction of these isomorphisms.

This theorem allows us to reduce either of the two horizontal operators (34) to the other.

The operators red, red$'$ have a global analogue: for any $X \in P_C$ there is a homomorphism

$$\text{Red} : \tilde{H}_*(\mathbf{C}^n, A \cup X) \to \tilde{H}_{*-2}(A \cap X) \tag{37}$$

commuting with the monodromy actions of the group $\pi_1(reg(P_C))$. It is defined just like the homomorphism (36) in which we take $\mathbf{C}^n$ instead of B; however, in this non-local situation is not necessarily an isomorphism.

Now we consider a similar reduction for the case when $a \in A \cap S$; only the situation when the hypersurface S is nonsingular close to a will be studied.

Denote the set $A \cap X \cap B$ by G, so that $\partial G = A \cap X \cap \partial B$.

Proposition 3. *Suppose that S is nonsingular close to the point $a \in A \cup S$, X is transversal to S, and each of the varieties $A \subset \mathbf{C}P^n$, $A \cap S \subset S$ is either nonsingular at this point or has an isolated singularity at it. Then there is a natural isomorphism*

$$\tilde{H}_*(B \setminus S, A \cup X \cup \partial B) \to \tilde{H}_{*-2}(G - S, \partial G) \tag{38}$$

and exact sequence

$$0 \to \tilde{H}_{*-3}(G \cap S, \partial G \cap S) \to \tilde{H}_{*-2}(G - S, \partial G) \to \tilde{H}_{*-2}(G, \partial G) \to 0 \tag{39}$$

containing the target group of (38). If, moreover, a is a nonsingular point of both A and $A \cap S$, and the plane Y (and hence also X) is transversal to A in B, then the term $\tilde{H}_{*-2}(G, \partial G)$ in (39) vanishes and there are natural isomorphisms forming the vertical arrows in the diagram

$$\begin{array}{ccc} \tilde{H}_*(B \setminus S, A \cup X \cup \partial B \setminus S) & \to & \tilde{H}_*(B \setminus S, A \cup X \setminus S) \\ \downarrow & & \downarrow \\ \tilde{H}_{*-3}(G \cap S, \partial G \cap S) & \to & \tilde{H}_{*-3}(G \cap S), \end{array} \tag{40}$$

the horizontal arrows in which are the variation operators (31), (31‴) defined by arbitrary loops in $D \setminus \operatorname{tang}(A \cup S)$; for any such loop this diagram is commutative.

The proof repeats essentially that of Pham's theorem from § 9, for instance the first nontrivial homomorphism in (39) is given by the Leray tube operation.

Example. Suppose that all the hypotheses of Proposition 3 are satisfied, and additionally $\dim A = n-1$ and a is a nonparabolic point of A. Then we are in the conditions of Pham's theorem with $r = 2$, $q = 1$, and the previous Proposition 3 follows from this theorem.

10.5. Monodromy and variation of hyperplane sections close to individual singularities of hypersurfaces.

In this subsection we consider the operators (32), (33) in the case when $\dim A = n-1$ and the point a of nontransversality of A and X is a stratum of a Whitney stratification of A.

In the two simplest cases – when either the singularity of A is isolated close to a or the singular locus of A close to a is one-dimensional – the calculation of the operators (32), (33) can be reduced to the standard Picard–Lefschetz theory. Indeed, in both cases, for a generic choice of the plane Y the restriction φ_0 of the function F (that distinguishes the divisor A) to the plane $Y \equiv X_0$ has an isolated singularity at the origin, and the family $\{\varphi_\xi\}, \xi \in \mathbf{C}^1$, formed by the restrictions of F to other planes X_ξ, is a deformation of this singularity.

First, suppose that the singularity of the divisor A at 0 is isolated. Then for generic choice of the plane Y the loop C belongs to the nondiscriminant set of this one-parameter deformation (i.e. the manifolds $A \cap X_\xi \cap B$ are nonsingular for $\xi \in C$). Therefore, the operator (32) defined by this loop can be calculated by the standard Picard–Lefschetz formulae.

Example 1. Let 0 be a Morse singular point of A, and Y a generic plane passing through 0; see Figure 18. Then φ_0 is a Morse function of $n-1$ variables, and, for $\xi \neq 0$, $\tilde{H}_{n-2}(A \cap X_\xi \cap B) \sim \tilde{H}_{n-2}(A \cap X_\xi \cap B, \partial B) \sim \mathbf{Z}$. Consider the map of the base $\mathbf{C}^1_\xi$ of the deformation φ_ξ into the base of the standard (one-dimensional) versal deformation of the function φ_0, inducing from it a deformation equivalent to φ_ξ. This map transposes the loop C into a loop which turns around the point 0

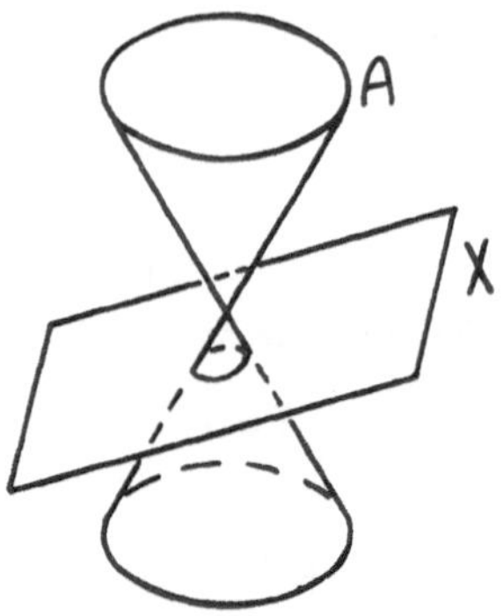

Fig. 18. Morse singularity of a hypersurface

twice. Hence, for even n the corresponding operator Var_C is trivial, and for odd n the operator Var_{kC} defined by the k-fold iteration of C takes a generator of the group $\tilde{H}_{n-2}(A \cap X_\xi \cap B, \partial B)$ into $2k$ times a generator of $\tilde{H}_{n-2}(X_\xi \cap A \cap B)$.

Example 2. Let $n = 2$, and let A be a union of s irreducible curves of multiplicities $m_1, \ldots, m_s$ at 0. Then the complex link consists of $m \equiv m_1 + \cdots + m_s$ points, and the monodromy operator acts by a cyclic permutation in any of these groups of points of cardinalities $m_1, \ldots, m_s$. Thus, the variation operator (32) defined by the loop C has an $(s-1)$-dimensional kernel which is generated by all these groups of m_i points modulo the union of all m points.

Now let the set $\mathrm{sing}(A)$ be one-dimensional close to 0. Then the computation of the operator Var_C can be reduced to a standard problem in the following way. Let $\nabla \in \tilde{H}_{n-2}(A \cap X_\xi \cap B, \partial B)$ be a class whose variation we wish to compute. We include the one-parameter deformation φ_ξ of the function φ_0 in an arbitrary multiparameter deformation, in which the nondiscriminant functions form a dense set. We approximate our loop C by a loop $\tilde{C}$ in the base of this deformation, contained in the nondiscriminant set and contracting onto C via a smooth homotopy whose image at all instants except for the last one lies outside the discriminant. Suppose that the initial point $\tilde{X}$ of this loop $\tilde{C}$ becomes the initial point X of the original loop C under this homotopy. The homotopy of the loop $\tilde{C}$ onto C (which implies a move of $\tilde{X}$ towards X) defines a homomorphism

$$\tilde{H}_{n-2}(A \cap \tilde{X} \cap B, \partial B) \to \tilde{H}_{n-2}(A \cap X \cap B, \partial B).$$

Proposition 4. *This homomorphism is epimorphic.*

Indeed, this is a special case of the following general fact (Proposition 4′). Let $F(x, \lambda)$ be a deformation of an isolated singularity $f : \mathbf{C}^m \to \mathbf{C}$, and λ' a discriminant value of the parameter, i.e. the variety $V_{\lambda'}$ is singular in B. Consider any smooth path in the parameter space $\mathbf{C}^m$ of this deformation, all points of which are nondiscriminant, except for the last point equal to λ'. This path defines a homomorphism of the group $\tilde{H}_{m-1}(V_{\lambda_0}, \partial V_{\lambda_0})$, corresponding to the initial point λ_0 of this path, to $\tilde{H}_{m-1}(V_{\lambda'}, \partial V_{\lambda'})$.

Proposition 4′. *This homomorphism is an epimorphism.*

Proof. First suppose that λ' is the "most discriminant" point 0 corresponding to the deformed singularity f itself. Consider the commutative diagram

$$\begin{array}{ccc} \tilde{H}_{m-1}(V_{\lambda_0}, \partial V_{\lambda_0}) & \to & \tilde{H}_{m-2}(\partial V_{\lambda_0}) \\ \downarrow & & \downarrow \\ \tilde{H}_{m-1}(V_{\lambda'}, \partial V_{\lambda'}) & \to & \tilde{H}_{m-2}(\partial V_{\lambda'}), \end{array}$$

in which both horizontal maps are boundary operators, and both vertical maps are defined by our path. The right vertical arrow is isomorphic because our singularity is isolated. The lower horizontal map is isomorphic by Theorem 1 of § 3. The upper horizontal map is epimorphic by the same theorem and the exact sequence of the pair $(V_{\lambda_0}, \partial V_{\lambda_0})$. Hence our assertion about the left vertical arrow follows.

Now let λ' be an arbitrary discriminant value of the parameter λ, and let β be the union of very small discs centred at all singular points of the corresponding variety $V_{\lambda'}$. Consider the exact sequences of triples $(V_\lambda, V_\lambda \setminus \beta, \partial V_\lambda)$ for $\lambda = \lambda_0$ and λ'. Our path defines a homomorphism of these exact sequences,

$$\begin{array}{ccc} \tilde{H}_{m-1}(V_{\lambda_0} \setminus \beta, \partial V_{\lambda_0}) & \xrightarrow{\text{iso}} & \tilde{H}_{m-1}(V_{\lambda'} \setminus \beta, \partial V_{\lambda'}) \\ \downarrow & & \downarrow \\ \tilde{H}_{m-1}(V_{\lambda_0}, \partial V_{\lambda_0}) & \longrightarrow & \tilde{H}_{m-1}(V_{\lambda'}, \partial V_{\lambda'}) \\ \downarrow & & \downarrow \\ \tilde{H}_{m-1}(V_{\lambda_0}, V_{\lambda_0} \setminus \beta) & \xrightarrow{\text{epi}} & \tilde{H}_{m-1}(V_{\lambda'}, V_{\lambda'} \setminus \beta) \\ \downarrow & & \downarrow \\ \tilde{H}_{m-2}(V_{\lambda_0} \setminus \beta, \partial V_{\lambda_0}) & \xrightarrow{\text{iso}} & \tilde{H}_{m-2}(V_{\lambda'} \setminus \beta, \partial V_{\lambda'}). \end{array}$$

The top and bottom horizontal arrows in it are isomorphic, since $V_{\lambda'}$ is nonsingular outside β and our path defines an isotopy there. The second from the top horizontal arrow is epimorphic by the above-considered special case applied to all singular points of $f_{\lambda'}$ separately. The general assertion of Proposition 4′ about the second from the bottom horizontal homomorphism now follows from the commutativity of this diagram.

Now let $\tilde{\nabla} \in \tilde{H}_{n-2}(\tilde{A} \cap X \cap B, \partial B)$ be an arbitrary element which goes into ∇ under this epimorphism. We calculate the action of the operator $\text{Var}_{\tilde{C}}$ on the

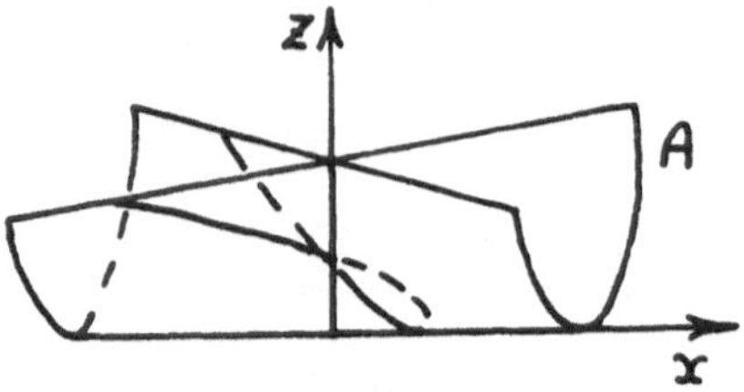

Fig. 19. Whitney umbrella

element $\tilde{\nabla}$: this is a problem in the standard Picard–Lefschetz theory. The image of the element $\mathrm{Var}_{\tilde{C}}(\tilde{\nabla})$ under the similar epimorphism

$$\tilde{H}_{n-2}(A \cap \tilde{X} \cap B) \to \tilde{H}_{n-2}(A \cap X \cap B)$$

of absolute homology groups is the desired element $\mathrm{Var}_C(\nabla)$.

Example 3. Suppose that A is the *Whitney umbrella*, defined in $\mathbf{C}^3$ by the condition $x^2z = y^2$, X_ξ is the plane distinguished by the equation $x + z = \xi$, and α is the only compact domain in $\mathbf{R}^3$ bounded by two surfaces A and X_ξ; see Figure 19. The group $\tilde{H}_3(B, X_\xi \cup A)$ is generated by the class of this domain, and the group $\tilde{H}_1(X \cap A \cup B)$ by the intersection of $X \cap A$ with the boundary of this domain. The obvious maps

$$\tilde{H}_3(B, A \cup X_\xi) \to \tilde{H}_3(B, A \cup X_\xi, \partial B),$$
$$\tilde{H}_1(A \cap X_\xi \cup B) \to \tilde{H}_1(A \cap X_\xi \cup B, \partial B)$$

are isomorphic, so that the latter two groups are also generated by the classes of these chains. In the manner just described, we immediately calculate that the monodromy operator acts on these elements as multiplication by -1, and hence the variation operator acts as multiplication by -2; cf. [Goryunov 1991].

§ 11. Stabilization of local monodromy and variation of hyperplane sections close to strata of positive dimension (stratified Picard–Lefschetz theory)

11.1. Stating the problem.

Let A be an analytic subvariety in $\mathbf{C}^n$ with some fixed analytic Whitney stratification, and $a \in A$ a point in the stratum σ of positive dimension k.

Definition 1. The point a is called a *generic point* of σ if the curvature form of σ is nondegenerate at this point.

Definition 2. A hyperplane $Y \in \text{tang}\,(\sigma)$ is *simply tangent* to the stratum σ at the point a if

a) for some disc $B \subset \mathbf{C}^n$ centred at a and some neighbourhood D of the point Y in P_C, all the planes $X \in D$ are transversal to all strata of our stratification except, maybe, for σ;

b) the restriction to σ of the linear function z distinguishing the plane Y in $\mathbf{C}^n$ is a Morse function on σ.

Of course, the stratum σ can have simple tangent hyperplanes only at its generic points.

Proposition 1. *For any stratum σ the hyperplanes Y satisfying condition a) of the previous definition for some disc B form a dense subanalytic subset in the space* tang (σ, a) *of all hyperplanes tangent to σ at the point a. Moreover, if a is a generic point of σ, then the simply tangent hyperplanes also form a dense subanalytic subset of* tang(σ, a). The proof is obvious.

Definition 3. A local affine coordinate system $\{z_1, \ldots, z_n\}$ with the origin at a generic point $a \in \sigma$ is called *adapted* if the tangent space to σ at a is spanned by the vectors $\partial/\partial z_1, \ldots, \partial/\partial z_k$ (so that the restrictions of the functions $z_1, \ldots, z_k$ constitute a local coordinate system on σ), the plane $Y = \{z | z_n = 0\}$ is simply tangent to σ, and the restriction of the function z_n to σ is expressed in the coordinates z_i as

$$z_n = z_1^2 + \cdots + z_k^2 + \phi_3(z_1, \ldots, z_k)$$

with $|\phi_3(z)| = O(|z|^3)$.

It follows from Proposition 1 and the normal form of quadratic forms that there exist adapted systems close to each generic point of any stratum.

In Figure 1b a real version of this situation is shown, where $n = 3$, the plane X is given by the linear equation $f \equiv -z_3 = c$, and z_1 is the coordinate along the stratum σ. The transversal slice of this picture by the plane $\{z_1 = 0\}$ is shown in Figure 1a.

Let us fix such a coordinate system. Define the one-parameter family of hyperplanes $X_\xi \subset \mathbf{C}^n$, $\xi \in \mathbf{C}^1$, by the condition

$$X_\xi = \{x | z_n(x) = \xi\}$$

and consider the loop C in the parameter space of this family that goes around the discriminant value $\xi = 0$: $\xi(t) = \delta e^{it}$, where δ is small positive and $t \in [0, 2\pi]$. This loop realizes a loop in $P_{\mathbf{C}} \setminus \operatorname{tang}(A)$ around the stratum $\operatorname{tang}(\sigma) \subset \operatorname{tang}(A)$; see § 10. Our present aim is to study the monodromy and variation operators defined by this loop.

Let B be the disc of small radius ε in $\mathbf{C}^n$ centred at a such that the plane Y is transversal in B to all strata of our stratification of A but σ, as well as to the intersections of all these strata (including σ) with the sphere ∂B.

Then for any ξ from a sufficiently small punctured disc in $\mathbf{C}^1$ the plane X_ξ is transversal to our stratification of A, and also to the intersections of these strata with the sphere ∂B. The pairs $(A \cap X_\xi \cap B, A \cap X_\xi \cap \partial B)$ form a locally trivial fibre bundle over our punctured disc, and the variation operator (31) (with empty S) is again well defined as well as its reduction (31′) (= the bottom horizontal operator in (34)). The similar operators, considered in § 10.5, are special cases of these, corresponding to the zero-dimensional stratum $\sigma = \{a\}$.

In the rest of this section we show that (and how) the calculation of these operators can be reduced to a similar problem corresponding to the special case from § 10.5: namely, to the calculation of similar operators (31′) for the transversal slice $A \cap \Xi \subset \Xi$, where $\Xi \subset \mathbf{C}^n$ is the transversal complex plane to σ of complementary dimension $n-k$. This reduction for the monodromy operator is quite easy, while the reduction of the operator Var_C (and even of the group $\tilde{H}_*(X \cap A \cap B, \partial B)$ on which this operator acts) is much more difficult; see Theorems 1, 3 below.

Recall the notations $X' = X \cap B, A' = A \cap B$. The desired variation and monodromy operators then relate the following objects:

$$\operatorname{Var} : \tilde{H}_*(A' \cap X', \partial B) \to \tilde{H}_*(A' \cap X'), \tag{41}$$

$$M : \tilde{H}_*(A' \cap X') \to \tilde{H}_*(A' \cap X'). \tag{42}$$

11.2. The reduction theorems.

Set $m = n-k$. Define the plane Ξ transversal to σ by the condition $z_1 = \cdots = z_k = 0$ and join it to $\mathbf{C}^n$ by a flag of planes

$$\Xi \equiv \Xi^m \subset \Xi^{m+1} \subset \cdots \subset \Xi^n \equiv \mathbf{C}^n, \tag{43}$$

where any plane Ξ^r is defined by the condition $z_1 = \cdots = z_{n-r} = 0$. The induction over this flag reduces the computation of operators (41), (42) to the computation of the analogous operators for the individual singularity $A \cap \Xi$; we now describe this reduction.

For any $r = n-k, n-k+1, \ldots, n$ the intersection of our picture with the plane Ξ^r defines the variation operator

$$\mathrm{Var}_r : \tilde{H}_*(A' \cap X' \cap \Xi^r, \partial B) \to \tilde{H}_*(A' \cap X' \cap \Xi^r),$$

in particular $\mathrm{Var}_n = \mathrm{Var}$.

Denote the groups participating in this operator as follows:

$$\bar{\mathcal{H}}_*(r) \equiv \tilde{H}_*(A' \cap X' \cap \Xi^r, \partial B),\ \mathcal{H}_*(r) \equiv \tilde{H}_*(A' \cap X' \cap \Xi^r). \tag{44}$$

For any $r = n-k, \dots, n-1$ we express the operator Var_{r+1} in terms of Var_r.

Theorem 1. *For any $r = n-k, \dots, n-1$ and any dimension i there are canonical isomorphisms*

$$\mathcal{H}_{i+1}(r+1) \simeq \mathcal{H}_i(r), \tag{45}$$

$$\bar{\mathcal{H}}_{i+1}(r+1) \simeq \mathcal{H}_i(r) \oplus [\mathcal{H}_i(r)/Im\,Var_r \bar{\mathcal{H}}_i(r)] \oplus Ker\,Var_r \bar{\mathcal{H}}_{i-1}(r); \tag{46}$$

the first of these isomorphisms exists for any coefficient group, while the second is valid for the coefficient ring with division by 2.

The exact construction of these isomorphisms will be specified in subsection 11.4 below.

For an arbitrary element $\alpha \in \mathcal{H}_i(r)$ denote by $\Sigma(\alpha)$ the element of the group $\mathcal{H}_{i+1}(r+1)$ that corresponds to it by the isomorphism (45), and by $[\alpha]$ the element of the group $\bar{\mathcal{H}}_{i+1}(r+1)$ that corresponds to the element α of the first summand on the right-hand side of (46).

Theorem 2. *The obvious map $\mathcal{H}_{i+1}(r+1) \to \bar{\mathcal{H}}_{i+1}(r+1)$ acts into the first term on the right-hand side of (46), and maps any element $\Sigma(\alpha)$ into $2[\alpha] + [Var_r(\alpha)]$.*

Theorem 3. *The operator Var_{r+1} is equal to zero on the second and the third terms of the right-hand side of (46), and maps any element $[\alpha]$ of the first term into the element $-\Sigma(\alpha)$.*

Corollary. *If $r > n-k$, then the second term in (46) is trivial.*

Definition 4. The operator stab_r is defined as the composite map

$$\bar{\mathcal{H}}_i(r) \to \mathcal{H}_i(r) \to \bar{\mathcal{H}}_{i+1}(r+1),$$

where the first arrow is the operator Var_r, and the second is $[\,\cdot\,]$, i.e. the inclusion into the first term in (46).

The previous theorems immediately imply the following assertions.

Corollary. *For any $r \geq m$,*

$$\mathrm{Var}_{r+1} \circ \mathrm{stab}_r \equiv -\Sigma \circ \mathrm{Var}_r.$$

Theorem 4. *Suppose that the operator*

$$\mathrm{Var}_r : \bar{\mathcal{H}}_*(r) \to \mathcal{H}_*(r)$$

is an isomorphism. Then for all $s \geq r+1$ *the operators* Var_s *are also isomorphisms, and the operators* stab_s *for* $s = r, \ldots, n-1$ *are isomorphisms of degree 1.*

Corollary (periodicity theorem). *Under the assumptions of the previous theorem, the two homomorphisms* $\Sigma^2 : \mathcal{H}_i(r) \to \mathcal{H}_{i+2}(r+2)$ *and* $\mathrm{stab} \circ \mathrm{stab} : \bar{\mathcal{H}}_i(r) \to \bar{\mathcal{H}}_{i+2}(r+2)$ *are isomorphisms and preserve the operators* $\mathrm{Var} : \bar{\mathcal{H}}_. \to \mathcal{H}_.$ *(which are also isomorphisms) and the obvious maps* $\mathcal{H}_. \to \bar{\mathcal{H}}_.$.

Note in proof. In fact, splitting (46) holds also for arbitrary coefficient group. If the homomorphism $\mathrm{Var}_r : \bar{\mathcal{H}}_i(r) \to \mathcal{H}_i(r)$ is epimorphic then all Theorems 1 – 5 of this section stay valid without changes, otherwise only the following change in Theorem 2 should be done. The homomorphism studied in this theorem acts in the sum of two first terms of (46) and maps $\Sigma(\alpha)$ into the element $2[\alpha] + [\mathrm{Var}_r(\alpha)]+$ the class of the element α in the second term.

In particular, all statements of Theorems 1–5 stay valid if $r > n-k$.

11.3. The first two reductions.

Let us change the coordinates $z_1, \ldots, z_k$ of the adapted coordinate system by a diffeomorphism $(\mathbf{C}^k, 0) \to (\mathbf{C}^k, 0)$ in such a way that the restriction of z_n to the stratum σ is exactly equal to $z_1^2 + \cdots + z_k^2$. (Of course, these new coordinates must be, in general, not affine, but it is not important for us now: it is sufficient that the coordinate z_n that distinguishes the planes X_ξ is a linear function in $\mathbf{C}^n$.)

Further, in our considerations we can change the disc B to the polydisc defined by these new coordinates $z_1, \ldots, z_n$. Namely, let $B' \subset B$ be the polydisc $\{z | |z_i| \leq \varepsilon/n\}$ (where z_i are the coordinates just defined and ε is the radius of B in the standard metric defined by these coordinates) and suppose that the number δ (participating in the definition of the loop $C \subset \mathbf{C}^1$) is sufficiently small with respect to ε. Denote by T_δ the disc bounded by this loop.

Proposition 2. *For every hyperplane* X_ξ *defined by the equation* $z_n = \xi$, $\xi \in T$, *the pair* $(A \cap X_\xi \cap B', A \cap X_\xi \cap \partial B')$ *is homotopy equivalent to the pair* $(A \cap X_\xi \cap B, A \cap X_\xi \cap \partial B)$, *and the corresponding homotopy equivalence of quotient spaces* $(A \cap X_\xi \cap B')/(A \cap X_\xi \cap \partial B') \to (A \cap X_\xi \cap B)/(A \cap X_\xi \cap \partial B)$ *is realized by the obvious factorization map contracting the whole complement of* B' *to one point. Moreover, for every* $r = n-k, \ldots, n$ *and every plane* X_ξ, *the pair* $(A \cap X_\xi \cap \Xi^r \cap B', A \cap X_\xi \cap \Xi^r \cap \partial B')$ *is also naturally homotopy equivalent to* $(A \cap X_\xi \cap \Xi^r \cap B, A \cap X_\xi \cap \Xi^r \cap \partial B)$.

Proof. We shall prove this proposition only for $r = n$ (when $\Xi^r = \mathbf{C}^n$): the proof for arbitrary r is obtained from this one by easily replacing $\mathbf{C}^n$ by Ξ^r, B by $B \cap \Xi^r$, etc.

First we prove our assertion for the plane $X_0 = Y$. Consider two metrics in $\mathbf{C}^n$, $|z| = (\sum |z_i|^2)^{1/2}$ and $\|z\| = \max |z_i|$.

Lemma 1. *If the radius ε of the disc B is sufficiently small, then there exists a homeomorphism of the analytic variety $A \cap X_0 \cap B$ onto the cone over its boundary $A \cap X_0 \cap \partial B$ which is identical on this boundary and maps the point a into the vertex of this cone, and is such that both functions $|z|$, $\|z\|$ strictly increase along the images of all segments of the cone.*

This is essentially a special case of Theorem 13 from § 8: the only additional requirement about the increase of $|z|$ and $\|z\|$ is ensured by the proof of this theorem.

The assertion of Proposition 2 for the plane X_0 follows immediately from this lemma.

Let $\mathcal{B}$ be an open disc centred at a and contained in B'. Since the plane X_0 was transversal to the stratified variety A everywhere outside a (in particular, in $B \setminus a$), then also any plane X_ξ for very small $|\xi|$ is transversal to A in $B - \mathcal{B}$. Hence, by Thom's isotopy theorem, all the varieties $A \cap X_\xi \cap (B - \mathcal{B})$ with sufficiently small $|\xi|$ are isotopic to $A \cap X_0 \cap (B - \mathcal{B})$; moreover, for every generating segment of the cone $A \cap X_0 \cap B \simeq C(A \cap X_0 \cap \partial B)$ the part of this segment belonging to $B - \mathcal{B}$ goes under this isotopy into a segment along which both functions $\|z\|$ and $|z|$ are again monotonic. This implies Proposition 2.

Thus, everywhere in the proof of Theorems 1 – 3 we can replace the disc B by the polydisc B'.

11.4. Realization of the formulae (45) and (46).

Denote the $2r$-dimensional polydisc $B' \cap \Xi^r$ by $B^{(r)}$ and the characteristic radius ε/n of all such polydiscs by ϵ. Denote the ϵ-disc $\{z_{n-r} \in \mathbf{C}^1 | |z_{n-r}| \leq \epsilon\}$ by Ω.

Consider the projection $z_{n-r} : A \cap X_\delta \cap B^{(r+1)} \to \Omega$.

Lemma 2. *If the radius ϵ of the polydisc B' is sufficiently small and δ is sufficiently small with respect to ϵ, then this projection z_{n-r} is a ramified fibre bundle, whose base is the disc Ω, the standard fibre is homeomorphic to the variety $A \cap X_\delta \cap B^{(r)}$, and the ramification set consists of two points $\sqrt{\delta}$ and $-\sqrt{\delta}$; over these two points the fibres of our bundle are homotopy trivial; see Figures 20, 1.*

This lemma follows directly from the construction, from Thom's isotopy theorem and from Theorem 13 of § 8.

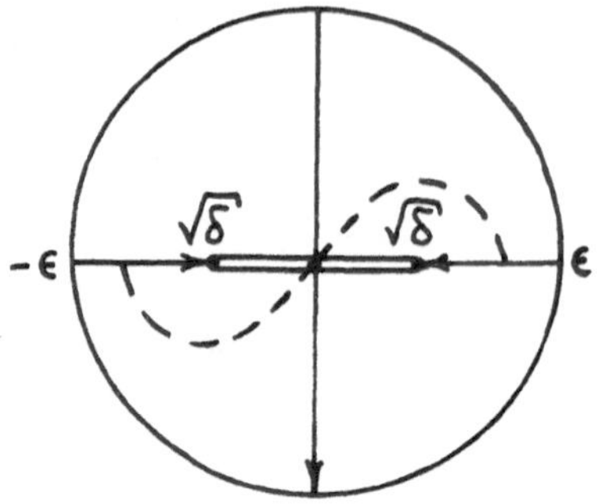

Fig. 20. The base Ω of the fibration z_{n-r}

For arbitrary $t \in \Omega$, denote by F_t the fibre $A \cap X_\delta \cap B^{(r+1)} \cap \{z|z_{n-r} = t\}$ of the projection $z_{n-r} : A \cap X_\delta \cap B^{(r+1)} \to \Omega$, and by ∂F_t its boundary $F_t \cap \partial B^{(r+1)}$. In particular, $F_0 \equiv A \cap X_\delta \cap B^{(r)}$, $\tilde{H}_*(F_0, \partial F_0) \equiv \bar{\mathcal{H}}_*(r)$ and $\tilde{H}_*(F_0) \equiv \mathcal{H}_*(r)$.

Each of the fibres $F_{\sqrt{\delta}}$, $F_{-\sqrt{\delta}}$ contains one singular point, the critical point of the restriction of z_{n-r} to the manifold $\sigma \cap X_\xi \cap \Xi^{r+1}$. (These points can however be nonsingular points of the whole $A' \cap X_\xi \cap \Xi^{r+1}$.) Let u be the union of two very small (with respect to ϵ and δ) discs in $X_\xi \cap \Xi^{r+1}$ around these points. Here is a small appendix to Theorem 1.

Proposition 3. *The group* $\tilde{H}_{i+1}(A' \cap X' \cap \Xi^{r+1} \setminus u, \partial B)$ *is naturally isomorphic to* $\bar{\mathcal{H}}_i(r)$.

Consider two variation operators $V_{+,-} : \tilde{H}_*(F_0, \partial F_0) \to \tilde{H}_*(F_0)$ defined by the simple loops in Ω corresponding to the segments $[0, \sqrt{\delta}]$ and $[0, -\sqrt{\delta}]$.

Lemma 3. *These two operators* V_+, V_- *are equal to each other and to the operator* $\mathrm{Var}_r : \bar{\mathcal{H}}_*(r) \to \mathcal{H}_*(r)$.

Indeed, all three operators are defined by homotopic loops in the space of $(r-1)$-dimensional complex planes parallel to $\Xi^r \cap X$ and transversal to A.

Let α be any element of the group $\mathcal{H}_i(r)$. Using the fibre bundle structure, we transport the cycle realizing α in F_0 over the S-shaped path in Figure 20 to the points of the segments $[\sqrt{\delta}, \epsilon]$ and $[-\epsilon, -\sqrt{\delta}]$. Then, transporting the resulting cycles over these segments, we sweep out two $(i+1)$-dimensional chains in Ξ^r. We orient them by using the pair of orientations, the first of which is the orientation of the base segment chosen as shown in Figure 20, and the second is induced by the original orientation of our cycle α over the basepoint 0. Since the fibres over the ramification points are homotopically trivial, we can add to these chains two chains in these singular fibres, which span their boundaries, so that the resulting two chains become relative cycles (modulo $(X_0 \cap A \cap \partial B^{(r)})$). Half the difference of these two cycles is exactly the cycle $[\alpha] \in \bar{\mathcal{H}}_{i+1}(r+1)$ corresponding to α in the first term of the right-hand side of (46). (If the basic coefficient group does not admit division by 2, we take for $[\alpha]$ just the first of these two cycles; see note in proof in subsection 11.2.)

The sum of these cycles is of course also a cycle. This cycle is homologous to zero if and only if $\alpha \in \mathrm{ImVar}_r \bar{\mathcal{H}}_i(r)$, and we get the second term of (46). If the relative cycle $\gamma \in \bar{\mathcal{H}}_{i-1}(r)$ belongs to the subgroup KerVar_r, then it can be transported continuously into all the fibres of our fibration over Ω, including the degenerate fibres; this two-dimensional family of $(i-1)$-dimensional cycles forms an $(i+1)$-dimensional cycle, which is included in the third term.

The class $\Sigma(\alpha) \in \mathcal{H}_{i+1}(r+1)$ is obtained from α by the suspension operation: it is swept out by an appropriate family of cycles obtained from α by transport over the segment $[-\sqrt{\delta}, \sqrt{\delta}]$ and subsequent contraction of the boundary of the resulting cycle inside the singular fibres over the endpoints $-\sqrt{\delta}$ and $\sqrt{\delta}$.

Theorem 5. *The operator* stab *(defined after the statement of Theorem 3) can be realized by the relative cycle swept out by the cycle α in transport over the imaginary axis, oriented downwards; see Figure 20.*

11.5. Proof of Theorem 1.

First we prove formula (45). Consider any smooth deformation retraction of the disc Ω onto the segment $[-\sqrt{\delta}, \sqrt{\delta}]$. Using the fibre bundle structure, we can lift this deformation to the space $A \cap X_\delta \cap B^{(r+1)}$, which is therefore homotopy equivalent to the preimage of this segment under the projection $z_{n-r} : A \cap X_\delta \cap B^{(r+1)} \to \mathbf{C}^1$. Since the preimages of the endpoints of this segment are contractible, and over the interior points of the segment our projection is locally trivial, this preimage is homotopy equivalent to the suspension of the fibre over any interior point, for instance of the fibre $A \cap X_\delta \cap B^{(r)}$. This implies formula (45).

To prove formula (46), consider the following cellular filtration $K \subset L \subset \Omega$ of the disc Ω mod $\partial\Omega$. The subset K consists of two points $\pm\sqrt{\delta}$, and the subset L consists of two segments $[\sqrt{\delta}, \epsilon]$ and $[-\epsilon, -\sqrt{\delta}]$, so that the set $\Omega \setminus L$ is a 2-cell. Lift this filtration onto the set $A \cap X_\delta \cap B^{(r+1)}$, and consider the homological spectral sequence $\{E^r_{p,q}\}$ generated by this filtration and calculating the relative (mod $\partial B^{(r+1)}$) homology of $A \cap X_\delta \cap B^{(r+1)}$. In the Table 9, representing this spectral sequence, F denotes the standard fibre $A \cap X_\delta \cap B^{(r)}$ of our ramified fibration $A \cap X_\delta \cap B^{(r+1)} \to \Omega$.

Table 9. Term E^1 of the main spectral sequence

....	....	....	....	0
$q=i+1$	$(\tilde{H}_i(\partial F))^2$	$(\tilde{H}_{i+1}(F,\partial F))^2$	$\tilde{H}_{i+1}(F,\partial F)$	0
$q=i$	$(\tilde{H}_{i-1}(\partial F))^2$	$(\tilde{H}_i(F,\partial F))^2$	$\tilde{H}_i(F,\partial F)$	0
$q=i-1$	$(\tilde{H}_{i-2}(\partial F))^2$	$(\tilde{H}_{i-1}(F,\partial F))^2$	$\tilde{H}_{i-1}(F,\partial F)$	0
....	....	....	....	0
	$p=0$	$p=1$	$p=2$	...

Proposition 4. *For any i, the three rows $q=i-1, i$, and $i+1$ of the term E^1 of our spectral sequence are as shown in Table 9.*

Indeed, the assertion about the column $\{p=0\}$ follows from the fact that the fibres over the points $\pm\sqrt{\delta}$ are contractible, but their boundaries (= intersections with $\partial B^{(r+1)}$) are isotopic to these for nonsingular fibres. The assertions about columns $\{p=1\}$ and $\{p=2\}$ follow immediately from the constructions.

We shall assume that a distinguished identification of the terms $E^1_{p,q}$, $p=1,2$, to the homology groups of $A\cap X_\delta\cap B^{(r)}$ is fixed: the second of them is given by the inclusion (and Künneth formula), and the first is realized additionally by the transportation over the S-shaped path in Figure 20; see subsection 11.4.

The involution $\Omega\to\Omega$ given by formula $z_{n-r}\to -z_{n-r}$ acts on this spectral sequence and splits it into invariant ($E^{r+}_{p,q}$) and antiinvariant ($E^{r-}_{p,q}$) subsequences, which we can investigate separately. (Here we use the assumption that the coefficient ring allows division by 2.)

It is easy to see that all the column $\{p=2\}$ of the term E^1 is invariant under this action, and the union of columns $\{p=0\}$ and $\{p=1\}$ splits into two parts, each of which coincides with the term E^1 of the spectral sequence calculating the homology mod $\partial B^{(r+1)}$ of the preimage of either of the two components of the set L. In particular, the antiinvariant and invariant parts of the term E^1 of our spectral sequence are as shown in Tables 10 and 11, respectively.

Let us study the antiinvariant part. The corresponding differential $\partial_1 : E^{1,-}_{1,i}\to E^{1,-}_{0,i}$ coincides with the boundary operator, hence the antiinvariant spectral sequence is nothing but the exact sequence of the pair $(F,\partial F)$ that calculates the absolute homology of F. This gives the first term in the right-hand part of formula (46). It is

Table 10. Term E^1 of the "antiinvariant" spectral sequence

....	....	....	0	0
$q=i+1$	$\tilde{H}_i(\partial F)$	$\tilde{H}_{i+1}(F,\partial F)$	0	0
$q=i$	$\tilde{H}_{i-1}(\partial F)$	$\tilde{H}_i(F,\partial F)$	0	0
$q=i-1$	$\tilde{H}_{i-2}(\partial F)$	$\tilde{H}_{i-1}(F,\partial F)$	0	0
....	....	....	0	0
	$p=0$	$p=1$	$p=2$	...

Table 11. Term E^1 of the "invariant" spectral sequence

....	....	....	....	0
$q=i+1$	$\tilde{H}_i(\partial F)$	$\tilde{H}_{i+1}(F,\partial F)$	$\tilde{H}_{i+1}(F,\partial F)$	0
$q=i$	$\tilde{H}_{i-1}(\partial F)$	$\tilde{H}_i(F,\partial F)$	$\tilde{H}_i(F,\partial F)$	0
$q=i-1$	$\tilde{H}_{i-2}(\partial F)$	$\tilde{H}_{i-1}(F,\partial F)$	$\tilde{H}_{i-1}(F,\partial F)$	0
....	....	....	....	0
	$p=0$	$p=1$	$p=2$	...

easy to see that the realization of this term is exactly the one indicated in subsection 11.4.

The invariant spectral sequence is naturally isomorphic to the one calculating the homology group of the space $A \cap X_\delta \cap B^{(r+1)}$ modulo the union of $\partial B^{(r+1)}$ and the half-space given by the condition $Re\ z_{n-r} \leq -\sqrt{\delta}/2$, in particular the final terms of both spectral sequences are adjoined to this homology. To calculate this group it will be convenient for us to subordinate our filtration as follows: $(K \subset L) \subset \Omega$. As above, the $(i+1)$-dimensional homology group of the preimage $z_{n-r}^{-1}(L^+)$ of the "positive" component of L mod $\partial B^{(r+1)}$ is isomorphic to $\tilde{H}_i(F)$, and we only have to calculate the boundary homomorphism

$$\tilde{H}_i(F, \partial F) \simeq$$

$$\simeq \tilde{H}_{i+2}(A \cap X_\delta \cap B^{(r+1)}, (\partial B^{(r+1)} \cup z_{n-r}^{-1}(L^+) \cup \{z | Re\ z_{n-r} \leq -\sqrt{\delta}/2\})) \to$$
$$\tilde{H}_{i+1}(z_{n-r}^{-1}(L^+), \partial B^{(r+1)}) \simeq \tilde{H}_i(F). \tag{47}$$

Lemma 4. *For the canonical choice of both isomorphisms involved in formula (47), the middle homomorphism in this formula coincides with the operator* Var_r.

Indeed, the first isomorphism assigns to an arbitrary relative homology class $\gamma \in \tilde{H}_i(F, \partial F)$ a chain $\gamma!$ swept out by cycles obtained from γ by transportation over arbitrary paths in the contractible domain $\Omega \setminus L$; we can choose this transportation in such a way that for any point t of the interval $(\sqrt{\delta}, \epsilon)$ two limit positions of these transported cycles from both sides of the complement of this interval coincide in a neighbourhood of ∂F_t. By Lemma 3, the difference of these two limit cycles is equal to the absolute cycle $\mathrm{Var}_r(\gamma_t)$, where γ_t is the cycle obtained from γ by transport over the S-shaped path in Figure 20. By definition of the boundary operator, the boundary of $\gamma!$ is the (absolute) chain in $z_{n-r}^{-1}(L^+)$ swept out by these differences. By the construction of the last isomorphism in (41), this is exactly the homology class corresponding to the element $\mathrm{Var}_r(\gamma) \in \tilde{H}_i(F)$.

Lemma 4 is proved, and formula (46) of Theorem 1 follows immediately from it.

11.6. Proof of Theorem 2.

The segment $[-\sqrt{\delta}, \sqrt{\delta}]$, considered as a relative cycle in $\Omega\,(\mathrm{mod}\, K \cup \partial\Omega)$, can be deformed in two ways into the union of two segments $[-\sqrt{\delta}, -\epsilon]$ and $[\epsilon, \sqrt{\delta}]$; see Figure 21.

Given a cycle $\alpha \in \tilde{H}_i(F)$, consider the corresponding cycle $\Sigma(\alpha) \in \mathcal{H}_{i+1}(r+1)$, divide it formally by 2, and deform these two halves in $A \cap X_\delta \cap B^{(r+1)}$ in correspondence with Figure 21: for any instant of this deformation, one of these

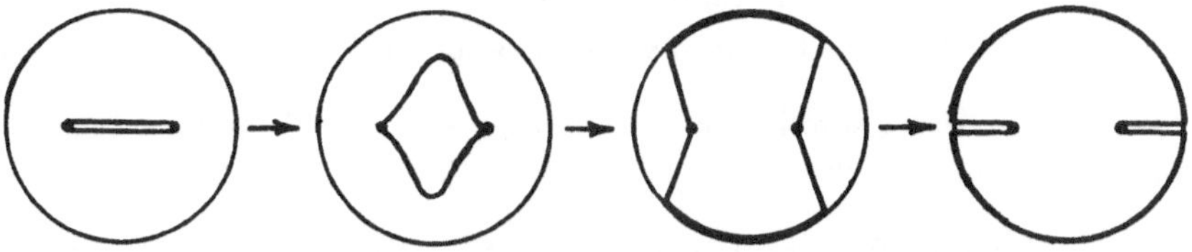

Fig. 21. Deformations of the segment $[-\sqrt{\delta}, \sqrt{\delta}]$

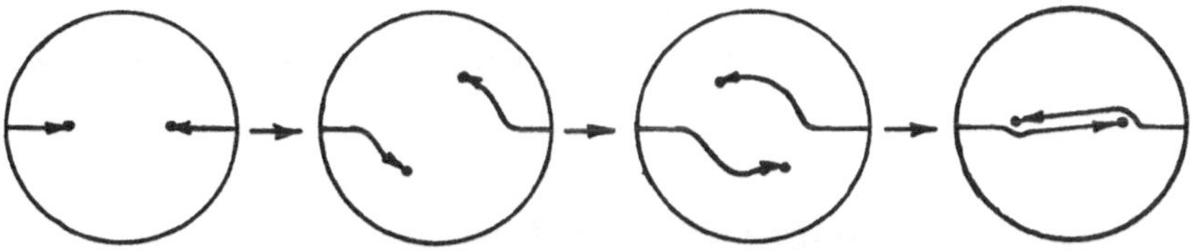

Fig. 22. Variation of the disc Ω

halves becomes a locally trivial bundle with fibres homeomorphic to α over the points of the upper curve obtained from the segment $[-\sqrt{\delta}, \sqrt{\delta}]$, and the other half is that over the lower curve. By construction, the resulting cycle is homologous to $\Sigma(\alpha)$ mod $\partial B^{(r+1)}$, and is equal to $[\alpha] + \overline{[\alpha]}$, where $\overline{[\alpha]}$ is obtained from α in the same way as $[\alpha]$ but by means of transportation over a path complex conjugate to the S-shaped path from Figure 20. By Lemma 3, $\overline{[\alpha]} \simeq [\alpha] + [\mathrm{Var}_r(\alpha)]$ and Theorem 2 is proved.

11.7. Proof of Theorem 3.

When ξ moves along the circle of radius δ around 0, the ramification points $\pm\sqrt{\delta} \subset \Omega$ move as shown in Figure 22. Let us also move the segments $[-\epsilon, -\sqrt{\delta}]$, $[\sqrt{\delta}, \epsilon]$ connecting the ramification points with "infinity" in such a way that they coincide with the parts of the real axis close to the boundary of Ω. Let us also move the cycle $[\alpha]$ in such a way that at any instant of this deformation it is a fibre bundle over the union of the corresponding two segments (except for their endpoints). At the final instant of the monodromy, each of these two parts of $[\alpha]$ will increase by the cycle $\Sigma(\alpha)/2$ swept by the initial cycle α in transport over the added part of resulting segment, so that $\mathrm{Var}_{r+1}([\alpha]) = \Sigma(\alpha)$.

The action of Var_{r+1} on the second term in (46) could be calculated in a similar way (where, however, one part increases by $\Sigma(\alpha)$ and the other decreases by exactly the same cycle). On the other hand, it follows directly from the fact that the second and the third terms in (46) are invariant under the action of the involution $\{z_{n-r} \to -z_{n-r}\}$ and the whole target group $\mathcal{H}_*(r+1)$ is antiinvariant, that the restriction of Var_{r+1} to these two terms is a zero operator. Theorem 3 is proved.

Theorem 4 is a direct corollary of Theorems 1 - 3, and Theorem 5 is proved in the same way as Theorem 2.

Proof of Proposition 3. A centrifugal vector field pushes any relative cycle $\alpha \in \tilde{H}_{i+1}(A' \cap X' \cap \Xi^{r+1} \setminus u, \partial B)$ out from the small neighbourhoods of the fibres $F_{\sqrt{\delta}}$, $F_{-\sqrt{\delta}}$, so that the resulting cycle intersects this neighbourhood only in points of ∂B^{r+1}, which can be ignored. Then, using the fibre bundle structure over the complement of the point $\pm\sqrt{\delta}$ in Ω, we push any relative cycle into the preimage of the imaginary axis. Since over this axis this bundle is trivializable, the proposition follows.

11.8. Proof of the usual Picard–Lefschetz formula.

The standard Picard–Lefschetz formula (Proposition 3 of § 2) describes the variation operator (41) in the special case when A is a smooth hypersurface in $\mathbf{C}^{n+1}$ (and $\sigma = A$); the adapted coordinates $z_1, \ldots, z_n$ identify the space $A \cap X$ with the Milnor fibre V_ξ, and hence the groups $\mathcal{H}_i(r)$ and $\bar{\mathcal{H}}_i(r)$ are the groups $\tilde{H}_i(V_\xi)$ and $\tilde{H}_i(V_\xi, \partial V_\xi)$ respectively for Morse functions of r variables; see § 2. In particular, these groups are trivial for $i \neq r-2$ and are free cyclic for $i = r-2$. In the case $n = 2$ the Picard-Lefschetz formula is trivial: the monodromy action is the permutation of two points.

Lemma 5. *The operators* $\Sigma : \mathcal{H}_i(r) \to \mathcal{H}_{i+1}(r+1)$ *and* stab : $\bar{\mathcal{H}}_i(r) \to \bar{\mathcal{H}}_{i+1}(r+1)$ *(where all homology groups are taken with integer coefficients) map the integer generators of the source groups into integer generators of the target groups.*

Indeed, it follows from the above realization of these operators that for any $\alpha \in \mathcal{H}_i(r)$, $\beta \in \bar{\mathcal{H}}_i(r)$, $|\langle \Sigma(\alpha), \mathrm{stab}(\beta)\rangle| = |\langle \alpha, \beta \rangle|$, thus the lemma follows from the Poincaré duality theorem.

Now the Picard–Lefschetz formula for Var_{r+1} follows from that for Var_r and Theorems 2–4.

Note in proof (see my paper "Stratified Picard–Lefschetz theory", to appear in Selecta Math.). Recently I have obtained the stabilization theorems similar to Theorems 1–5 for the intersection homology groups (both absolute and relative) of spaces $A \cap X \cap \Xi^r \cap B$ and the intersection pairings of these groups. The sum of two last terms in (46) can be expressed in a simpler form $\tilde{H}_i(A' \cap \Xi^r \cap \partial B)$.

11.9. Important example.

Under the assumptions of Theorems 1 – 3 let the codimension $m \equiv n - k$ of the stratum σ be equal to 2. Then the transversal slice $A' \cap \Xi^m$ of A' at a point a is a germ of a complex curve, and the set $A' \cap X' \cap \Xi^m$ consists of several points; the number of them is the local multiplicity of this plane curve, and the monodromy and variation operators for this slice are described in Example 2 of § 10.5.

Let γ_1 be any of these points, and u the local multiplicity at a of the local irreducible component of the curve $A' \cap \Xi^m$ that contains γ_1. Denote by $\gamma_2, \ldots, \gamma_u$ the points obtained consecutively from γ_1 by the iterations of the monodromy of the plane X along the circle C. Obviously, all these u points are different, and the monodromy operator acts on them by cyclic permutation.

For any $j = 1, \ldots, u$, denote by Γ_j the element $\text{stab}^{n-2}(\gamma_j) \in \bar{\mathcal{H}}_{n-2}(n)$.

In this subsection we calculate explicitly the action on these elements of the variation operator defined by the loop C studied in Theorem 3 and multiples of this loop.

Proposition 5. *If n is even, then*

$$\text{Var}_{u \cdot C}(\Gamma_j) = 0,$$

i.e. the u-fold iteration of the monodromy along the loop C takes any element Γ_i into itself and adds nothing to the group $\mathcal{H}_{n-2}(n)$.

If n and u are odd, then the same is true for the $2u$-fold iteration of the monodromy:

$$\text{Var}_{2u \cdot C}(\Gamma_j) = 0.$$

If n is odd and u is even, then for any $j = 1, \ldots, u$

$$\text{Var}_{u \cdot C}(\Gamma_j) = \Sigma^{n-2}((\gamma_{j+1} - \gamma_j) + (\gamma_{j+3} - \gamma_{j+2}) + \ldots + (\gamma_{j-1} - \gamma_{j-2}) \qquad (48)$$

(where the lower indices are considered as residues mod u).

The cycle (48) defines a zero element of the relative homology group $\bar{\mathcal{H}}_{n-2}(n)$.

This is a direct corollary of Theorems 2 and 3.

Proposition 6. *Under the assumptions of Proposition 5, suppose that the curve $A' \cap \Xi^2$ is locally irreducible at the point a, so that its multiplicity is equal to u. Then for any n the group $\bar{\mathcal{H}}_{n-2}(n)$ is isomorphic to $\mathbf{Z}^{u-1}$ and is generated by the elements Γ_j, $j = 1, \ldots, u$, subject to the relation $\Gamma_1 + \cdots + \Gamma_u = 0$. The group $\mathcal{H}_{n-2}(n)$ is also isomorphic to $\mathbf{Z}^{u-1}$ and is generated by the elements $\Sigma^{(n-2)}(\gamma_{j+1} - \gamma_j)$ subject to the obvious relation.*

This is a direct corollary of Theorems 1, 3 and their realization given in subsection 11.4.

Definition 5. Let the number u be even. Then the *harm* of an element $\alpha_1\Gamma_1 + \cdots + \alpha_u\Gamma_u \in \bar{\mathcal{H}}_{n-2}(n)$ is the number $|\alpha_1 - \alpha_2 + \cdots - \alpha_u|$. This element is *harmful* if its harm is not equal to zero.

Obviously this number does not depend on either the choice of the point γ_1 or the representation of the element in terms of the generators Γ_j: the harm of the relation element $\Gamma_1 + \cdots + \Gamma_u$ is equal to zero.

By formula (48), if n is odd and u is even, then for any element $\Gamma \in \bar{\mathcal{H}}_{n-2}(n)$ its variation $\mathrm{Var}_{u \cdot C}(\Gamma)$ is a nontrivial element in $\mathcal{H}_{n-2}(n)$ if and only if Γ is harmful.

§ 12. Homology of local systems. Twisted Picard–Lefschetz formulae

12.1. Local systems.

Definition 1. A *local system* on a topological space M is a covering over M whose fibres are endowed with the structure of a fixed Abelian group, which depends continuously on the fibre (i.e., the group structure extends to the set of local sections over any small domain in the base).

Examples. 1. For any Abelian group A the direct product $M \times A$ is a local system called the *trivial system with fibre* A and denoted simply by A.

2. Let $p: M' \to M$ be a τ-fold covering and A an Abelian group. Then there is a local system $p_!A$ on M with the fibre over any point $x \in M$ isomorphic to A^τ and consisting of all possible A-valued functions on the fibre $p^{-1}(x)$. This local system is called the *direct image* of the trivial local system with fibre A on M'.

3. On the set of local systems on M there are obviously defined operations of direct sums, tensor products, Hom, factorization of a local system by its subsystem, etc.

4. Let M be a manifold. The *orientation sheaf* on M is a local system with fibre $\mathbf{Z}$, such that the transportation over a closed path in M sends every sheet to itself or to its opposite, depending on whether the orientation of M is preserved along this path or not. In terms of the previous examples this local system can be described as the quotient system of the direct image of the trivial $\mathbf{Z}$-system on the orientation two-fold covering of M by the trivial $\mathbf{Z}$-system on M.

In general, to any vector bundle over a topological space M there corresponds its orientation sheaf with fibre $\mathbf{Z}$: it is isomorphic to the trivial local system if and only if the bundle is orientable.

An *isomorphism of local systems* is an isomorphism of the coverings preserving the group structure in fibres.

To any local system with fibre A over a path-connected space M there corresponds a representation $\pi_1(M) \to \mathrm{Aut}(A)$ defined by the Gauss–Manin connection. Two local systems over the same space are isomorphic if and only if these representations coincide up to a conjugation.

12.2. Homology of local systems.

To each local system $L \to M$ a chain complex $C_*(M, L)$ is related. Its elements are formal sums of singular simplices in the space of the covering L with the following relations: the sum of a simplex Δ in a sheet U of the covering L and the similar simplex having exactly the same projection onto M as Δ and lying in a sheet U' is identified with a similar simplex in the sheet $U + U'$; any simplex in the zero sheet is equal to 0.

The boundary operator of this complex is defined in the standard way; its homology is called the *homology of M with coefficients in the system L* or, more briefly, the *homology of the system L.*

The dual construction defines *cohomology groups of the system L.*

As usual, it is possible to define the homology of L by using finite and locally finite chains. (A locally finite chain with coefficients in L is a sum of simplices in the space of the corresponding covering whose projection onto M is a locally finite chain in M.) These homology groups are denoted by $H_*(M, L)$ and $H_*^{lf}(M, L)$, respectively.

Examples. 1. The homology and cohomology of M with coefficients in a trivial local A-system are just the usual homology and cohomology of M with coefficients in the group A: a singular simplex in M, taken with coefficient a, can be regarded as a simplex in the sheet $M \times \{a\}$.

2. For any k-dimensional vector bundle $E \to B$ we have the *Thom isomorphism*

$$H^i_{lf}(E, \mathbf{Z}) \simeq H^{i-k}_{lf}(B,\ \mathrm{Or}(E)),$$

where $\mathrm{Or}(E)$ is the orientation sheaf of the bundle.

Let M be a simply-connected complex n-dimensional manifold, and ϕ a ramified holomorphic n-form on it. Suppose that the ramification set S of ϕ consists of d components, $S_1, \ldots, S_d$, and the analytic continuation of ϕ along any closed path l

in $M \setminus S$ multiplies ϕ by

$$\exp\left(\sum_{i=1}^{d} \alpha_i \{l, S_i\}\right),$$

where α_i are some constants, and $\{l, S_i\}$ is the linking number of l and S_i. Then a local system L_α over $M \setminus S$ with fibre $\mathbf{C}^1$ exists, such that the integrals of ϕ along the homology classes of this system are well defined. Namely, we consider the Riemann surface of ϕ, i.e., the covering over $M \setminus S$, consisting of pairs {a point $z \in M \setminus S$, one of the values of the form ϕ at z} : thus, ϕ can be considered as a single-valued form on this covering. Then we define the group of singular chains of the local system L_α as the group of chains of this Riemann surface factorized through the following condition: if Δ and Δ' are two simplices whose projections into $M \setminus S$ coincide, and l is an arbitrary closed path in M which is the projection of some path in the covering space, connecting the corresponding points of these simplices, then we identify Δ' and $\exp(-\sum_{i=1}^{d} \alpha_i \{l, S_i\}) \cdot \Delta$.

If all the numbers α_i are nonzero real numbers (respectively, nonzero rational numbers, respectively, are equal to ± 1), then the similar local system with fibre $\mathbf{R}$ (respectively, $\mathbf{Q}$, respectively, $\mathbf{Z}$ or $\mathbf{Z}_q$) is defined in exactly the same way.

Like the homology groups with constant coefficients, the homology groups of local systems may be calculated using the cellular decompositions or triangulations of M: over any cell or simplex we fix a basis section (or a set of such sections corresponding to all generators of the fibre group) and calculate their incidence coefficients; in the case of finite cell complexes this method reduces the homology calculation to a study of a finitely generated complex.

12.3. Dual local systems and Poincaré duality in their homology.

Let A be one of the groups $\mathbf{Z}$, $\mathbf{R}$ or $\mathbf{C}$, and L a local system on M with fibre A.

Definition 2. The system $\check{L}$ *dual to* L is the local system with fibre $A \equiv \mathrm{Hom}(A, A)$ whose fibres are dual to the corresponding fibres of L.

The representation $\pi_1(M) \to \mathrm{Aut}(A)$ given by the system $\check{L}$ is dual to the representation defined by the original system L.

Theorem 1 (Poincaré duality theorem). *For any oriented n-dimensional manifold M and any local system L with one-dimensional fibre on it, there is a canonical isomorphism*

$$H^i(M, L) \simeq H_{n-i}^{lf}(M, \check{L}). \tag{49}$$

In particular, there is a nondegenerate pairing

$$H_i(M, L) \otimes H^{lf}_{n-i}(M, \check{L}) \to A; \tag{50}$$

as in the case of homology of trivial systems, it is given by intersection indices.

12.4. Twisted vanishing homology groups of singularities.

Let $f : (\mathbf{C}^n, 0) \to (\mathbf{C}, 0)$ be an isolated function singularity, and F a deformation of it, $F : (\mathbf{C}^{n+k}, 0) \to (\mathbf{C}, 0)$. Let $\lambda \in \mathbf{C}^k$ be any nondiscriminant value of the parameter of F. Then, for any $\alpha \in \mathbf{C}$, over $B \setminus V_\lambda$ a local system L_α with fibre $\mathbf{C}$ is defined such that the corresponding representation of $\pi_1(B \setminus V_\lambda)$ maps any path l into multiplication by $\exp\,(-2\pi i\alpha \cdot \{l, V_\lambda\}) \cdot \Delta$. For instance, this system can be defined as in subsection 12.2 above starting from the differential form $(f_\lambda)^\alpha dz_1 \wedge \ldots \wedge dz_n$. All four homology groups from the upper and lower rows of the diagram (7′) from § 3.6 can be generalized to the homology groups with coefficients in this local system: we consider the groups

$$\tilde{H}_n(B \setminus V_\lambda, L_\alpha), \tilde{H}_n(B \setminus V_\lambda, \partial B; L_\alpha), \tilde{H}^{lf}_n(B \setminus V_\lambda, L_\alpha), \tilde{H}^{lf}_n(B \setminus V_\lambda, \partial B; L_\alpha). \tag{51}$$

(If the system L_α is nontrivial, here all $\tilde{H}$ can be replaced by H.)

Proposition 1 (see [Givental 88]). *For any α, all four groups (51) are isomorphic to $\mathbf{C}^{\mu(f)}$.*

Proposition 2. *For any $\alpha \notin \mathbf{Z}$, the obvious homomorphisms*

$$\tilde{H}_n(B \setminus V_\lambda, L_\alpha) \to \tilde{H}^{lf}_n(B \setminus V_\lambda, L_\alpha), \tag{52}$$

$$\tilde{H}_n(B \setminus V_\lambda, \partial B; L_\alpha) \to \tilde{H}^{lf}_n(B \setminus V_\lambda, \partial B; L_\alpha)$$

are isomorphisms.

The group $\tilde{H}^{lf}_n(B \setminus V_\lambda, L_\alpha)$ for any α is generated again by the Lefschetz thimbles, i.e., by the cycles with boundary in V_λ spanning the integral vanishing $(n-1)$-dimensional cycles; see § 3.6. By Proposition 2, in the case of noninteger α these cycles can also be realized by some *compact* cycles which generate the group $\tilde{H}_n(B \setminus V_\lambda, L_\alpha)$. These generators can be chosen as follows (cf. [Pham 65]).

First let f be the standard Morse singularity, $f = z_1^2 + \cdots + z_n^2$, and $f_\lambda = f - \epsilon$. If $n = 1$, then the vanishing cycle $\kappa_\epsilon \equiv \kappa_{\epsilon,\alpha}$ is represented by the curve in $\mathbf{C}^1 \setminus \{\pm\sqrt{\epsilon}\}$ shown in Figure 23, and lifted into the Riemann surface of the form $(f - \epsilon)^\alpha dz$ in such a way that the positively directed real line segment of the curve lies on the leaf where this form is a negative real form. In the case of arbitrary n, this cycle is swept by the $(n-1)$-parameter family of those curves that lie in the complexifications of all

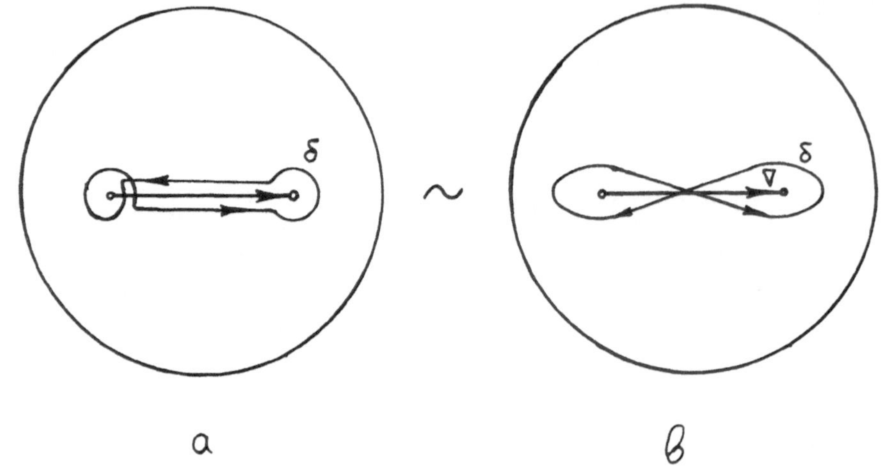

Fig. 23. Vanishing cycle and vanishing cell of a Morse singularity

real lines through the origin in $\mathbf{R}^n$. The distinguished "real" segments sweep a real disc which almost coincides with the corresponding "thimble" $\delta_\epsilon \equiv \delta_{\epsilon,\alpha}$ generating the group $\tilde{H}_n^{lf}(B \setminus V_\lambda, L_\alpha)$: δ_ϵ is the real open disc $\{x | \|x\| < \sqrt{\epsilon}\}$ lifted to the leaf of the Riemann surface where $(f-\epsilon)^\alpha$ takes real negative values. Obviously, the homomorphism (52) maps κ_ϵ into $(1-e^{-2\pi i\alpha})\delta_\epsilon$.

Now let f be an arbitrary (maybe not Morse) singularity, and f_λ a nondiscriminant morsification of it. As in the case of nontwisted homology, the bases in the groups $\tilde{H}_n(B \setminus V_\lambda, L_\alpha)$, $\tilde{H}_n^{lf}(B \setminus V_\lambda, L_\alpha)$ are provided by *vanishing cycles*, which are obtained in the standard way (i.e., by Gauss–Manin transportation over some distinguished paths) from the just described elements κ and δ of the similar groups $\tilde{H}_n(B \setminus V_{\lambda_i}, L_\alpha)$, $\tilde{H}_n^{lf}(B \setminus V_{\lambda_i}, L_\alpha)$, where V_{λ_i}, $i = 1, \ldots, \mu(f)$, are the levels of some μ noncritical values close to all critical values of the morsification f_λ.

Let $s(\lambda)$ be the set of critical values of f_λ; then the monodromy representation of the group $\pi_1(\mathbf{C}^1 \setminus s(\lambda))$ in any of the groups (51) is well defined. This representation is determined by the *twisted Picard–Lefschetz formulae*; see [Pham 65], [Pham 67], [Givental 88].

Namely, consider the local system $\check{L}_\alpha \simeq L_{\bar{\alpha}}$ dual to the system L_α. There is a nondegenerate Poincaré duality

$$\tilde{H}_n(B \setminus V_\lambda, \partial B; L_\alpha) \otimes \tilde{H}_n^{lf}(B \setminus V_\lambda, \check{L}_\alpha) \to \mathbf{C}. \tag{53}$$

Proposition 3. *Suppose that* $\alpha \notin \mathbf{Z}$. *Then the intersection index* $\langle \kappa_{i,\alpha}, \delta_{i,\bar{\alpha}} \rangle$ *of the basic cycles* $\kappa_{i,\alpha} \in \tilde{H}_n(B \setminus V_\lambda; L_\alpha)$ *and* $\delta_{i,\bar{\alpha}} \in \tilde{H}_n^{lf}(B \setminus V_\lambda, L_{\bar{\alpha}})$ *constructed above, that vanish along the same path and are defined by the same canonical coordinate* *system close to the corresponding Morse critical point, is equal to* $(-1)^{n(n-1)/2}(1+e^{-2\pi i\alpha})$.

The proof follows immediately from the construction; see especially Figure 23b.

Theorem 2 (see [Pham 65]). *Suppose that* $\alpha \notin \mathbf{Z}$. *Then the variation of any relative cycle* $\nabla \in \tilde{H}_n(B\setminus V_\lambda, \partial B; L_\alpha)$, *defined by the simple loop going around some critical value of* f_λ, *is equal to*

$$(-1)^{n(n+1)/2}\langle \nabla, \delta_{i,\bar{\alpha}}\rangle \kappa_{i,\alpha}\ , \tag{54}$$

where $\delta_{i,\bar{\alpha}}$ *and* $\kappa_{i,\alpha}$ *are the above-described basic cycles, vanishing over the same critical value along the path defining our simple loop.*

For an illustration (which essentially contains the main idea of the proof) look from the side of the left margin at the first four pictures of Figure 5.

Remark. Pham also proved similar generalized Picard–Lefschetz formulae for all $n+1$ standard degenerations of the ramification set of the differential form (see § 9). The above-described Morse case corresponds to degeneration 1. The situation from the opposite margin of the list of degenerations, when $n+1$ smooth manifolds meet at the same point, plays a very important role in the theory of generalized hypergeometric functions and quantum groups; see [Gelfand 86], [VGZ], [Schechtman & Varchenko 90].

The monodromy action of the group $\pi_1(\mathbf{C}^1 \setminus s(\lambda))$ can obviously be lowered to a representation of some quotient group of it, the fundamental group of the complement of the discriminant variety of any versal deformation of the singularity f; cf. § 3.4.

12.5. Twisted Picard–Lefschetz formulae and stabilization of singularities.

The case when α is a half-integer, and hence any loop going once around V_λ acts on the fibres of the system L_α as multiplication by -1, is especially important. First of all, in this case this local system is selfdual: $\check{L}_\alpha = L_\alpha$.

Moreover, A.B. Givental [Givental 88] remarked that in this case the twisted Picard–Lefschetz formula (54) for a morsification f_λ coincides with the standard Picard–Lefschetz formula which controls the monodromy of vanishing cycles of the stabilized function

$$f_\lambda^{(1)} \equiv f_\lambda(z_1,\ldots,z_n) + z_n^2;$$

see § 3.3. In particular, for any versal deformation F of f, the monodromy representation $\pi_1(\mathbf{C}^\mu \setminus \Sigma(F)) \to \mathrm{Aut}\ \tilde{H}_n(B \setminus V_\lambda, L_{1/2})$ is isomorphic to the standard (untwisted) monodromy representation of the group $\pi_1(\mathbf{C}^\mu \setminus \Sigma(F^{(1)}))$ after obvious identification of bases of the deformation F and of its stabilization $F^{(1)}$.

§ 13. Singularities of complete intersections and their local monodromy groups

The notion and theory of singularities of complete intersections are direct generalizations of those of holomorphic functions.

13.1. First definitions.

Let $f : (\mathbf{C}^n, 0) \to (\mathbf{C}^p, 0)$ be a holomorphic map, $n \geq p$, and X_f the set $f^{-1}(0)$.

Definition 1. X_f is a *complete intersection* if codim $X_f = p$. The map f (and X_f) has an *isolated singularity* at 0 if the rank of the Jacobian matrix of f is equal to p everywhere in $X_f \setminus \{0\}$ close to 0.

Let f be a complete intersection of codimension p in $\mathbf{C}^n$ with isolated singularity at 0. As in the case of smooth functions, we take a small disc $B \subset \mathbf{C}^n$ centred at 0 and a very small (with respect to B) disc $T \subset \mathbf{C}^p$ centred at 0. All varieties $f^{-1}(t)$, $t \in T$, intersect ∂B transversally. The varieties $f^{-1}(t) \cap B$ will be denoted by V_t and called the *Milnor fibres* of f. For almost all $t \in T$ these fibres are nonsingular.

Theorem 1 (see [Hamm 71]). *If f defines a complete intersection with isolated singularity at* 0, *then any its nonsingular Milnor fibre is homotopy equivalent to the wedge of finitely many $(n-p)$-dimensional spheres.*

The number of these spheres is called the *Milnor number* of the singularity f at 0.

The generators of the group $\tilde{H}_{n-p}(V_t)$ can be obtained as follows (see for example Chapter I, § 2 in [AVGL 89]).

Let $(f_1, \ldots, f_p)$ be the coordinate representation of the map f. Suppose that the first $p-1$ functions f_i, $i = 1, \ldots, p-1$, also define a complete intersection $f' : (\mathbf{C}^n, 0) \to (\mathbf{C}^{p-1}, 0)$ (this condition can always be ensured by a small linear change of coordinates in $\mathbf{C}^p$). Let X' be a fixed nonsingular level manifold of f', $X' = \{z | f'(z) = c\} \cap B$. Consider the restriction of the function f_p to this level. After changing (if necessary) f_p by some small perturbation of it, we can asume that all critical points of this restriction $f_p|_{X'}$ are Morse, and the critical values $t^1, \ldots, t^\nu$

are distinct. The nonsingular fibre $\{f_p = t^0\} \cap X'$ of this restriction is diffeomorphic to the nonsingular fibre $V_\lambda \subset \mathbf{C}^n$ of the map f.

Choose a distinguished system of ν disjoint smooth paths in $\mathbf{C}^1$, going from the point t^0 to all points $t^1, \ldots, t^\nu$ (see Definition 3 in § 3). When t moves along the i-th of them, an $(n-p)$-sphere in the fibre $\{f_p = t\} \cap X'$ contracts into a point, the vanishing cycle Δ_i. The corresponding "thimble" swept by this sphere in transport over our path will be denoted by $\check{\delta}_i$.

Proposition 1 (see [Hamm 71]). *The thimbles $\check{\delta}_i$ form a basis of the relative homology group $\tilde{H}_{n-p+1}(X', V_\lambda) \simeq \mathbf{Z}^\nu$. The only nontrivial fragment in the exact homology sequence of the pair (X', V_λ) is as follows:*

$$0 \to \tilde{H}_{n-p+1}(X') \to \tilde{H}_{n-p+1}(X', V_\lambda) \to \tilde{H}_{n-p}(V_\lambda) \to 0. \qquad (55)$$

In particular, the cycles Δ_i generate the $(n-p)$-dimensional homology group of the nonsingular fibre of f, while the elements of the group $\tilde{H}_{n-p+1}(X')$ generate the relations between these vanishing cycles: $\nu = \mu(f) + \mu(f')$.

The Milnor numbers $\mu(f)$ of *quasihomogeneous* complete intersections were calculated explicitly in [Greuel & Hamm 78]. Later we shall need the following very special case of this calculation.

Theorem 2. *Let $f = (f_1, f_2)$ be a complete intersection of codimension 2 in $\mathbf{C}^n$ with isolated singularity at 0, while the functions f_1, f_2 are homogeneous of degrees a and b respectively. Then the Milnor number of f is equal to*

$$\frac{1}{a-b}((a-1)^n b - (b-1)^n a)$$

if $a \neq b$, and $(a-1)^n(an - a + 1)$ if $a = b$.

13.2. Deformations and discriminants.

Definition 2. A *deformation* of a complete intersection

$$f : (\mathbf{C}^n, 0) \to (\mathbf{C}^p, 0)$$

is any map $F : (\mathbf{C}^n \times \mathbf{C}^k) \to \mathbf{C}^p$ such that $F(\cdot \times 0) \equiv f$.

Definition 3. Two germs f and g of such complete intersections are *$\mathcal{K}$-equivalent* if there exists a germ h of a diffeomorphism $(\mathbf{C}^n, 0) \to (\mathbf{C}^n, 0)$ and a germ M at $0 \in \mathbf{C}^n$ of a nondegenerate functional $p \times p$-matrix such that $f(x) \equiv M(x)g(h(x))$.

(In the special case of functions, when $p = 1$, this definition becomes the following one: two function germs are $\mathcal{K}$-equivalent if they can be obtained from one another by a diffeomorphism and multiplication by a smooth function that is not equal to 0 at 0.)

By the definition, the germs at 0 of varieties X_f, X_g defined by $\mathcal{K}$-equivalent complete intersections are ambient diffeomorphic.

The notions of $\mathcal{K}$-equivalent, induced and $\mathcal{K}$-versal deformations are defined in exactly the same way, as the similar notions for function-singularities were defined in § 3: we only need to replace the (pseudo) group of diffeomorphisms acting on the space of functions by that of $\mathcal{K}$-equivalences.

Definition 4. The smallest possible dimension of the parameter space of a $\mathcal{K}$-versal deformation of a singularity f is called its *Tyurina number*, $\tau(f)$. The $\tau(f)$-parameter versal deformation is called a *miniversal* deformation.

Proposition 2 (see [Greuel 80], [Looijenga 84]). *If $n > p$, then $\tau(f) \leq \mu(f)$.*

Let $F = F(x, \lambda)$ be a versal deformation of a complete intersection singularity f. For any $\lambda \in \mathbf{C}^k$, denote by f_λ the map $F(\cdot, \lambda)$, and by V_λ the set $f_\lambda^{-1}(0) \cap B$. The *discriminant* $\Sigma(F)$ of the deformation F is the set of those λ that V_λ is a singular variety.

13.3. Monodromy group and Picard–Lefschetz formulae.

The manifolds V_λ define a locally trivial fibre bundle over $\mathbf{C}^k \setminus \Sigma(F)$, hence also the $\mu(f)$-dimensional homology vector bundle and the Gauss–Manin connection in it are well defined. The corresponding monodromy group, i.e. the image of $\pi_1(\mathbf{C}^k \setminus \Sigma(F))$ under this representation in Aut $\tilde{H}_*(V_\lambda)$, can be described by the Picard–Lefschetz formulae in the same way as in the case of hypersurface singularities. Namely, consider a generic line l in the base $\mathbf{C}^k$ of F. The line l intersects $\Sigma(F)$ transversally at a discrete set of points. The set X' of pairs $\{\lambda \in l, x \in V_\lambda\}$ is a smooth manifold, which is homotopy equivalent to a wedge of several $(n-p+1)$-dimensional spheres. The number of these spheres is denoted by $\tilde{\mu}(f)$. (In subsection 13.1 we have essentially considered a special case of this construction, when the line l consists of all maps $\{f_1 - c_1, \ldots, f_{p-1} - c_{p-1}, f_p - t\}$, c_i are some "generic" constants, and t runs over all numbers). Any path s in $l \setminus \Sigma(F)$ connecting a distinguished nondiscriminant point λ_0 with some point of intersection of l with $\Sigma(F)$ defines a vanishing cycle $\Delta_s \in \tilde{H}_{n-p}(V_{\lambda_0})$ and a thimble $\breve{\delta}_s \in \tilde{H}_{n-p+1}(X', V_{\lambda_0})$. The self-intersection indices of the cycles Δ_s are again determined by Proposition 2.2 (where $n-1$ should be replaced by $n-p$: namely, $\langle \Delta_s, \Delta_s \rangle$ is equal to 0 if $n-p$ is odd, to 2 if $n-p$ is a multiple of 4, and to -2 if $n - p \equiv 2(mod\ 4)$). The monodromy of $\tilde{H}_{n-p}(V_{\lambda_0})$ along the simple loop in l, defined by our path s, is given by the Picard–Lefschetz formula

$$\mathrm{Var}_l(\nabla) = (-1)^{(n-p+1)(n-p+2)/2} \langle \nabla, \Delta_s \rangle \Delta_s. \qquad (56)$$

Thus, the calculation of intersection indices of vanishing cycles in V_{λ_0} becomes very important. These indices define a bilinear form on $\tilde{H}_{n-p}(V_{\lambda_0})$, which is symmetric if $n-p$ is even, and skew-symmetric if $n-p$ is odd.

For a classification of complete intersections see [Giusti 77], [Giusti 83], [Dimca & Gibson 83], [Wall 83], [Ebeling 87], [AVGL 89]; the calculation of intersection forms and corresponding monodromy groups for many such singularities was accomplished in [Ebeling 87], [Janssen 83], and other publications.

13.4. Twisted vanishing homology of complete intersections.

Using Givental's philosophy, we can define a symmetric bilinear form of a complete intersection even in the case when $n-p$ is odd: this is the intersection index in the twisted homology group

$$\tilde{H}_{n-p+1}(X' \setminus V_\lambda, L_{1/2}). \tag{57}$$

(Unlike the case of function singularities, this bilinear form cannot be immediately interpreted as the usual (untwisted) intersection form on the vanishing homology of a stabilized singularity; see [Pickl 85].)

More generally, for any complex numbers $\alpha \notin \mathbf{Z}$ and c_p, define the local system L_α with fibre $\mathbf{C}$ in $B \setminus f_p^{-1}(c_p)$ such that any loop whose linking number with $f_p^{-1}(c_p)$ is equal to m acts on the fibre of L_α as multiplication by $e^{-2\pi i m\alpha}$. As in subsection 13.1, define X' as a generic level manifold of $(f_1, \ldots, f_{p-1})$, and V_c as a generic level submanifold in X' distinguished by the additional condition $f_p = c_p$. Restrict L_α to $X' \setminus V_c$. The vanishing cycles in the group (57) (and in all similar groups with coefficients in systems L_α) are defined in exactly the same way as for the function-singularities; see § 12.4. The self-intersection indices of these cycles are also equal to $(-1)^{(n-p+1)(n-p)/2}(1+e^{-2\pi i\alpha})$, cf. Proposition 12.3, and the monodromy action on these groups, given by closed paths in the set of noncritical values of $f|_{X'}$, is given by the twisted Picard–Lefschetz formula (54), in which n must be replaced by $n-p+1$. Indeed, all these formulae describe local phenomena, and hence "do not know" if they concern hypersurfaces in X' or in $\mathbf{C}^n$.

Let us consider again the special case when α is a half-integer, and hence rotation around V_c in X' acts as multiplication by -1.

Theorem 3. *The obvious homomorphism*

$$\tilde{H}_{n-p+1}(X' \setminus V_\lambda, L_{1/2}) \to \tilde{H}^{lf}_{n-p+1}(X' \setminus V_\lambda, L_{1/2}) \tag{58}$$

is an isomorphism, and the dimensions of both groups (58) are equal to $\nu(f) \equiv \mu(f) + \mu((f_1, \ldots, f_{p-1}))$. The right group (58) is freely generated by the Lefschetz *thimbles specified by some distinguished system of paths connecting the noncritical value of $f_p|_{X'}$ with all critical values, and lifted into an appropriate nonzero leaf of the system $L_{1/2}$. The left group (58) is generated by the cycles vanishing over the same paths.*

Proof. The fact that the map (58) is isomorphic is a general algebraic fact, which is true for all local systems L_α, $\alpha \notin \mathbf{Z}$: it follows from the comparision of the *Leray spectral sequences* (see for example [Griffiths & Harris 78], § III.5) calculating the indicated homology groups and applied to the imbedding $X' \setminus V_\lambda \to X'$.

Denote by $\pm\mathbf{Z}$ the local system over $X' \setminus V_\lambda$ with fibre $\mathbf{Z}$, such that the simplest loop around V_λ also acts on its fibres as multiplication by -1. In particular, $L_{1/2} \equiv \pm\mathbf{Z} \otimes \mathbf{C}$, and for any i

$$\dim \tilde{H}_i^{lf}(X' \setminus V_\lambda, L_{1/2}) = \text{rank } \tilde{H}_i^{lf}(X' \setminus V_\lambda, \pm\mathbf{Z}). \tag{59}$$

Lemma 1. *For any $i \neq n - p + 1$ the rank of the group*

$$\tilde{H}_i^{lf}(X' \setminus V_\lambda, \pm\mathbf{Z}) \tag{60}$$

is equal to zero, and the $\mathbf{Z}_2$-torsion of these groups is trivial for all i.

Indeed, by Proposition 1, $\tilde{H}_i^{lf}(X' \setminus V_\lambda, \mathbf{Z}_2) = 0$ for $i \neq n-p+1$. But $\pm\mathbf{Z}\otimes\mathbf{Z}_2 \equiv \mathbf{Z}_2$, and Lemma 1 follows from the formula for universal coefficients.

Also from Proposition 1 we see that the Euler characteristic of X'/V_λ is equal to $\nu(f)$, hence the free part of the group $\tilde{H}_{n-p+1}^{lf}(X' \setminus V_\lambda, \pm\mathbf{Z})$ is $\nu(f)$-dimensional. Therefore the free part of the group $\tilde{H}_{n-p+1}^{lf}(X' \setminus V_\lambda, \pm\mathbf{Z})$ is generated by any set of $\pm\mathbf{Z}$-cycles that forms a basis of the group $\tilde{H}_{n-p+1}^{lf}(X' \setminus V_\lambda, \mathbf{Z}_2)$ after reduction mod 2; the thimbles mentioned in Theorem 3 provide such system a of generators. In particular they also form a basis in $\tilde{H}_{n-p+1}^{lf}(X' \setminus V_\lambda, L_{1/2})$.

The assertion that the right group (58) is generated by the vanishing cycles follows from the fact that (58) is an isomorphism and under this isomorphism the vanishing cycles go into the corresponding thimbles with coefficient $(1 - e^{-2\pi i \cdot (1-1/2)}) = 2 \neq 0$.

Definition 5. A symmetric bilinear form on a space $\mathbf{R}^\mu$ is called *elliptic* if it is positive or negative definite (or, equivalently, among its inertia indices μ_-, μ_0, μ_+ only one, μ_- or μ_+, is nonzero); it is called *parabolic* if it is semidefinite (i.e. either μ_- or μ_+ is zero).

Definition 6. The *trivial extension* of a map $f : (\mathbf{C}^n, 0) \to (\mathbf{C}^p, 0)$ is the map $(f, \text{Id}|_{\mathbf{C}^r}) : (\mathbf{C}^{n+r}, 0) \to (\mathbf{C}^{p+r}, 0)$.

Theorem 4 (see [Ebeling 87]). *Suppose that the singularity of a complete intersection $f : \mathbf{C}^n \to \mathbf{C}^p$ is not $\mathcal{K}$-equivalent to a trivial extension of an isolated* *function singularity* $\mathbf{C}^{n-p+1} \to \mathbf{C}^1$. *Then the symmetric intersection form of the singularity* f *(defined on the standard vanishing homology group if* $n - p$ *is even, and on the group (58) if* n *is odd) is never elliptic, and it is parabolic if and only if* f *is* $\mathcal{K}$*-equivalent to a trivial extension of a singularity* $(\mathbf{C}^{n-p+2}, 0) \to (\mathbf{C}^2, 0)$ *given by a generic pair of quadratic forms (i.e. by two quadratic forms having no multiple mutual eigenvalues).*

In the case of even $n - p$ this theorem is formulated and proved explicitly in the work [Ebeling 87], and in the case of odd $n - p$ and twisted homology it follows directly from the results of this work.

Chapter II. NEWTON'S THEOREM ON THE NONINTEGRABILITY OF OVALS

Any compact domain in $\mathbf{R}^n$ defines a two-valued function on the space of affine hyperplanes: the volume cut off by a hyperplane from the domain. This function (in the planar case) was considered by Newton in connection with the study of the motion of bodies in the gravitational field. Using essentially the concepts of monodromy, Newton proved that this function cannot be algebraic. In this chapter we extend this result to the case of domains in all even-dimensional spaces, and show that the algebraicity of the volume function defined by an odd-dimensional solid ellipsoid (discovered by Archimedes) is an exceptional phenomenon. Indeed, the analytic properties of the volume functions (say, their algebraicity and asymptotic behaviour of the analytic continuation) depend on the monodromy action on the homology groups defined by the complexifications of the hyperplane and the boundary of the body, hence these properties can be investigated by the methods of the previous chapter.

§ 1. Stating the problems and the main results

1.1. Main definitions.

Definition 1. An *algebraic body* in $\mathbf{R}^n$ is a compact connected component of the set $\{x | f(x) < 0\}$, where f is a polynomial. An algebraic body is *nonsingular* if $df \neq 0$ at the points of its boundary.

Denote by P the space of all affine hyperplanes in $\mathbf{R}^n$.

Any algebraic body $K \subset \mathbf{R}^n$ defines a two-valued function on P, the volumes of the two parts of K separated by the hyperplane.

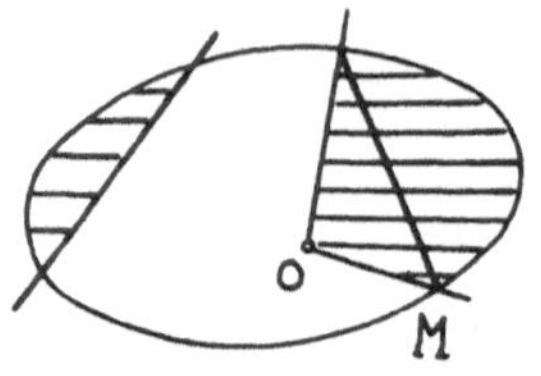

Fig. 24. Volume function of a line and a ray

Denote this function by V_K.

Definition 2. The body K (and the surface bounding it) is called *algebraically integrable* if the function V_K is algebraic (i.e., there exists a polynomial $S = S(a_1, \ldots, a_n, b, V)$ such that if the plane $X \in P$ is given by the equation $a_1x_1 + \cdots + a_nx_n = b$, then $S(a_1, \ldots, a_n, b, V_K(X)) = 0$ for both values of $V_K(X)$).

1.2. Newton's theorem.

Theorem 1 (see [Newton 1687], Lemma XXVIII of Book I). *There exist no algebraically integrable convex nonsingular algebraic bodies in* $\mathbf{R}^2$.

Newton's proof. Let us fix a point O inside the oval bounding the body, and a ray with origin O; see Figure 24. Consider the function on the oval (or, equivalently, on the space of all rays issuing from O) whose value at the point M is equal to the area of the sector bounded by the fixed ray, the radius OM and the oval. If the oval is integrable, this function is algebraic. Indeed, a sector consists of a segment and a triangle, and the areas of both of them depend algebraically on the point M: this follows from the algebraicity and the integrability of the oval. Let us move the point M along the oval. After any complete cycle, the area of the sector increases by the area bounded by the oval. In particular, for any point M this function has infinitely many values, which contradicts its algebraicity.

In the statement of Theorem 1, four conditions on the body are imposed: the body should be convex, nonsingular, algebraic and two-dimensional. In the next items of this section we show that most of these conditions are unnecessary. Now we can remark only that the third condition (the algebraicity) is: if a bounded nonsingular body K in $\mathbf{R}^n$ is algebraically integrable, then it is algebraic. Indeed, consider the envelope of the body K, i.e., the set of hyperplanes tangent to its boundary. This set belongs to the zero level of the function V_K, and hence is semialgebraic. Thus, the boundary of K (which is projectively dual to this set) is also a semialgebraic hypersurface.

The general problem on the existence and classification of integrable bodies was posed by V.I.Arnold by analogy with Newton's result; see [Arnold 87, 87′, 89′], [Arnold & Vassiliev 89].

1.3. Archimedes' example.

Theorem 2 (Archimedes). *The standard disc in* $\mathbf{R}^3$ *is algebraically integrable.*

Moreover, any ellipsoid in any odd-dimensional space is algebraically integrable. Indeed, the volume cut from the unit disc in $\mathbf{R}^{2k+1}$ by a hyperplane lying at a distance $t < 1$ from the origin is a polynomial in t; it is also obvious that the algebraic integrability is a property invariant under affine transformations.

First Main Conjecture. *If n is odd, then any irreducible smooth algebraically integrable body in* $\mathbf{R}^n$ *is an ellipsoid.*

1.4. Newton's theorem for convex bodies in even-dimensional spaces.

Theorem 3. *There are no smooth convex algebraically integrable bodies in* $\mathbf{R}^{2k}$, $k \geq 1$.

Proof. Denote by P_C the space of all complex affine hyperplanes in $\mathbf{C}P^n$. The space P can be considered as a subspace in P_C. Indeed, to any real plane there corresponds a complex plane, its complexification.

Lemma 1. *For any smooth hypersurface $A \subset \mathbf{R}^n$ there exists a linear function $L : \mathbf{R}^n \to \mathbf{R}$ whose restriction to A is a Morse function.*

Indeed, consider the Gauss map $A \to \mathbf{R}P^{n-1}$ which assigns to any point of A the subspace parallel to the tangent space of A at this point. Choose a noncritical value of this map (which exists by Sard's lemma); then we can take for the desired function L a linear function whose zero set is the corresponding hyperplane in $\mathbf{R}^n$.

Let L be such a function for the surface A of the convex body K. Let m and M be the minimal and maximal values of the restriction of L to A. For any $t \in [m, M]$, define the number $W(t)$ as the volume of the domain $K \cap \{x | L(x) \leq t\}$, in particular $W(m) = 0, W(M) = \mathrm{Vol}(K)$. Let θ be the oriented path in the complexification $\mathbf{C}^1$ of the line of values of the function L, which goes once around the segment $[m, M]$ (see Figure 25), and its initial point t_0 lies on the real segment of θ in which $\mathrm{Re}\, t$ increases.

Lemma 2. *For n even, the increment along the path θ of the analytic continuation of the function $W(t)$ from the point t_0 is twice as large as the volume of the body K. For n odd, this increment is equal to zero.*

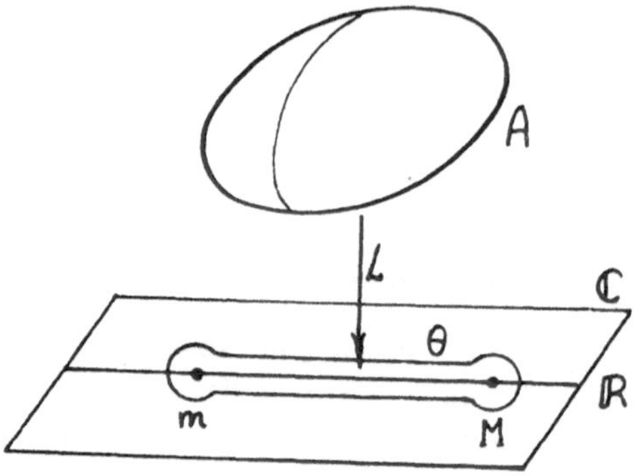

Fig. 25. Analytic continuation of the volume function

Theorem 3 follows immediately from this lemma, which is based on the following fact.

Lemma 3. *For n even, the analytic continuation of the function $W(t)$ along the small circles centred at the points m, M is equal to the functions $-(W(t) - W(m))$ and $W(M) - W(t)$ respectively. For n odd, this analytic continuation keeps the function $W(t)$ unchanged.*

In other words, the power series of the function $W(t)$ in the neighbourhoods of the points m, M contain only half-integer (but not integer!) powers of $t - m$, $M - t$ if n is even, and only integer powers if n is odd.

This lemma follows from the Picard–Lefschetz formula for the fibre bundle of pairs (B, V_t) described in § 2.2 of Chapter I (and especially from the fact that the monodromy around a Morse critical value of a function of k variables maps the corresponding vanishing cycle into itself if k is even and into minus itself if k is odd). For the precise reduction of our problem to the Picard–Lefschetz theory see §§ 2, 3 below.

1.5. Newton's theorem for nonconvex ovals in $\mathbf{R}^2$.

Second Main Conjecture. *For n even there exist neither convex nor nonconvex smooth algebraically integrable bodies in $\mathbf{R}^n$.*

Here are some partial confirmations of this conjecture.

Theorem 4. *For any d and any even n, almost all algebraic bodies of degree d in $\mathbf{R}^n$ are not algebraically integrable (i.e., the algebraically integrable bodies of degree d in $\mathbf{R}^n$, if they exist, belong to a proper algebraic subset of the space of all such bodies).*

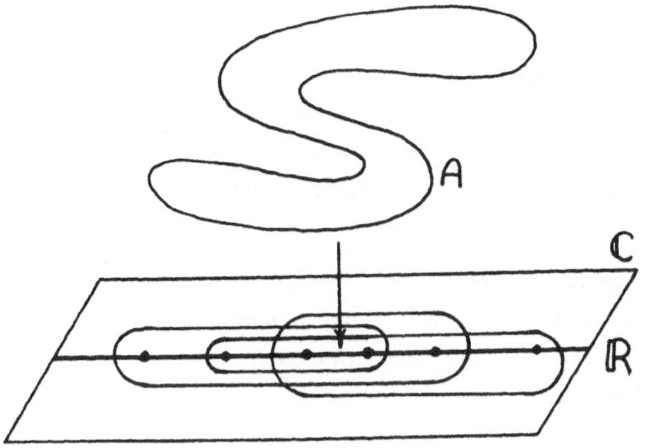

Fig. 26. Obstruction to the algebraicity of the area function

Indeed, the global monodromy group, which controls the ramification of the volume function (which we shall specify precisely in § 2), is defined by the geometry of the complexification of the surface of the body, and is the same for almost all bodies of the same degree. Thus the totality of values of the analytic continuation of the volume function at any given point of P_C depends analytically on the body. Being infinite for an open subset in the space of such bodies (consisting of the convex bodies) it is thus infinite almost everywhere.

Theorem 5. *The Second Main Conjecture is true if* $n = 2$.

The proof is similar to that of Theorem 3 and is based on the same Lemma 3. We choose a more complicated path in the complex line of values of L; see Figure 26. This path turns around the critical values of the restriction of L to the oval, consecutively in the order of the corresponding critical points on the oval. When t returns to its initial value along this path, the analytic continuation of the volume function increases by twice the area bounded by the oval.

The Second Main Conjecture for arbitrary even n could follow in a similar way from the next conjecture.

Second′ Main Conjecture. *In the case of any even* n, *a path with this property exists for any smooth algebraic body.*

One more obstruction to the integrability of (nonconvex) bodies in even-dimensional spaces will be described in subsection 1.7.

1.6. Obstructions to the integrability of odd-dimensional bodies.

In this subsection we describe some obstructions to the integrability of bodies in odd-dimensional spaces, formulated in the terms of the local geometry of the complexifications of their boundaries; these obstructions are strong enough to prove that almost all algebraic bodies of any fixed degree in $\mathbf{R}^{2k+1}$ are nonintegrable. Here is a (quite weak) consequence of these results.

Theorem 6. *Suppose that the closure in $\mathbf{C}P^{2k+1}$ of the complexification of the boundary of an algebraic body $K \subset \mathbf{R}^{2k+1}$ is a nonsingular manifold, and the degree of the polynomial defining this manifold is more than 2. Then the body K cannot be algebraically integrable.*

Obviously the hypotheses of this theorem are satisfied for almost all bodies of any fixed degree $d > 2$ in $\mathbf{R}^{2k+1}$.

Definition 3. A nonsingular point a of a smooth hypersurface $A \subset \mathbf{C}P^n$ is called *parabolic* if the second fundamental form of A (calculated in an arbitrary affine chart) has nonmaximal (less than $n-1$) rank at this point. The point a is called *degenerate* if the hyperplane in $\mathbf{C}P^n$ that is tangent to A at this point is also tangent to A at all points of a set of positive dimension close to a.

For any polynomial $f : \mathbf{C}^n \to \mathbf{C}$, denote the divisor $\{z \in \mathbf{C}^n | f(z) = 0\}$ by A_f.

Let f be a real polynomial $(\mathbf{C}^n, \mathbf{R}^n) \to (\mathbf{C}, \mathbf{R})$, and K_f a nonsingular algebraic body bounded by a component of $A_f \cap \mathbf{R}^n$.

Theorem 7. *If n is odd and the algebraic body K_f is integrable, then the divisor A_f does not have nondegenerate parabolic points in $\mathbf{C}^n$.*

Theorem 6 is a consequence of this one. Indeed, let A be a smooth (in particular, irreducible) algebraic hypersurface of degree d in $\mathbf{C}P^n$. The parabolic points of A are exactly the singular points of the Gauss map $A \to P_C \equiv (\mathbf{C}P^n)^*$ which sends any point of A to the hyperplane tangent to A at this point.

Lemma 4. *The class in $H^2(A)$ dual to the set of parabolic points is equal to the class dual to the hyperplane section of A, taken with multiplicity $(n+1)(d-2)$. In particular, if $d \geq 3$, then this class is nontrivial.*

Indeed, the parbolic points in A_f are distinguished by the additional homogeneous equation $\text{Hess}(f) = 0$ in $\mathbf{C}^{n+1}$, whose degree is equal to $(n+1)(d-2)$.

Corollary. *The set of parabolic points of a smooth algebraic hypersurface of degree $d \geq 3$ in $\mathbf{C}P^n$ is a nonempty subvariety of complex codimension 1 in A.*

Lemma 5 (see [Zak 87]). *A smooth projective algebraic hypersurface has no degenerate parabolic points.*

Lemma 6. *All the parabolic points of a smooth projective hypersurface cannot be contained in its hyperplane section.*

Indeed, suppose that all the parabolic points belong to a hyperplane section. If the tangent planes at all these points coincide with this secant plane, we get a contradiction with Lemma 5. Hence this plane is transversal to the hypersurface at most of these points, in particular the multiplicity of the intersection at the generic points is equal to 1.

If $d \geq 3$ and $n > 0$, this contradicts Lemma 4.

Taking for this secant plane the improper plane in $\mathbf{C}P^n$, we deduce that A has nondegenerate parabolic points in the affine space $\mathbf{C}^n$, and Theorem 6 follows from Theorem 7 (which will be proved in § 4).

So the complexification of any algebraically integrable surface of degree ≥ 3 must have singular points in $\mathbf{C}P^n$: otherwise case it has nondegenerate parabolic points. The (singular) hypersurfaces of arbitrarily high degrees, having no parabolic points, actually exist: such surfaces can be constructed by means of the notion of projective duality.

Recall the notation P_C for the space of all hyperplanes in $\mathbf{C}P^n$; this space is called the *space dual to* $\mathbf{C}P^n$.

Obviously, this space is again isomorphic to $\mathbf{C}P^n$, and its dual space can be naturally identified with the original space $\mathbf{C}P^n$.

Let A be an analytic hypersurface in $\mathbf{C}P^n$.

Definition 4. The closure in the space P_C of the set of all tangent hyperplanes at all nonsingular points of A is called the *projective dual hypersurface of* A and is denoted by A^*.

In terms of Chapter I, § 10, A^* is the component tang(reg A) of tang(A), where reg A is the open stratum in A.

Proposition 1 (see for example [Arnold 1978′]). *The operation of projective duality is an involution ($A^{**} = A$) provided that the set of nonparabolic points is dense in A.*

It is easy to see that the projective duality map sends the nonparabolic smooth points of A to nonparabolic smooth points of the corresponding local irreducible branch of A^*, while the parabolic points of A correspond to the singular points of A^*.

Now consider any nonsingular algebraic hypersurface A^* in P_C. The corresponding dual surface $A^{**} \equiv A$ in $\mathbf{C}P^n$ will have no parabolic points, as desired.

Nevertheless, we cannot get an example of an integrable body in this way: the hypersurface A thus constructed has cuspidal edges (which correspond to the generic parabolic points of A^*) and, as we show in the next theorem, the cuspidal edges (satisfying certain nondegeneracy conditions) are again obstructions to integrability.

Theorem 8. *If n is odd and K is an algebraically integrable smooth body in $\mathbf{R}^n$, then the complexification of the boundary of K cannot have generic cuspidal edges.*

(The singular point of a hypersurface belongs to its *generic cuspidal edge* if the projective dual surface is smooth at the corresponding point and has there the simplest (of type A_2) parabolic point, or, which is the same, if our hypersurface is locally projectively equivalent to a wave front of type A_2; see [AVG 82], [Arnold 83]. We recall this definition explicitly in § 5 before the proof of Theorem 8; see Definition 1 there.)

Unlike Theorems 3–5, Theorems 7 and 8 are based on the study of the *local monodromy*: if $Y \in P_C$ is a plane that is tangent to A_f at a nondegenerate parabolic point, or is tangent to a generic cuspidal edge, then the analytic continuation of the function V_K has a logarithmic ramification in an arbitrary neighbourhood of the point Y.

Conjecture. *The obstructions formulated in Theorems 7 and 8 are strong enough to prove the First Main Conjecture from subsection 1.3 (at least for $n = 3$). That is, any algebraic irreducible hypersurface of degree ≥ 3 bounding a compact body and nonsingular at the points of its boundary, must have points of one of the classes forbidden in these theorems.*

1.7. Obstructions to integrability arising from the asymptotic behaviour of complexified boundaries of algebraic bodies.

Let $\bar{A}_f \subset \mathbf{C}P^n$ be the projective compactification of the divisor A_f, and let S be the "infinite" hyperplane in $\mathbf{C}P^n$.

Definition 5. A point $a \in \bar{A}_f \cap S$ is *trivial* if in a neighbourhood of a, $\bar{A}_f$ intersects S transversally and a is not a parabolic point of either the divisor $\bar{A}_f \subset \mathbf{C}P^n$ or the divisor $\bar{A}_f \cap S \subset S$.

Theorem 9. *If the nonsingular algebraic body $K_f \subset \mathbf{R}^{2k}$ is algebraically integrable, then the set $\bar{A}_f \cap S$ does not have trivial points.*

For the proof see § 6.

§ 2. Reduction of the integrability problem to the generalized Picard–Lefschetz theory

All the theorems stated in § 1 are proved by means of monodromy theory: to prove that the volume function is nonalgebraic it is sufficient to prove that its analytic continuation is infinite-valued (which Newton actually did), and the ramification of this analytic continuation is controlled by some monodromy group related to the divisor A_f. In the rest of this chapter we study these groups and reduce the previous nonintegrability theorems to the suitable Picard–Lefschetz formulae of Chapter 1.

2.1. Integrability and monodromy.

Here is a more general problem of integral geometry, containing Newton's problem as a special case. Let ω be a meromorphic n-form in $\mathbf{C}P^n$, S the set of its poles, and A an arbitrary divisor in $\mathbf{C}P^n$. Let X be a hyperplane in $\mathbf{C}P^n$ and $\theta(X)$ an n-dimensional contour of integration (= a compact singular chain with smooth simplices) in $\mathbf{C}P^n - S$, whose boundary belongs to $A \cup X$.

Lemma 1. *The value of the integral*

$$\int_{\theta(X)} \omega \tag{1}$$

depends only on the class of $\theta(X)$ in the homology group

$$\mathcal{K}(X) \equiv \tilde{H}_n(\mathbf{C}P^n - S, A \cup X). \tag{2}$$

This follows immediately from Stokes' theorem and from the fact that the restriction of the holomorphic n-form ω to the $(n-1)$-dimensional complex variety $A \cup X \setminus S$ is zero.

If X is a plane in general position with respect to $A \cup S$, then a small perturbation X' of it moves the chain $\theta(X)$ weakly, and the corresponding class $\{\theta(X')\} \in \mathcal{K}(X')$ is well defined. The integral of the form ω along this class is a function of X', and our problem is to investigate this function.

For example, in Newton's case ω is the volume form

$$dz_1 \wedge \ldots \wedge dz_n, \tag{3}$$

S is the "infinite" plane, A is the projective closure of A_f, and $\theta(X)$ (for the distinguished plane X which is supposed to be a complexification of a real plane) is the domain in $\mathbf{R}^n$ cut out from the body K_f by the hyperplane X.

It is natural to solve this problem by the methods of Chapter I, § 10; let us recall them. We fix some analytic Whitney stratification of the divisor $A \cup S$. The hyperplanes that are in general position with respect to all strata of this stratification form a Zariski open subset in the space $P_C \equiv (\mathbf{C}P^n)^*$ of all hyperplanes in $\mathbf{C}P^n$; we denote this set by $\operatorname{reg}(A \cup S)$, and its complement by $\operatorname{tang}(A \cup S)$. By Thom's isotopy theorem, for all $X \in \operatorname{reg}(A \cup S)$ the groups $\mathcal{K}(X)$ are isomorphic to each other. This isomorphism for two planes X, X' is noncanonical: it is specified by an arbitrary path in $\operatorname{reg}(A \cup S)$ connecting X and X' and depends on the homotopy type of this path. The exact definition of these isomorphisms is as follows.

Define the space E as the direct product $\operatorname{reg}(A \cup S) \times (\mathbf{C}P^n - S)$ and consider a subset K of E, consisting of pairs of the form {a plane $X \in \operatorname{reg}(A \cup S)$, a point in $(X \cup A - S)$}.

Lemma 2. *The obvious projection $E \to \operatorname{reg}(A \cup S)$ defines a locally trivial fibre bundle of pairs $(E, K) \to \operatorname{reg}(A \cup S)$.*

This follows immediately from Thom's isotopy theorem.

By this lemma, we are in the situation described in Chapter I, § 1. In particular, the Gauss–Manin connection on the bundle of groups $\mathcal{K}(X), X \in \operatorname{reg}(A \cup S)$, is well defined, and hence also the monodromy representation

$$\pi_1(\operatorname{reg}\,(A \cup S)) \to \operatorname{Aut} \mathcal{K}(X). \tag{4}$$

Let X_0 be a distinguished point in $\operatorname{reg}(A \cup S)$, and α a class in $\mathcal{K}(X_0)$. Then on $\operatorname{reg}(A \cup S)$ close to the point X_0 an analytic function $V_{\alpha,\omega,X_0}(X)$ is defined: its value at a point X is equal to the integral of the form ω over the class in $\mathcal{K}(X)$ obtained from α by translation via the Gauss–Manin connection over any path joining X_0 and X and lying close to X_0.

The analytic continuation of this function into the whole of $\operatorname{reg}(A \cup S)$ is a multivalued analytic function, whose ramification is defined by the transportations of α along different paths in $\operatorname{reg}(A \cup S)$.

Example. In the Newton's case the function $V_{\alpha,\omega,X_0}(X)$ coincides on an open set in $\operatorname{reg}(A \cup S) \cap P_R$ with the volume function V_K, and hence the condition that K is integrable implies that the function $V \equiv V_{\alpha,\omega,X_0}(X)$ is finite-valued.

Theorem 1. *Let the differential n-form ω be rational in $\mathbf{C}P^n$ (as, for instance, the volume form from Newton's problem). Then the above-defined function $V_{\alpha,\omega,X_0}(X)$ is algebraic if and only if it is finite-valued.*

Indeed, it is easy to prove that close to the points of tang($A \cup S$) the function $|V_{\alpha,\omega,X_0}(X)|$ grows at most as a negative power of the distance between X and tang($A \cup S$), hence our theorem follows from the generalized existence theorem of Riemann (see [Grauert & Remmert 58], [Hartshorne 77]).

2.2. Some preliminary remarks about the monodromy action on the groups $\mathcal{K}(X)$.

The space $P_C \equiv (\mathbf{C}P^n)^*$ is simply-connected, therefore the group $\pi_1(\text{reg}(A \cup S))$ is generated by a finite set of simple loops going around the set tang($A \cup S$) close to some of its nonsingular points. (Recall that a simple loop going around a divisor at a nonsingular point Y of it is a path in the complement of this divisor, going from the base point to the point Y, then around the divisor close to this point in the positive direction, and returning along its old path to the base point; see Figure 9.

The set tang($A \cap S$) is reducible: a component tang(σ) of it corresponds to each stratum σ of the chosen stratification of $A \cup S$; this component is the closure of the set of planes not transversal to σ.

Remark. Of course, only the components tang(σ) of dimension $n-1$ give a contribution to the group $\pi_1(\text{reg}P_C)$; we mention also that to generate this group it is, in general, not sufficient to take only one loop for each component.

Thus the calculation of the monodromy representation of this group reduces to the calculation of the monodromy along the loops going around the simplest points of the set tang($A \cup S$), i.e., the points corresponding to planes having a unique *simple tangency* with some stratum of $A \cup S$ at its *generic point*; see Definitions 1, 2 of Chapter I, § 11. In reality it turns out to be useful to consider more complicated points of the set tang($A \cup S$), for example, points corresponding to planes that have nonsimple tangencies with the strata of $A \cup S$.

In §§ 10, 11 of Chapter I we have studied explicitly the action (4) corresponding to loops going around different strata tang(σ) of tang($A \cup S$). In §§ 4–6 below we consider the integrals of the volume form along the cycles on which this monodromy acts, and obtain the results on the ramification of these integrals. In particular we prove that for all strata σ mentioned in Theorems 1.7, 1.8, 1.9, this action is sufficiently nontrivial to ensure the logarithmic ramification of the (analytic continuation of the) volume function. Also Lemma 1.3 (and hence Theorems 1.3 - 1.5) follow immediately from these considerations.

In all cases, the monodromy action corresponding to a simple loop can be localized, i.e. it can be reduced to a variation operator defined in terms of the local behaviour of the set $A \cup S$ close to the point at which the plane $Y \in$ tang($A \cup S$) is

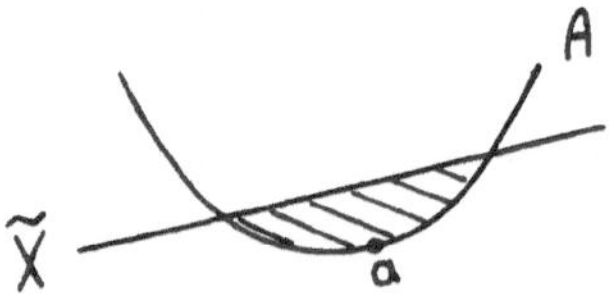

Fig. 27. Element "cap"

nontransversal to some stratum of it; see Chapter I. In particular, the increment of the analytic continuation of the volume function along a loop C close to a ramification point is equal to the integral of the volume form (3) along the cycle $\mathrm{Var}_C(\theta)$, where θ is the original integration cycle.

§ 3. The element "cap"

Here we describe the element of the group $\tilde{H}_n(\mathbf{C}^n, A \cup X)$ participating in Newton's problem.

Let a be a nonsingular and nonparabolic point of an irreducible divisor $A \subset \mathbf{C}^n$, B a small disc centred at a, and $\tilde{X}$ a hyperplane that is not tangent to A but very close to the tangent plane at a. See Figs. 27 and 15. Then the group $\tilde{H}_n(B, A \cup \tilde{X})$ is isomorphic to $\mathbf{Z}$. Its generator is called the *cap* corresponding to the plane $\tilde{X}$, and the corresponding element of the group $\mathcal{K}(\tilde{X}) \equiv \tilde{H}_n(\mathbf{C}^n, A \cup \tilde{X})$ is denoted by $\kappa(\tilde{X}, a)$. The reduction homomorphism (I.37) maps the cap into the *vanishing cycle* that generates the group $\tilde{H}_{n-2}(A \cap \tilde{X} \cap B)$.

Suppose that $X \in \mathrm{reg}(A \cup S)$. We shall say that an element $\alpha \in \mathcal{K}(X)$ is *representable by a cap* if there is a path in $\mathrm{reg}(A \cup S)$ that joins $\tilde{X}$ and X in such a way that α is obtained from $\kappa(\tilde{X}, \alpha)$ by the Gauss–Manin transportation over this path. An element of the group $\tilde{H}_{n-2}(A \cap X)$ is *representable by a vanishing cycle* if it is obtained by the homomorphism (I.37) from some element that is representable by a cap.

Remarks. *The set of elements that are representable by caps or by vanishing cycles is invariant with respect to the Gauss–Manin connection. The definition of this set does not depend on the choice of a and $\tilde{X}$, since the set of nonparabolic points is path-connected in A.*

The significance of this element "cap" for Newton's problem is determined by the following fact. Consider the multivalued function I_A on the set $\mathrm{reg}(A \cap S)$ whose values at the point X are equal to all the integrals of the volume form (3) over all elements in $\mathcal{K}$ representable by the caps.

Theorem 1. *An algebraic surface in* $\mathbf{R}^n$ *is algebraically integrable if and only if the function* I_A *defined by the complexification* A *of this surface is an algebraic function.*

In particular, to prove that the body K is nonintegrable it is sufficient to show that integrals of the volume form along the cycles representable by caps take infinitely many values.

The proof of Theorem 1 follows from the fact that the set $P_{\mathbf{R}} \setminus \mathrm{tang}(A \cap S)$ consists of a finitely many connected components, and from the following lemma.

Lemma 1. *a) For any* $X \in \mathrm{reg}(A \cup S) \cap P_{\mathbf{R}}$, *each of the two homology classes* $\theta(X)$ *defined by the real domains cut off the algebraic body by* $\mathrm{Re}\, X$ *is equal to the sum of several elements representable by caps. In particular, in a neighbourhood of such* X *in* $P_{\mathbf{R}}$ *the volume function is equal to the sum of a finitely many leaves of the function* I_A.

b) There exists $X \in P_{\mathbf{R}}$ *such that one of the corresponding classes* $\theta(X)$ *is representable by a cap, and hence the corresponding volume function coincides with (some leaf of)* I_A.

Proof. It is sufficient to prove this lemma for the case when the plane X is distinguished by a real linear equation $L = 0$ such that the restriction of the linear function L to the manifold Re A is a strictly Morse function: indeed, such X constitute a dense subset in $\mathrm{reg}(A \cup S) \cap P_{\mathbf{R}}$. Suppose for example that we count the volume of the part of the body contained in the half-space $\{L < 0\}$. Consider the system of paths in $\mathbf{C}^1$ connecting the value 0 with all negative critical values of the function $L|_{\mathrm{Re}\, A}$ and going along the real axis in the domain $\{\mathrm{Im} > 0\}$, as in Figure 11. Any such path defines a path in $\mathrm{reg}(A \cup S)$ consisting of complex planes X_t parallel to X that are distinguished by the equations $L \equiv t$, where t runs along the former path in $\mathbf{C}^1$. The final positions of these paths, corresponding to the Morse critical values of L_A, are the simply tangent planes to A at its real points. Now Lemma 1 can be made more precise as follows.

Lemma 2. *The class* $\theta(X)$ *is equal to the sum of (suitable oriented) caps corresponding to all negative critical values of the restriction of* L *to* Re A *and transported by the Gauss–Manin connection over the indicated paths.*

This lemma is proved by induction when ReX moves parallel to itself from the position $\{L \approx -\infty\}$ to $\{L = 0\}$. The basis of the induction is obvious. Let us prove the inductive step, i.e., the fact that for any critical value t of $L|_A$ two elements in $\tilde{H}_n(\mathbf{C}^n, A \cap X_{t+\varepsilon})$, one of which is $\theta(X_{t+\varepsilon})$ and the other is the class obtained by the Gauss–Manin transportation from $\theta(X_{t-\varepsilon})$, differ by the cap located close to the point of tangency of A and X_t.

Outside a small neighbourhood B of this point the family of varieties $A \cup X_\tau$, $\tau \approx t$, does not degenerate, hence the difference between these two elements can actually be realized by a cycle lying in this small neighbourhood. Therefore (and by Theorem I.10.2) it is sufficient to prove the "reduced" version of this assertion: the elements in $\tilde{H}_{n-2}(A \cap X_{t+\varepsilon})$, obtained from the two considered above by the reduction operator (I.37), differ by the vanishing cycle corresponding to the critical value t. This fact will be proved in Chapter V, § 3; see Theorem 2 there.

§ 4. Ramification of integration cycles close to nonsingular points. Generating functions and generating families of smooth hypersurfaces

Let a be a nonsingular point of a hypersurface A that is nondegenerate in the sense of Definition 1.3, and Y a hyperplane in $\mathbf{C}P^n$ tangent to A at the point a. In this section we show that in this case the ramification of cycles from the group $\mathcal{K}(X)$ (and hence the corresponding integrals) for X close to Y is described precisely by the Picard–Lefschetz theory of isolated singularities of functions, considered in Chapter I, § 3. In particular, applying the results of this theory to Newton's problem, we get Lemmas 2 and 3 (and hence also Theorem 3) from § 1, as well as Theorems 4, 5, 6 and 7 from the same section. All these results are consequences of Theorem 4.1, which will be formulated a little later. Here is a new notion which we need to formulate it: the generating function of a smooth hypersurface.

Choose an affine coordinate system $\{z_1, \ldots, z_n\}$ in $\mathbf{C}P^n$ with centre $a \in A$, such that the hyperplane Y tangent to A at a is distinguished by the equation $z_n = 0$. Then close to the point a the surface A can be distinguished by a condition of the form $z_n = \varphi(z_1, \ldots, z_{n-1})$, where the function φ is holomorphic and has a singularity at a.

This function φ is called the *generating function* of the divisor A at the nonsingular point a.

It is easy to see that the point a is parabolic if and only if the generating function has a non-Morse singularity at a, and a is degenerate if and only if this singularity is nonisolated.

Of course, the generating function depends on the choice of the coordinate system $\{z_i\}$, but all such functions defined by different affine coordinate systems at the same point of A are obviously $\mathcal{K}$-equivalent to each other; see Definition 3 in Chapter I, § 13.

Theorem 1. *Let $a \in \mathbf{C}P^n - S$ be a nonsingular and nondegenerate point of an* *irreducible divisor A, Y the tangent plane to A at the point a, D a small disc in P_C centred at the point Y, and ω a holomorphic n-form in $\mathbf{C}P^n - S$ that does not vanish at a. Let V be an analytic (in general, multivalued) function on $D \setminus \operatorname{tang}(A \cap S)$, whose values at the point X are equal to the integrals of the form ω over all possible cycles in $\mathcal{K}(X)$ representable by caps which lie entirely in a small neighbourhood of a and are obtained from a fixed cap by the Gauss–Manin connection over a path lying in $D \setminus \operatorname{tang}(A \cap S)$. Then*

a) for odd n the function V has an infinite-sheeted ramification close to Y if a is a parabolic point, and V is regular close to Y if a is nonparabolic;

b) for even n the function V has a finite-sheeted ramification close to Y if the singularity of the generating function φ at a is simple (i.e., belongs to one of the classes A_k, D_k, E_6, E_7, E_8, see Chapter I, § 5). In the special case when a is non-parabolic in the sense of Definition 1.3, this function V has exactly two sheets, whose sum is identically equal to zero.

To reduce this theorem to the standard Picard–Lefschetz theory, we use one more notion: the *generating family of a hypersurface* (see [Hörmander 71], [Zakalyukin 76]), which in our special case can be defined as follows.

Again let $a \in A$, Y, let the coordinate system $\{z_1, \dots, z_n\}$ be the one from the definition of the generating function, and let $\phi = \phi(z_1, \dots, z_{n-1})$ be the corresponding generating function of the germ of A at the point a.

Any hyperplane $X \in P_C$ sufficiently close to Y can be distinguished by an equation of the form

$$z_n = \lambda_1 z_1 + \cdots + \lambda_{n-1} z_{n-1} + \lambda_n \tag{5}$$

in the same coordinate system. The coefficients $\lambda_1, \dots, \lambda_n$ of this equation are determined uniquely by the point $X \in P_C$, and they can be considered as local coordinates on P_C in a neighbourhood of the point Y. It is easy to see that this coordinate system in P_C is affine.

Definition 1. The *projective generating family* of the germ (A, a) defined by the coordinates $z_1, \dots, z_n$ is the deformation $F(z; \lambda_1, \dots, \lambda_n)$ of the generating function φ determined by the equation

$$F(z_1, \dots, z_{n-1}; \lambda_1, \dots, \lambda_n) =$$

$$= \varphi(z_1, \dots, z_{n-1}) - \lambda_n - \lambda_1 z_1 - \cdots - \lambda_{n-1} z_{n-1}, \tag{6}$$

where the parameter $\lambda = \{\lambda_1, \dots, \lambda_n\}$ of the deformation runs over a small disc D in P_C centred at the point Y.

The set $\operatorname{tang}(A \cup S) \cap D$ can, in general, consist of several components corresponding to different points of nontransversality of Y and A. Denote by $\Sigma(a)$ the component of it consisting of planes X tangent to A at points close to a.

Proposition 1. *The set $\Sigma(a)$ coincides with the discriminant variety of the deformation (6); see Chapter I, § 3.4.*

The proof is immediate.

For any $\lambda \in D$ consider the corresponding variety

$$W_\lambda \equiv B \cap A \cap X_\lambda, \tag{7}$$

where X_λ is the plane given by (5) and B is a small disc around a (while the disc D is assumed to be very small even with respect to B). The local coordinates $z_1, \ldots, z_{n-1}$ on $B \cap X_\lambda$ locally identify this variety with the local level variety $\{z | F(z, \lambda) = 0\}$ of the deformation (6). In particular, for $\lambda \in D \setminus \Sigma(a)$ this variety is a smooth complex $(n-2)$-dimensional manifold with boundary, which is homotopically equivalent to a wedge of $\mu(\varphi)$ $(n-2)$-dimensional spheres. Thus, the groups $\tilde{H}_{n-2}(A' \cap X'_\lambda, \partial B)$, $\tilde{H}_{n-2}(A' \cap X'_\lambda)$ from the Reduction Theorem I.10.2 are just the groups $\tilde{H}_{n-2}(W_\lambda, \partial W_\lambda)$ and $\tilde{H}_{n-2}(W_\lambda)$ participating in the classical monodromy theory of isolated singularities of functions, see Chapter I, § 3; the operator $\mathrm{Var}_C : \tilde{H}_{n-2}(A' \cap X'_\lambda, \partial B) \to \tilde{H}_{n-2}(A' \cap X'_\lambda)$ defined by any loop C in $D \setminus \Sigma(a)$ is thus determined by the standard Picard–Lefschetz formula.

The generators in the groups $\tilde{H}_{n-2}(A' \cap X')$ and $\pi_1(D \setminus s(Y))$ can be chosen in the following way. Let X_0 be the distinguished point in $\mathrm{reg}(A \cup S) \cap D$, and $\lambda^0 = \{\lambda_1^0, \ldots, \lambda_n^0\}$ its coordinate description. Consider the complex line $l(X_0)$ through X_0 in P_C that consists of points whose coordinates $\lambda_1, \ldots, \lambda_{n-1}$ are the same as for the point X_0 (so that λ_n is a coordinate on this line).

Lemma 1. *For almost any X_0 the corresponding line $l(X_0)$ has in D exactly $\mu(\varphi)$ intersection points with the set $\Sigma(a)$. The group $\tilde{H}_{n-2}(A' \cap X'_0)$ is generated by $\mu(\varphi)$ cycles, which vanish at these points along any distinguished system of paths connecting these points with λ^0 in the line $l(X_0)$, and the group $\pi_1(D \setminus \Sigma(a))$ is generated by the simple loops (see Figs. 8, 9) in this line that correspond to the paths of this system.*

Indeed, X_0 satisfies this assertion if the function $\varphi_{\lambda^0} \equiv F(\cdot\,; \lambda^0)$ is a nondiscriminant strict morsification of φ (see Chapter I, § 3); thus we is only need to prove that a strict morsification of φ can be always chosen in the form (6) with certain λ. This is exactly the assertion of Lemma 1 in Chapter I, § 3.

The assertion (b) of Theorem 1 now follows from Proposition I.5.5 (and from the fact that the monodromy group of a Morse singularity in an odd-dimensional space consists of two elements acting on the (one-dimensional) vanishing homology group as multiplications by 1 and -1). To prove assertion a) in the case when the point a is parabolic, let us choose an arbitrary pair of vanishing cycles $\Delta, \Delta' \in$

$\tilde{H}_{n-2}(A' \cap X_0)$ such that their intersection index $\langle \Delta, \Delta' \rangle$ is not equal to zero (such a pair exists by the connectedness of the Dynkin diagram; see Proposition I.3.4). We act on the cycle Δ by successive transportations over the simple loop in $l(X_0)$, corresponding to the path along which the cycle Δ' vanishes. If n is odd, then any such transportation adds the vanishing cycle Δ', taken with the nonzero coefficient $(-1)^{n(n-1)/2}\langle \Delta, \Delta' \rangle$, to the vanishing cycle Δ. In particular, any such transportation increases the corresponding volume function by the integral of the volume form (3) along the "cap" element corresponding to Δ' taken with this coefficient. Since the volume of any cap element is not identically zero on P_C, the volume function V has infinitely many values in D.

The assertion of item a) concerning nonparabolic points follows easily from Hartogs' lemma. Indeed, the volume function is single-valued in D outside $\Sigma(a)$ (because in this case the unique cycle vanishing in D is stable under the local monodromy); moreover, from the fact that the form ω is holomorphic it folows that this function can be extended to a continuous function on the whole of D (and this function is equal to zero on $\Sigma(a)$: the integral of ω along a zero cycle vanishes).

§ 5. Obstructions to integrability arising from the cuspidal edges. Proof of Theorem 1.8

Now, possessing the notions of the generating function and projective duality, we are in a position to define the notion of generic cuspidal edge used in Theorem 1.8.

Definition 1. A point a of a divisor $A \subset \mathbf{C}P^n$ belongs to the *generic cuspidal edge* of A if A is locally irreducible at a, the germ of the dual surface $A^* \subset (\mathbf{C}P^n)^*$ at the corresponding point a^* is smooth, and the generating function of A^* at this point has a singularity of type A_2; see Chapter I, § 5.

It is easy to verify that the germ of the surface projectively dual to the smooth divisor at its point of type A_2 actually has a cuspidal edge, in particular, its singular set is $(n-2)$-dimensional and smooth, and the transversal slice of this germ by a 2-plane is the curve having in some affine coordinates the parametric representation $\{x = t^2, y = t^3 + O(t^4)\}$.

The proof of Theorem 1.8 consists of two steps. First we consider a wide class of singularities, including the generic cuspidal edges as a very special case (and included as a special case in the class considered in Chapter I, § 11.9), and prove a conditional version of Theorem 1.8 for it: if a certain local homology class, defined close to such a singularity, is representable by caps, then the surface is not algebraically integrable.

Then we prove that at least for the generic cuspidal edges this additional condition is surely satisfied.

5.1. Stating the results.

Definition 2. An analytic submanifold $\sigma \subset \mathbf{C}P^n$ of arbitrary dimension is τ-*degenerate* if the set of all hyperplanes in $\mathbf{C}P^n$ tangent to it (i.e. containing its tangent planes) has dimension less than $n-1$.

The germ of any irreducible analytic curve in $\mathbf{C}^2$ can be represented parametrically in some affine coordinate system in the form

$$x = t^u, \quad y = t^{b_1} + \lambda_2 t^{b_2} + \cdots, \tag{8}$$

where $u < b_1 < b_2 < \ldots$ and the only common divisor of $u, b_1, b_2, \ldots$ is 1. Then u is equal to the local multiplicity of the curve, and the pair of numbers u, b_1 does not depend on the choice of suitable affine coordinates x, y. For instance, if the curve has a standard cuspidal singularity, then $u = 2, b_1 = 3$.

Definition 3. The germ of a locally irreducible plane curve is called *obstructing* if in its representation (8) the number u is even, and $2b_1/u$ is an odd integer. A germ of a locally reducible analytic curve is called obstructing if at least one of its irreducible components is obstructing.

In particular, the standard cusp point is obstructing.

Let A be a hypersurface in $\mathbf{C}^n$ supplied with some analytic Whitney stratification, and σ a $(n-2)$-dimensional stratum of this stratification. This stratum is called obstructing if the intersection of A with some (and then any) two-dimensional complex affine plane Ξ^2 transversal to σ is an obstructing curve in this plane close to the point of its intersection with σ.

Lemma 1. *The generic cuspidal edge is obstructing and* τ*-nondegenerate.*

The proof is elementary; see for example subsection 5.3 below.

Suppose that a divisor $A \subset \mathbf{C}^n$ has an obstructing τ-nondegenerate stratum σ. Then the set of generic points of this stratum (see Definition I.11.1) is dense in it. Close to any such generic point a of σ we are in the situation described in Chapter I, § 11.9. In particular, if B is a small disc around a in $\mathbf{C}^n$, and X a hyperplane transversal to A and very close to a generic tangent plane to σ at a, then the corresponding vanishing homology groups $\bar{\mathcal{H}}_{n-2}(n), \mathcal{H}_{n-2}(n)$ are defined by (I.44) and are isomorphic to $\mathbf{C}^{u-1}$, where u is the exponent for the transversal slice (8).

Theorem 1. *Suppose that n is odd, and the affine complexification $A_f \subset \mathbf{C}^n$ of the boundary of the algebraic body $K_f \subset \mathbf{R}^n$ has an obstructing τ-nondegenerate $(n-2)$-dimensional stratum σ. Let a be a generic point of this stratum, B a small disc around a in $\mathbf{C}^n$, and X a hyperplane transversal to A and very close to a generic tangent plane to σ at a. Suppose that the local relative homology group $\bar{\mathcal{H}}_{n-2}(n) \equiv \tilde{H}_{n-2}(A_f \cap X \cap B, \partial B)$ contains a harmful element (see Definition I.11.5) which is a reduction modulo the complement of B of some element in $\tilde{H}_{n-2}(A_f \cap X)$ representable by a vanishing cycle (see § 3). Then the body K_f is not algebraically integrable.*

Theorem 2. *For a generic cuspidal point of the divisor $A_f \subset \mathbf{C}^n$ the vanishing homology group $\bar{\mathcal{H}}_{n-2}(n) \simeq \mathbf{C}^1$ is generated by an element representable by a vanishing cycle.*

Theorem 1.8 is a direct consequence of these two theorems: indeed, in the case of odd u the generator of the group $\bar{\mathcal{H}}_{n-2}(n)$ for a generic cuspidal edge is obviously harmful.

5.2. Proof of Theorem 1.

Let X be the hyperplane participating in Theorem 1, and C the loop in regP_C with beginning and end at the point X, the monodromy along which was investigated in Chapter I, § 11.9. Let $\kappa(X) \in \tilde{H}_n(\mathbf{C}^n, A_f \cup X)$ be an element representable by a cap and such that applying to it successively 1) the reduction map Red : $\tilde{H}_n(\mathbf{C}^n, A_f \cup X) \to \tilde{H}_{n-2}(A_f \cap X)$ defined in Chapter I, § 10.4, 2) reduction modulo the complement of B, and 3) the excision isomorphism, we get the harmful element of the group $\bar{\mathcal{H}}_{n-2}(n)$ mentioned in Theorem 1. By Proposition I.11.6, for any natural number r the $(r \cdot u)$-fold iteration of the monodromy of the plane X along the loop C adds to the cycle Red$(\kappa(X))$ r times the element (I.48) taken with a nonzero multiplicity (equal to $\pm$ the harm of our element in $\bar{\mathcal{H}}_{n-2}(n)$).

Denote by $\mathcal{I}(X)$ the element of the group $\tilde{H}_n(B, A_f \cup X)$ obtained by the isomorphism red^{-1} (described in Theorem I.10.2) from the element (I.48) with $j = 1$, and by $\mathcal{I}_1(X)$ the class obtained by red^{-1} from the first term $\Sigma^{n-2}(\gamma_2 - \gamma_1)$ of this element.

Since the operations red$'$, red commute with the monodromy (see Theorem I.10.2), this $(r \cdot u)$-fold rotation adds to the volume function $V_\kappa(X)$ r times the integral of the volume form (3) along the cycle $\mathcal{I}(X)$ multiplied by the same nonzero coefficient. Thus, to prove the nonalgebraicity of the volume function it remains to prove that this integral is not equal to zero. (In fact, the last assertion is true for all $n \geq 2$, not necessarily odd.)

We shall estimate this integral together with all similar integrals of the volume form along the cycles $\mathcal{I}(X_\zeta)$ defined by all planes X_ζ distinguished (in the adapted coordinate system choosen as in Chapter I, § 11.1) by the equation $z_n = \zeta$, $|\zeta|$ sufficiently small. Denote this integral by $V[\zeta]$, and the similar integral along the cycle $\mathcal{I}_1(X_\zeta)$ by $V_1[\zeta]$.

Lemma 2. *If u and b_1 are the exponents from formula (8) for the curve $A_f \cap \Xi^2$ in the transversal slice of a τ-nondegenerate stratum σ, and u is even, then*

$$a) \quad V_1[\zeta] =_{\zeta\to 0} \zeta^{(n/2+b_1/u)}(c+o(1)), \tag{9}$$

where the constant c is equal to zero if and only if u divides b_1;

b) if u does not divide b_1, then

$$\lim_{\zeta\to 0}\frac{V[\zeta]}{V_1[\zeta]} = \sum_{j=0}^{u/2-1}[\exp(2\pi i\frac{2b_1}{u})]^j. \tag{10}$$

Corollary. *If under the assumptions of Lemma 2 the stratum σ is obstructing, then $V[\zeta] \neq 0$ for sufficiently small $|\zeta| \neq 0$.*

Indeed, in this case the right-hand side of (10) is equal to $u/2 \neq 0$.

Thus Theorem 1 follows immediately from this lemma.

Proof of Lemma 2. Throughout this proof, the symbol $\approx$ means "equal up to terms of higher degree in the small parameter".

First, consider the case when $n = 2$, in particular, A_f is a plane curve given locally by the parametric equation $x = t^u$, $y \approx at^{b_1}$. Then the coordinate $z_n \equiv z_2$ of the adapted system satisfies $\partial z_2/\partial x \neq 0$, and, using a dilation of x and t, we can assume that $z_2 = x + \alpha y$. So the lines X_ζ are given by $x + \alpha y = \zeta$ with certain fixed α. See Figure 28.

The coordinates (x, y) of all the u points $A_f \cap X_\zeta$ are

$$(\approx \zeta, \approx \zeta^{b_1/u}). \tag{11}$$

Let us choose one of the roots $\zeta^{1/u}$ of ζ so that (11) is the equation of the point γ_1; see Chapter I, § 11.9. Then the coordinates of the other points γ_l, $l = 2, \ldots, u$, are equal to $(\approx \zeta, \approx \exp(2\pi i\frac{l-1}{u})\zeta^{b_1/u})$. The integral of the volume form $dx \wedge dy \equiv dz_2 \wedge dy$ along the cycle $\mathrm{red}^{-1}(\gamma_{2j} - \gamma_{2j-1})$ (shown in Figure 28 by the shadowed triangle) is

$$\approx \exp\Big(2\pi i\frac{2(j-1)}{u}\Big)\frac{[\exp(2\pi i\frac{b_1}{u}) - 1]}{\frac{b_1}{u}+1}\zeta^{\frac{b_1}{u}+1}. \tag{12}$$

This proves Lemma 2 in the case $n = 2$.

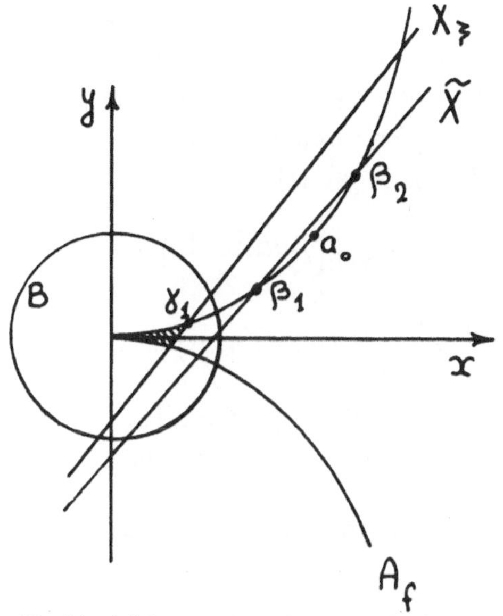

Fig. 28. Vanishing cycles close to a cuspidal point

In the case of arbitrary n, the cycles $\mathrm{red}^{-1}\Sigma^{n-2}(\gamma_{2j} - \gamma_{2j-1})$ are constructed as follows. As in Chapter I, § 11.3, we introduce the local coordinates $z_1, \ldots, z_n$ close to a in such a way that the coordinates z_{n-1}, z_n coincide with the last two coordinates of an adapted system, while $z_1, \ldots, z_{n-2}$ differ from the first $n-2$ coordinates of this system by terms of degree ≥ 2, and the stratum σ is given in these coordinates by the equation $z_n = z_1^2 + \cdots + z_{n-2}^2$. Since the functions $V[\zeta], V_1[\zeta]$ are obviously analytic, it is sufficient to prove Lemma 2 for real positive ζ. For such ζ the group $\tilde{H}_{n-2}(\sigma \cap B, \sigma \cap X_\zeta \cap B)$ is generated by the vanishing disc distinguished in local coordinates $z_1, \ldots, z_{n-2}$ by the conditions

$$\operatorname{Im} z_1 = \cdots = \operatorname{Im} z_{n-2} = 0, \; z_1^2 + \cdots + z_{n-2}^2 \leq \zeta. \tag{13}$$

Denote this disk by Θ_ζ. For any interior point x of this disc, consider the two-dimensional plane $\Xi^2_{(x)}$ through x along which all the coordinates $z_1, \ldots, z_{n-2}$ are constant. In the intersection with this plane, A_f and X_ζ form a configuration considered in the above-described two-dimensional situation (with one change: ζ should be replaced by $\zeta - z_1^2 + \cdots + z_{n-2}^2$ so that for the marginal values of x the shadowed triangle in Figure 28 contracts to a single point). Fix the point γ_1 of the set $A \cap X_\zeta \cap \Xi^2_{(0)}$; then the points $\gamma_2, \ldots, \gamma_u$ of this set also become well defined, as also the cycles $\mathrm{red}^{-1}(\gamma_{2j} - \gamma_{2j-1})$, $j = 1, \ldots, u/2$, in the group

$\tilde{H}_2(\Xi^2_{(0)}, A_f \cap X_\zeta)$; see Figure 28. Transport this cycle by the Gauss–Manin connection in similar groups over all other points x of the vanishing disc (13). The desired class $\mathrm{red}^{-1}(\Sigma^{n-2}(\gamma_{2j} - \gamma_{2j-1})) \in \tilde{H}_n(B, A_f \cap X_\zeta)$ is realized by the cycle swept out by this family of two-dimensional cycles over all points x of our disc.

Since all estimates of smaller terms, assumed in the symbol $\approx$ in formula (12), are uniform for all planes $\Xi^2_{(x)}$ for x from our disc Θ, the integral of the volume form along the cycle $\mathrm{red}^{-1}(\Sigma^{n-2}(\gamma_{2j} - \gamma_{2j-1}))$ is

$$\approx \exp\Big(2\pi i \frac{2(j-1)}{u}\Big) \times$$

$$\times \int_\Theta \frac{[\exp(2\pi i \frac{b_1}{u}) - 1]}{\frac{b_1}{u} + 1} J(x)(\zeta - \|x\|^2)^{\frac{b_1}{u}+1} dz_1 \wedge \ldots \wedge dz_{n-2}, \tag{14}$$

where Θ is a disc of radius $\sqrt{\zeta}$ in $\mathbf{R}^{n-2}$, $\|x\|^2 = z_1^2(x) + \cdots + z_{n-2}^2(x)$, and $J(x) \approx 1$ is the Jacobian determinant of the coordinate system $z_1, \ldots z_{n-2}$ with respect to the original adapted affine system.

Lemma 2 follows immediately from (14), thus Theorem 1 is also proved.

5.3. Proof of Theorem 2.

Main example: let $n = 2$, and let the curve $A_f^{(2)}$ be given by the equation $y^2 = x^3$; see Figure 28. A vanishing cycle $\{\beta_1 - \beta_2\} \in \tilde{H}_0(\tilde{X} \cap A_f^{(2)})$ arises at the point a_0. The Gauss–Manin connection over the segment connecting the planes $\tilde{X}$ and X_ζ and the subsequent reduction modulo $\mathbf{C}^2 \setminus B$ obviously takes this cycle to the element $\{\gamma_1\} \in \tilde{H}_0(X_\zeta \cap A_f^{(2)} \cap B, \partial B)$ that generates this group.

Now let n be arbitrary. First we consider the model cuspidal edge whose projectively dual manifold has the generating function

$$\phi(\zeta_1, \ldots, \zeta_{n-1}) = \frac{1}{3}\zeta_1^3 + \frac{1}{2}\zeta_2^2 + \cdots + \frac{1}{2}\zeta_{n-1}^2, \tag{15}$$

i.e. is distinguished by the equation $\zeta_n = \phi(\zeta_1, \ldots, \zeta_{n-1})$ in some local affine coordinates.

It is easy to calculate that the germ of the projectively dual surface in $\mathbf{C}^n$ is given parametrically by $(\zeta_1, \ldots, \zeta_{n-1}) \to (\zeta_1^2, \zeta_2, \ldots, \zeta_{n-1}, \frac{2}{3}\zeta_1^3 + \frac{1}{2}\zeta_2^2 + \cdots + \frac{1}{2}\zeta_{n-1}^2)$, or, equivalently, in some affine coordinates in $\mathbf{C}^n$ it can be defined by the equation

$$x_n^3 = (x_{n-1} - x_1^2 - \cdots - x_{n-2}^2)^2. \tag{16}$$

In particular, the singular stratum σ is given by the equations $x_n = 0, x_{n-1} = x_1^2 + \cdots + x_{n-2}^2$, and the whole surface A_f is swept out by a $(n-2)$-parameter family of singular curves lying in the 2-planes along which $x_1, \ldots, x_{n-2}$ are constant,

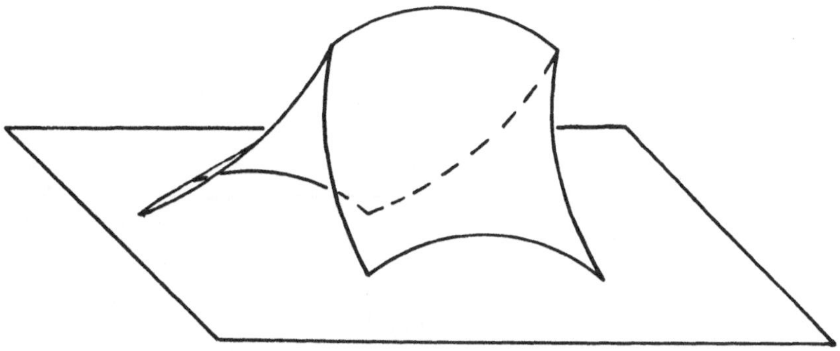

Fig. 29. The model cuspidal edge

and are obtained from the single curve $\{x_n^3 = x_{n-1}^2, x_1 = \cdots = x_{n-2} = 0\}$ by parallel transportations. The intersection of this surface with the real space $\mathbf{R}^n$ (in the case $n = 3$) is shown in Figure 29. This intersection obviously consists of two smooth parts, one of which is convex. Identify the plane of the main example with the transversal plane $\Xi^2 = \{x | x_1 = \cdots = x_{n-2} = 0\}$ by the correspondence $(x, y) \leftrightarrow (x_n, x_{n-1})$. Then the curve $A_f^{(2)}$ of the above two-dimensional example goes into the curve $A_f \cap \Xi^2$, while the branch of the real curve on which $y > 0$ goes into the convex part of $\operatorname{Re} A_f$. To any affine line $X \subset \mathbf{C}^2$ we assign the hyperplane $\mathbf{X} \subset \mathbf{C}^n$ such that $X = \mathbf{X} \cap \Xi^2$ and the equation of $\mathbf{X}$ does not contain the coordinates $x_1, \ldots, x_{n-2}$. Then again the plane $\tilde{\mathbf{X}}$ contains a vanishing cycle given by the intersection of $\operatorname{Re} \tilde{\mathbf{X}}$ with the convex part of $\operatorname{Re} A_f$, and the group $\tilde{H}_{n-2}(A_f \cap \mathbf{X}_\zeta \cap B, \partial B)$ also contains an element representable by a vanishing cycle, namely, the intersection of $\operatorname{Re} \mathbf{X}_\zeta$ with this part and with B. By the construction, this cycle coincides with the cycle $\operatorname{stab}^{n-2}(\gamma_1)$ generating this group, and Theorem 2 is proved for the model cuspidal edge.

In the dual space $\mathbf{C}P^n$ the path connecting $\tilde{\mathbf{X}}$ with $\mathbf{X}_\zeta$ looks like this. The variety $\operatorname{tang}(A_f)$ corresponding to the considered piece of A_f consists of two components, the manifold $(A_f)^*$ described by the generating function (15) and the set of planes tangent to σ. For the model two-dimensional situation, these two components look as in Figure 30a. The plane $\tilde{\mathbf{X}}$ is very close to the first component and relatively far from the second, and the plane $\mathbf{X}_\zeta$ is close to the second component and relatively far from the first one.

Finally, we consider a divisor A'_f with arbitrary generic cuspidal edge.

In an appropriate affine coordinate system the generating function ϕ' of its dual variety is equal to the function ϕ given by (15) plus the terms of higher quasihomo-

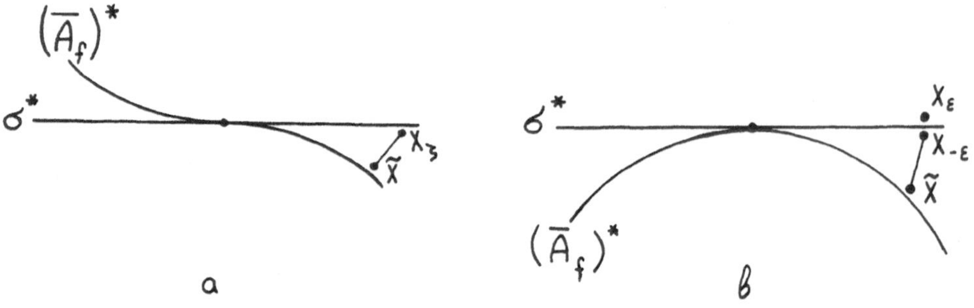

Fig. 30. Sets of nongeneric lines with respect to a cuspidal point and to a trivial point of $A_f \cap S$

geneous degrees (see Definition I.5.3).

Consider simultaneously a similar close surface A_f whose dual manifold is given in the same coordinates by the model generating function (15). Two components of $\text{tang}(A_f)$ are very close to these for $\text{tang}(A'_f)$, moreover the one-parameter family $\{\tau\phi + (1-\tau)\phi',\ \tau \in [0,1]\}$ realizes an isotopy between these pairs of manifolds. Indeed, it is easy to verify that at any instant of this isotopy these two components of the current variety $\text{tang}(\cdot)$ are smooth in the investigated domain, intersect along a smooth submanifold of dimension $n-2$, and have tangency of order 2 at the points of this submanifold. During this isotopy, the segment connecting the planes $\tilde{\mathbf{X}}$ and $\mathbf{X}_\zeta$ also moves to a close path in $\text{reg}(A'_f)$ connecting a plane $\tilde{\mathbf{X}}'$ close to $\tilde{\mathbf{X}}$ and almost tangent to $(A'_f)^*$ with a plane $\mathbf{X}'_\zeta$ close to $\mathbf{X}_\zeta$ and almost tangent to the cuspidal edge of A'_f. Since all the above constructions of the vanishing cycle $\text{stab}^{n-2}(\gamma_1)$ for A_f are localized in a small neighbourhood of the investigated cusp point (while the disc B could even be chosen much smaller than this neighbourhood), the cycle in $A'_f \cap \mathbf{X}'_\zeta$ obtained from the vanishing cycle $A'_f \cap \tilde{\mathbf{X}}'$ is well defined and again generates the group $\tilde{H}_{n-2}(A'_f \cap \mathbf{X}'_\zeta \cap B, \partial B)$. This proves Theorem 2.

§ 6. Obstructions to integrability arising from the asymptotic hyperplanes. Proof of Theorem 1.9

Throughout this section we use the notations of § 1.7, in particular, S is the "infinite" hyperplane in $\mathbf{C}P^n$ and $\bar{A}_f$ the projective closure of A_f.

The proof of Theorem 1.9 is very similar to the proofs of Theorems 1.7 and 1.8 given in §§ 4 and 5 respectively. We consider a family of manifolds localized close to the trivial point of $\bar{A}_f \cap S$ and parametrized by the points of a punctured disc. Then we prove that a) a certain relative homology class in the (arbitrarily chosen) distinguished manifold of this family is representable by caps, b) the r-fold rotation along the generator of the fundamental group of the base of our family adds to this class r times a generator of the absolute homology group of this manifold, and c) the integral of the volume form along this absolute cycle is nonzero.

6.1. Main lemmas.

Let $a \in \bar{A}_f \cap S$ be a nonparabolic point of the divisor $\bar{A}_f \cap S \subset S$, B a small disc around a in $\mathbf{C}P^n$, and Y a hyperplane through a in $\mathbf{C}P^n$ which is transversal to both S and $\bar{A}_f$ but has a simple tangency with $\bar{A}_f \cap S$. Then the triple of manifolds $(\bar{A}_f, S, Y)$ realizes close to a the standard degeneration of type P_3 from Chapter I, § 9; see Figures 16 and 31b.

Let $l = 0$ be the linear equation of Y in some affine chart close to a. Consider the one-parameter family X_ζ, $\zeta \in \mathbf{C}^1$, of hyperplanes distinguished by the equations $l = \zeta$. Let T' be a small punctured disc in $\mathbf{C}^1$ around 0. Consider the locally trivial fibre bundle over T' whose fibre over a point ζ is the pair $(B \setminus S; (\bar{A}_f \cup X_\zeta) \cap (B \setminus S))$.

Lemma 1. *If the disc T' is sufficiently small, then for any $\zeta \in T'$ the groups*

$$\tilde{H}_n(B \setminus S; (\bar{A}_f \cup X_\zeta \cup \partial B) \setminus S)\,, \tag{17}$$

$$\tilde{H}_n(B \setminus S; (\bar{A}_f \cup X_\zeta) \setminus S) \tag{18}$$

are isomorphic to $\mathbf{Z}$.

If n is even, $n > 2$, then the operator Var *in our fibre bundle of pairs, defined by r times the generator of the group $\pi_1(T')$, maps the generator of the group (17) into r times a generator in (18).*

This is just a special case of Theorem I.9.1 or Proposition I.10.3.

Lemma 2. *If a is a trivial point of $\bar{A}_f \cap S$, then the generator of the group (17) is obtained by the obvious reduction homomorphism $\tilde{H}(\mathbf{C}^n, \bar{A}_f \cup X_\zeta) \to \tilde{H}_n(B \setminus S; \bar{A}_f \cup X_\zeta \cup \partial B \setminus S)$ from an element representable by a cap.*

Lemma 3. *For any $n \geq 3$ the integral of the volume form (3) along the generator of the group (18) is nonzero.*

Theorem 1.9 follows immediately from these three lemmas.

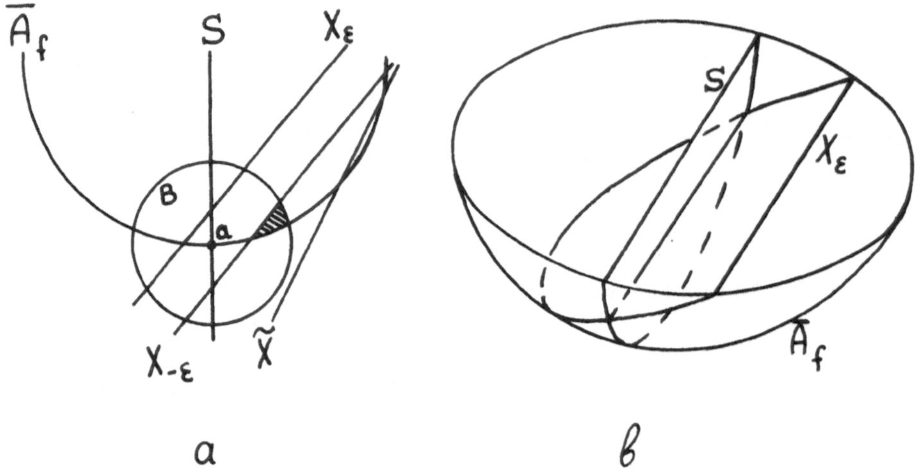

Fig. 31a. Vanishing cycles close to a standard point in $\mathbf{C}P^2$
Fig. 31b. The model standard point of $A_f \cap S$ in $\mathbf{C}P^3$

6.2. Proof of Lemma 2.

This proof is very similar to that of Theorem 5.1. The *main example* is again two-dimensional and is shown in Figure 31a: we suppose that in some local affine coordinates x and y with origin at a, S is given by the equation $x = 0$, $\bar{A}_f$ by $y = x^2$, and X_ζ by $y = \alpha x + \zeta$, $\alpha \neq 0$.

For $\alpha > 0$ and $\zeta = -\varepsilon < 0$ the group (17) is generated by the shadowed triangle in Figure 31a bounded by $X_{-\varepsilon}$, $\bar{A}_f$ and ∂B. It is obvious from the diagram that this triangle is representable by a cap modulo the complement of B. Note however that this model example is not typical from the viewpoint of Lemma 1 (which does not hold for $n = 2$).

The *model example* for arbitrary $n \geq 2$ is obtained from the previous one by a suspension procedure as in § 5.3: we introduce affine local coordinates $x, y, z_1, \ldots, z_{n-2}$ with origin at a, and assume that S is given locally by the equation $x = 0$, X_ζ by $y = \alpha x + \zeta$, where $\alpha > 0$ is comparatively far from 0, and $\bar{A}_f$ by $y = x^2 + z_1^2 + \cdots + z_{n-2}^2$; see Figure 31b.

By Pham's Theorem I.9.1 (and the explicit construction of the isomorphisms assumed there), the statement of Lemma 2 is equivalent to the fact that the generator of the group

$$\tilde{H}_{n-2}(B \cap \bar{A}_f \cap X_\zeta \setminus S; \partial B \setminus S) \tag{19}$$

can be obtained by the reduction homomorphism $\tilde{H}_{n-2}(\bar{A}_f \cap X_\zeta \setminus S) \to \tilde{H}_{n-2}(B \cap$

$\bar{A}_f \cap X_\zeta \setminus S; \partial B \setminus S)$ from an element representable by a vanishing cycle.

Obviously, for small negative ζ the relative cycle

$$B \cap \operatorname{Re} \bar{A}_f \cap \operatorname{Re} X_\zeta (\operatorname{mod} \partial B)$$

is obtained from such an element, and the fact that this cycle generates the group (19) follows from Corollary 1 of Proposition I.2.1 (in which n should be replaced by $n-2$, $\mathbf{C}^n$ by $\bar{A}_f \cap X_\zeta$, B by $B \cap \bar{A}_f \cap X_\zeta$, and f by the equation of S in $\bar{A}_f \cap X_\zeta$, which has the standard Morse form in appropriate coordinates).

Finally, in the general case the plane S and the family X_ζ have in appropriate affine coordinates the same form as in the above model case, while the equation of $\bar{A}_f$ is equal to that for the model situation plus terms of degree ≥ 3. The entire situation is topologically the same. In particular, the set of planes $X \in P_C$ nontransversal to $\bar{A}_f \cup S$ close to a consists of two components, the manifold $(\bar{A}_f)^*$ projectively dual to $\bar{A}_f$ and the set of planes having tangency with $\bar{A}_f \cap S$. These two components are smooth and simple tangent to each other along a smooth $(n-2)$-dimensional submanifold of both; see Figure 30b. Arguing as in § 5.3, we see that also for the pair $(S, \bar{A}_f)$ with deformed $\bar{A}_f$ and appropriate X_ζ the generator of the group (19) is representable by a vanishing cycle. Q.e.d.

6.3. Proof of Lemma 3.

In the local affine coordinates chosen above close to a, the volume form is equal (up to the constant Jacobian factor) to $\frac{1}{x} dx \wedge dy \wedge dz_1 \wedge \ldots \wedge dz_{n-2}$ (where x is the equation of S). By Theorem I.9.1, the generator of the group (17) is realized by the Leray tube around an arbitrary cycle in S generating the group

$$\tilde{H}_{n-1}(S \cap B, (\bar{A}_f \cup X_\zeta) \cap B). \tag{20}$$

Therefore the investigated integral is equal (up to the factor $2\pi i$) to the Euclidean volume of the latter cycle (i.e. the integral of the residue form $dy \wedge dz_1 \wedge \ldots \wedge dz_{n-2}$ along it). In the model example (see Figure 31b) with $\zeta > 0$ this generating cycle is realized by the segment in $\operatorname{Re} S$ bounded by (the real parts of) $\bar{A}_f$ and X_ζ; the volume of it is obviously nonzero. Finally, consider the general case when the local equation of $\bar{A}_f$ differs from the model one by terms of degree ≥ 3. Then by the Morse lemma there exists a local diffeomorphism of S that preserves the function y (so that $X_\zeta \cap S$ is again given by the equation $y = 0$), and changes the coordinates $z_1, \ldots, z_{n-2}$ in such a way that $\bar{A}_f \cap S$ is given by the above model equation $y = z_1^2 + \ldots + z_{n-2}^2$. The generator of the group (20) is then given again by the segment in $S \cap \{\operatorname{Im} y = \operatorname{Im} z_1 = \cdots = \operatorname{Im} z_{n-2} = 0\}$ bounded by $\bar{A}_f$ and X_ζ, and the desired volume is equal to the integral along this segment of the volume form $dy \wedge dz_1 \wedge \ldots \wedge dz_{n-2}$ multiplied by the Jacobian determinant $J(z_1, \ldots, z_{n-2})$ of our diffeomorphism. The values that this determinant takes close to a lie in a small neighbourhood of the point 1 in $\mathbf{C}^1$, therefore the entire integral is again nonzero.

§ 7. Several open problems

A. The First and Second Main Conjectures of § 1.3 and § 1.5; see also the last conjecture of § 1.6.

B. Instead of the planes X we could consider arbitrary algebraic surfaces. Are the First and Second Main Conjectures true at least in this extension? This problem allows us to ignore the geometrical properties of body K_f (like convexity) and to deal only with its topological structure. Probably the version of the Second′ Main Conjecture for this problem can be solved by methods like those of [Smale 62], [Milnor 65]; see also Lemmas 3.1 and 3.2.

C. Instead of the volume function $V_K(X)$ we can consider the function on the space of k-dimensional oriented planes whose value on a plane X is equal to the k-dimensional volume of the intersection of X with K. The corresponding monodromy group acts on the spaces $\tilde{H}_k(X, A_f \cap X)$; in the case of hyperplanes the monodromy group (4) considered previously is a central extension of it.

Chapter III.
NEWTON'S POTENTIAL OF ALGEBRAIC LAYERS

By two celebrated theorems of Newton, a homogeneous spherical layer in Euclidean space does not attract bodies inside the sphere, and exterior bodies are attracted by it to the centre of the sphere as by the pointwise particle whose mass is equal to the mass of the entire sphere.

Ivory [Ivory 1809] generalized these theorems to the Newton–Coulomb attraction of ellipsoids: again, the standard layer on an ellipsoid does not attract the interior points, and the exterior attraction force is the same for all confocal ellipsoids.

Arnold [Arnold 82] extended the first of these assertions to the attraction of arbitrary hyperbolic [1] layers: the particle inside the hyperbolicity set is not attracted by the standard layer distributed on the hyperbolic hypersurface of degree d; moreover, the same is true for the attraction by the standard layer multiplied by any "density" polynomial of degree $\leq d-2$. Givental [Givental 84] generalized this fact to the potential functions of arbitrary linear homogeneous elliptic operators with constant coefficients, and proved that if the hyperbolic surface is compact, then for *arbitrary* degrees of the density polynomial and the operator, the potential is a polynomial of a certain degree in the hyperbolicity set.

In this chapter we recall these theorems and study the behaviour of the Newtonian potentials (and the attraction forces defined by them) outside the hyperbolicity set. The main new results are as follows.

Main Theorem 1. *For any strictly hyperbolic polynomial F of degree d in $\mathbf{R}^2$,*

[1] In this chapter we use a notion of hyperbolicity that is slightly different from the standard one considered in Chapter IV and is essentially a "projectivization" of it. A (nonhomogeneous) polynomial F of degree d in $\mathbf{R}^n$ is called here strictly hyperbolic with respect to a point x if any real line in $\mathbf{R}^n$ through x intersects its zero level set $M_F \equiv \{F=0\}$ at d different real points (one of which can be infinitely distant), in other words, if the corresponding homogeneous polynomial in $\mathbf{R}^n \oplus \mathbf{R}^1$ is strictly hyperbolic in the sense of Chapter IV, § 0 with respect to the vector $(x, 1)$. The set of all points x with respect to which F is hyperbolic is called the hyperbolicity set of F.

the attracting force defined by the Newton–Coulomb charge ω_F *of the curve* $\{F = 0\}$ *is an algebraic vector-function outside this curve.*

Moreover, for any polynomial (respectively, holomorphic single-valued) function P *in* $\mathbf{R}^2$*, the attraction force of the polynomial (respectively, holomorphic) charge* $P \cdot \omega_F$*, distributed on a compact hyperbolic curve* M_F*, is also an algebraic (respectively, analytic finite-valued) vector-function. If* $\deg P \leq d - 2$*, the same is also true for noncompact hyperbolic curves.*

Recall that the Newton attraction force is equal to minus the gradient of a function, the potential function of the attraction.

Main Theorem 2. *The Newton potential function in* $\mathbf{R}^n, n \geq 3$*, defined by the standard charge distributed on any ellipsoid or two-component hyperboloid, coincides with an algebraic function outside this surface. In the case of an ellipsoid the potential function defined by the standard charge multiplied by any polynomial (respectively, holomorphic single-valued) function is an algebraic (respectively, finite-valued analytic) function.*

Main Theorem 3. *Let* F *be a generic hyperbolic polynomial* F *of degree* $d \geq 3$ *in* $\mathbf{R}^n$*,* $n \geq 3$*, and* $n+d \geq 8$*. Then the potential function of the standard layer of the surface* $\{F = 0\}$ *does not coincide with an algebraic function in any component of the set* $\{F \neq 0\}$ *other than the hyperbolicity set; moreover, some arbitrarily large partial derivatives of this potential function are also not algebraic in such components.*

Main Conjecture. *The assumption* $n + d \geq 8$ *in Theorem 3 is unnecessary.*

Here is a refinement of Theorem 1, which estimates the number of leaves of the analytic continuation of the attraction force. Let us order the components of the complement of any strictly hyperbolic surface in $\mathbf{R}^n$ (we call these components *zones*) starting from the hyperbolicity set (which has number 0): thus, the k-th zone is the union of points such that all paths in $\mathbf{R}^n$ connecting them with the hyperbolicity set have at least k intersections with M_F, and there is a path having exactly k intersections.

Proposition 1. *The gradient vector field of the potential function of the standard charge, distributed on a hyperbolic curve* $\{F = 0\}$ *of degree* d *in the Euclidean plane, coincides in the* k*-th zone with the sum of two algebraic vector-functions, the number of leaves of each of which does not exceed* $\binom{d-s}{k}$*, where* s *is the number of factors* $x^2 + y^2$ *in the decomposition of the principal (degree* d*) homogeneous part of the polynomial* F *into the simplest real factors. In particular, the attraction force defined by this charge is always an* m*-valued algebraic vector function with* $m \leq \binom{d}{k}^2$*. The same assertion is true for the standard charge with density* P*, where* P *is a polynomial of degree* $\leq d - 2$*.*

If the hyperbolic curve $\{F = 0\}$ is compact and the density function P is holomorphic, then the corresponding attraction force coincides in the k-th zone with an analytic finite-valued (and even algebraic if P is a polynomial) vector-function, the number of leaves of which is majorized by the same number $\binom{d-s}{k}^2$.

Corollary. *If d is even and the principal part of F is equal to $(x^2+y^2)^{d/2}$, then the attraction force coincides with a rational vector-function in the "most nonhyperbolic" $(d/2)$-th zone.*

Also the numbers of leaves of analytic functions from Theorem 2 can be estimated by the cardinalities of the orbits of certain classical finite reflection groups; see § 3.

Example. In the plane Newtonian case, when the attracting surface is a circle, the attraction force coincides with a single-valued vector-function outside the circle, in particular, the upper bound $\binom{2}{1}^2 = 4$ is not attained. This is due to the fact that the equation of the circle is not generic: its asymptotic directions $(1, i)$ and $(1, -i)$ coincide with the singular lines of the Green function $\ln(x^2+y^2)$ of the potential. For any ellipse with different eigenvalues, the estimate 4 of the number of leaves is exact.

In a similar way, the "genericity" condition in Theorem 3 implies that the "infinite part" of the complexification of the hyperbolic surface (i.e. the set of its limit points on the improper hyperplane $\mathbf{C}P^{n-1} \subset \mathbf{C}P^n$) should be smooth and transversal to the standard quadric given by the homogeneous equation $x_1^2 + \cdots + x_n^2 = 0$.

These results depend on the monodromy action on the integration cycle defining the Newton–Coulomb potential. This cycle (called the *Arnold cycle*) is defined for any point of the complement of the hyperbolic surface. As a geometrical object it does not depend on the component of this complement, but such cycles for points from different components are considered as classes in different homology groups.

In the two-dimensional case the monodromy group in question is finite even if our curve is nonhyperbolic; therefore similar indices (the cardinalities of orbits of certain cycles, given by the set of real points and considered as elements of certain homology groups depending on the components of the complement of the algebraic curve) provide a strong system of invariants of rigid isotopy type of the curve; see [Gudkov 74], [Rokhlin 78], [Kharlamov 86].

We also give an interpretation of Ivory's theorem on the attraction of confocal ellipsoids and explain why it has no generalizations to the case of surfaces of higher degrees.

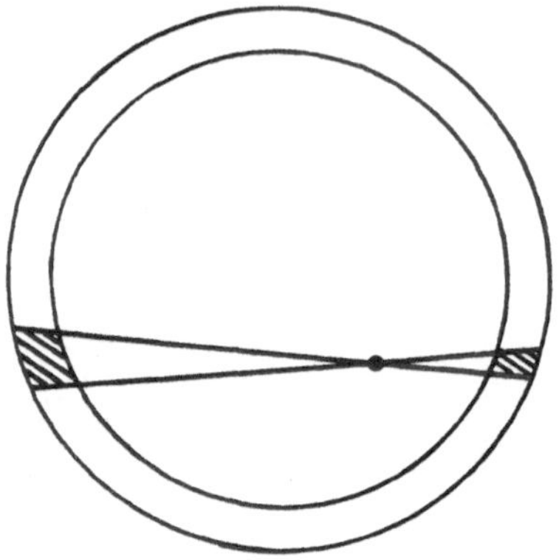

Fig. 32. Annihilation of attraction forces

§ 1. Theorems of Newton and Ivory

Theorem 1 (see [Newton 1687], Theorem XXX of Book 1). *"If toward the individual points of a spherical surface* [2] *are directed forces, decreasing inversely proportional to the squared distances from these points, then a particle inside this surface is not attracted to any side".*

Indeed (see Figure 32) for any infinitely narrow cone with its wedge at this particle the intersections of the sphere with the opposite parts of the cone attract the particle with equal forces, since the areas of these intersections are proportional to the squares of distances.

Theorem 2 ([Newton 1687], Theorem XXXI). *"With the same assumptions, I affirm that a particle outside a spherical surface is attracted to the centre with the force inversely proportional to its squared distance from the centre".*

Proof (due apparently to Laplace; see [Arnold 89′]). A spherically symmetric noncompressible (of divergence zero) vector field decreases inversely proportional to the square of the distance to the centre (since its flows through all the spheres are the same). The attraction field of any particle is noncompressible. Hence the attraction field of any body is noncompressible outside this body. Thus the attraction field of any sphere is noncompressible outside the sphere. Being obviously spherically symmetric, it coincides with the attraction field of a particle at the centre.

These theorems together with the corresponding proofs hold in the n-dimensional case if the attraction force is inversely proportional to the $(n-1)$-th power of the distance.

---

[2] in $\mathbf{R}^3$

A generalization of these two theorems to the case when the attracting surface is arbitrary ellipsoid was given by Ivory [Ivory 1809].

Let F be a smooth function in Euclidean space $\mathbf{R}^n$, and M_F the hypersurface given by $F = 0$. Suppose that grad $F \neq 0$ at the points of M_F, so that M_F is smooth. Denote by dV the volume differential form in $\mathbf{R}^n$, i.e. the form $dx_1 \wedge \ldots \wedge dx_n$ in the Euclidean positively oriented coordinates $x_1, \ldots, x_n$.

Denote by r the Euclidean norm function in $\mathbf{R}^n$, $r = (x_1^2 + \cdots + x_n^2)^{1/2}$, and by C_n the area of the unit sphere in $\mathbf{R}^n$.

Definition 1. The *elementary Newton–Coulomb potential function*, or, which is the same, the *standard fundamental solution of the Laplace operator* in $\mathbf{R}^n$, is the function equal to $\frac{1}{2\pi}\ln r$ if $n = 2$, and to $\frac{-1}{(n-2)C_n} r^{2-n}$ if $n \geq 3$. Denote this function by G.

It is easy to verify that this function is actually a fundamental solution of the Laplace operator. This function can be interpreted as the potential of the force of attraction by a particle of unit mass placed at the origin, i.e., the attraction force of this particle is equal to $-$grad G.

The attraction force with which a body K with density distribution P attracts a particle of unit mass placed at the point $x \in \mathbf{R}^n$ is equal to minus the gradient of the corresponding *potential function*, whose value at the point x is equal to the integral over K of the differential form

$$G(x - y)P(y)dV(y) \tag{1}$$

(if such an integral exists).

Definition 2. The *standard charge* ω_F on the surface M_F is the differential form dV/dF, i.e. the $(n-1)$-form such that for any tangent frame $(l_2, \ldots, l_n)$ of M_F and a transversal vector l_1 the product of the values $\omega_F(l_2, \ldots, l_n)$ and (dF, l_1) is equal to the value $dV(l_1, \ldots, l_n)$. The *natural orientation* of the surface M_F is the orientation defined by this differential form, i.e., the orientation such that the n-frame {grad F; a tangent $(n-1)$-frame positively oriented with respect to this orientation} is positive with respect to the form dV.

In particular, the value at a point $x \notin M_F$ of the limit of potential functions of homogeneous (with density $1/\epsilon$) distributions of charges between the surfaces $F = 0$ and $F = \epsilon$ is equal to the integral of the standard charge form

$$G(x - y)\omega_F(y) \tag{2}$$

along the naturally oriented surface M_F.

In a similar way, any function P on the surface M_F defines the charge $P \cdot \omega_F$, which is called the *standard charge with density* P; the potential at the point x of this charge is equal to the integral of the form

$$G(x-y)P(y)\omega_F(y) \tag{3}$$

along the naturally oriented surface M_F. The attraction force of this charge is equal to minus the gradient of this potential function.

In these terms, Theorems 1 and 2 have the following form: the potential function of the standard charge of a spherical surface (given by the equation $x_1^2+\cdots+x_n^2=c$) is constant inside the sphere and equal to G outside it.

The theorems of Ivory in these terms are as follows.

Theorem 3 (see [Ivory 1809]). *The potential of the standard charge of the ellipsoid in $\mathbf{R}^n$ given by a canonical equation (i.e. by a polynomial F of degree 2) is equal to a constant inside the ellipsoid, while outside the ellipsoid it coincides with the potential function defined by any other ellipsoid confocal to ours (and such that the point where the potential function is evaluated is also exterior for this ellipsoid) multiplied by the constant equal to the ratio of the whole mass distributed on these ellipsoids.*

A proof of the first assertion is almost the same as that of Theorem 1; see the next section for a general form of all these statements. For an explanation of the second assertion, see § 3.5.

§ 2. Potentials of hyperbolic layers are polynomial in the hyperbolicity domains (after Arnold and Givental)

Definition 1. A polynomial $F : \mathbf{R}^n \to \mathbf{R}$ of degree d (and the surface M_F defined by it) is called *hyperbolic with respect to a point* $x \notin M_F$ if any line through x in the projectivization $\mathbf{R}P^n$ of $\mathbf{R}^n$ intersects the closure of the surface M_F in exactly d real points (counted with multiplicities). F and M_F are *strictly hyperbolic with respect to* x if additionally for any such line all these intersection points are distinct.

Lemma 1. *If a surface M_F is hyperbolic with respect to a point x, then it is also hyperbolic with respect to any point in the same component of the complement of M_F.*

This is a "projectivized" form of Theorem 2 of § 0 in Chapter IV below.

Definition 2. The *hyperbolicity domain* of a surface M_F is the union of points x such that M_F is hyperbolic with respect to x.

Example. If M_F is an ellipsoid, then the hyperbolicity domain consists of one component, its interior part. If M_F is a two-component hyperboloid (for example, a hyperbola in $\mathbf{R}^2$), then the hyperbolicity domain consists of two components.

All the smooth projective closures $\bar{M}_F$ of hyperbolic surfaces of a given degree d are situated topologically in the same way: if d is even, then $\bar{M}_F$ consists of $d/2$ surfaces diffeomorphic to a sphere and is ambient isotopic to $d/2$ concentric spheres in an affine chart in $\mathbf{R}P^n$; if d is odd, then there is one more "odd" component homeomorphic to a projective hyperplane, so that the whole surface $\bar{M}_F$ is ambient (and even rigid, see Theorem 1 of Chapter IV, § 0) isotopic to $[d/2]$ concentric affine spheres plus the improper plane. The hyperbolicity domain consists of the proper (not infinitely distant) interior points of the "most interior" spheroid. This spheroid is always convex in $\mathbf{R}P^n$, in particular, the hyperbolicity domain in $\mathbf{R}^n$ may consist of at most two connected components.

Lemma 2. *The set of all strictly hyperbolic surfaces $\bar{M}_F$ of given degree d in $\mathbf{R}P^n$ is path-connected, and its closure coincides with the set of all hyperbolic surfaces of the same degree.*

The set of all strictly hyperbolic compact surfaces $M_F \subset \mathbf{R}^n$ is contractible.

The first statement of this lemma is a direct corollary of a theorem of Nuij (see Theorem IV.0.1), and the second follows immediately from its proof; see [Nuij 68].

Given a strictly hyperbolic polynomial F, let us fix some component of its hyperbolicity domain in $\mathbf{R}^n$ and number the components of M_F starting from the boundary of this component (which becomes number 1), its neighbouring component gets number 2, etc.

Definition 3. The *Arnold cycle* of F is the manifold $\bar{M}_F$, oriented in such a way that in the restriction to its finite part M_F all odd components are taken with the natural orientation (see Definition 1.2), while all even components are taken with the reverse orientations.

The *hyperbolic potential* (respectively, *hyperbolic potential with density P*, where P is any function on M_F) of the surface M_F at a point $x \notin M_F$ is the integral of the form (2) (respectively, (3)) along the Arnold cycle.

As usual, the attraction force defined by these potentials is equal to minus the gradient of the potential functions.

Lemma 3. *This definition of the Arnold cycle is correct, i.e. the orientations of different noncompact components of M_F thus defined are the restrictions of the same orientation of the corresponding components of $\bar{M}_F$.*

The proof is immediate.

Theorem 1 (see [Arnold 82]). *The hyperbolic potential of the surface M_F (and moreover any hyperbolic potential with density P, where P is a polynomial of degree $\leq d-2$) is constant inside the hyperbolicity domain.*

(In other words, the points of the hyperbolicity domain are not attracted by the standard charge on M_F taken with sign 1 or -1 depending on the parity of the number of the component on which this charge is distributed.)

The proof follows Newton's original proof: for any infinitesimally narrow cone centred at the point x, whose direction is not asymptotic for the surface M_F, the forces of attraction to the pieces of M_F cut by the cone are annihilated. Indeed, let us restrict the polynomial F to the line L in $\mathbf{R}^n$ through x contained in this cone; then this attraction force is equal to the solid angle of our cone multiplied by the sum over all zeros A_i of the polynomial $F|_L$ in the numbers $\frac{P(A_i)}{F'(A_i)}$. The last sum is zero because it is the sum of the residues of a rational function over all its complex poles (note that in our assumptions on deg P this function is regular at infinity).

Givental generalized this theorem in two ways: he replaced the Newton–Coulomb potential by the fundamental solution of an arbitrary linear elliptic differential operator with constant coefficients in $\mathbf{R}^n$, and proved that the potential function in the hyperbolicity domain of a compact surface inherits some good properties even if the degree of the density polynomial P is arbitrarily large: it coincides there with a polynomial function.

The fundamental solutions of constant elliptic operators are described by the following theorem.

Definition 4. A homogeneous polynomial $L(\xi)$ of n variables with complex coefficients is called *elliptic* if $L(\xi) \neq 0$ for all $\xi \in \mathbf{R}^n \setminus 0$. A homogeneous differential operator with constant coefficients $L(-i\partial/\partial x_1, \ldots, -i\partial/\partial x_n)$ is called elliptic if its characteristic polynomial $L(\xi_1, \ldots, \xi_n)$ is an elliptic polynomial.

The most famous example of an elliptic polynomial is the polynomial $x_1^2 + \cdots + x_n^2$ corresponding to the Laplace operator.

Proposition 1 (see [Hörmander 83, vol. I]). *Let L be an elliptic homogeneous polynomial in $\mathbf{R}^n$ of degree s. Then the operator $L(-i\partial/\partial x_1, \ldots, -i\partial/\partial x_n)$ has a fundamental solution $E \equiv E(L)$ of the form*

$$E_0 - Q(x) \log|x|, \tag{4}$$

where E_0 is a positively homogeneous distribution of degree $s-n$ in $\mathbf{R}^n$, which coincides with an analytic function in $\mathbf{R}^n \setminus 0$ and satisfies the condition

$$E_0(-x) = (-1)^s E_0(x), \tag{5}$$

while Q is a polynomial of degree $s - n$, in particular, $Q \equiv 0$ if $s < n$.

Let G be any function in $\mathbf{R}^n \setminus 0$ of the form (4), where E_0 is a C^{s-1}-smooth positively homogeneous function of degree $s-n$ satisfying (5), and Q a polynomial of degree $s-n$; let F be a hyperbolic polynomial of degree d in $\mathbf{R}^n$. The corresponding hyperbolic G-potentials and G-potentials with density P of the surface M_F are defined in $\mathbf{R}^n \setminus M_F$ in exactly the same way as the potentials of the fundamental solutions of the Laplace operator were defined in Definition 3: their values at the point x are equal to the integral of the form (3) along the Arnold cycle.

Theorem 2 (see [Givental 84]). *If the hyperbolic surface M_F defined by the hyperbolic polynomial F is compact, then for any G as above, and any polynomial P of degree p, the hyperbolic G-potential with density P of the surface M_F coincides in the hyperbolicity domain with a polynomial of degree $\leq p - d + s$.*

If the degree p of P is at most $d - s$, then the hyperbolic G-potential with density P of the surface M_F is locally constant in the hyperbolicity domain even if the surface M_F is noncompact.

The proof again follows the scheme of that for Theorem 1.1. We consider the integral that defines some partial derivative of order $s-1$ of the potential at a point x in the hyperbolicity domain. The deposit of the charges from a narrow cone with vertex x in this integral is equal to the sum of the residues of the meromorphic form $P/F dz$ on the axis of the cone. The $(p-d+2)$-th partial derivatives of this form on the parameter x are holomorphic at infinity. Hence, by the theorem on the sum of residues, all $(p-d+s+1)$-th partial derivatives of the potential function at the point x are zeros.

§ 3. Proofs of Main Theorems 1 and 2

Here we prove the Main Theorems 1 and 2 of the introduction to this chapter, and reduce the general problem to the study of certain reflection groups: as in the previous chapter, the ramification of the potential functions depends on the structure of the orbits of certain characteristic homology classes under this action.

Since the main method of proof is the Picard–Lefschetz theory, the two cases of even and odd n need to be considered separately; nevertheless the algebraic picture for all dimensions is almost the same. This phenomenon depends on the fact that in the case of odd n the complex continuation of the integration form (3) is a two-valued ramified form, and hence the ramification of the corresponding integral is controlled by the twisted Picard–Lefschetz theory, which is very similar to the untwisted one for even n; see Chapter I, §§ 12, 13.

3.1. Arnold homology classes.

For any point $x \in \mathbf{C}^n$, $x = (x_1, \dots, x_n)$, denote by $S(x)$ the cone in $\mathbf{C}^n$ given by the equation

$$(z_1 - x_1)^2 + \cdots + (z_n - x_n)^2 = 0. \tag{6}$$

Denote by $@ \equiv @(y)$ a local system over $\mathbf{C}^n \setminus S(x)$ with fibre $\mathbf{Z}$ and such that the corresponding representation $\pi_1(\mathbf{C}^n \setminus S(x)) \to \mathrm{Aut}(\mathbf{Z})$ maps the loops whose linking numbers with $S(x)$ are odd (respectively, even) in multiplication by -1 (respectively, the identity operator).

We specify this local system in such a way that integrals of the form $r(\cdot - x)dz_1 \wedge \ldots \wedge dz_n$ along the $(n-1)$-dimensional cycles with coefficients in it are well defined; see Chapter I, § 12.2. Namely, we consider the two-fold covering over $\mathbf{C}^n \setminus S(x)$ on which this form is single-valued, then take the trivial $\mathbf{Z}$-bundle over it, and the direct image in $\mathbf{C}^n \setminus S(x)$ of this bundle under the obvious projection of this covering. The trivial $\mathbf{Z}$-bundle over $\mathbf{C}^n \setminus S(x)$ is naturally included in this direct image as a subbundle; the desired local system is the quotient bundle of these two local systems. Obviously, integrals of the form $r(\cdot - x)dz_1 \wedge \ldots \wedge dz_n$ (and of its products by all single-valued functions) along the piecewise smooth n-chains with coefficients in this local system are well-defined, and if these chains are cycles, they depend only on their homology classes.

Let $F : (\mathbf{C}^n, \mathbf{R}^n) \to (\mathbf{C}, \mathbf{R})$ be a fixed real hyperbolic polynomial of degree d, and W_F the set of its complex zeros, so that $M_F \equiv W_F \cap \mathbf{R}^n$. Let $\bar{W}_F$ be the closure of W_F in $\mathbf{C}P^n$. We assume that W_F is smooth. For such F the *Arnold cycle* was defined in § 2: see Definition 2.3.

Definition 1. If the hyperbolic surface M_F is compact, then for any $x \in \mathbf{R}^n \setminus M_F$ the *Arnold class* $A(x)$ is the class of the Arnold cycle in the group

$$\tilde{H}_{n-1}(W_F \setminus S(x)) \equiv \tilde{H}_{n-1}(W_F \setminus S(x), \mathbf{Z}) \tag{7}$$

in the case of even n, and in the group

$$\tilde{H}_{n-1}(W_F \setminus S(x),\ @(x)) \tag{8}$$

in the case of odd n.

For any $x \in \mathbf{C}^n \setminus W_F$ denote by $\mathcal{H}(x)$ the group (7) in the case of even n and the group (8) if n is odd.

Similarly, denote by $\mathcal{PH}(x)$ the group

$$\tilde{H}_{n-1}(\bar{W}_F \setminus \bar{S}(x)) \equiv \tilde{H}_{n-1}(\bar{W}_F \setminus \bar{S}(x), \mathbf{Z}) \tag{7'}$$

in the case of even n, and the group

$$\tilde{H}_{n-1}(\bar{W}_F \setminus \bar{S}(x),\ @(x)) \tag{8'}$$

in the case of odd n.

Definition 1′. For an arbitrary hyperbolic surface M_F and for any $x \in \mathbf{R}^n \setminus M_F$ the *projective Arnold class* $PA(x)$ is the class of the Arnold cycle in the group $\mathcal{PH}(x)$.

In the case of odd n, integrals of the form (3) along $(n-1)$-chains in $W_F \setminus S(x)$ with coefficients in $@(x)$ are well defined, and the values of these integrals along the cycles depend only on their homology classes in the group (8). Moreover, if $\deg P \leq d-2$, and hence the form (3) is regular at infinity, then it can be integrated along the chains in $\bar{W}_F \setminus \bar{S}(x)$, and the integrals along the cycles depend only on their classes in the group (8′).

In the case of even $n > 2$ the form (3) is single-valued, and no problems with the definition of similar integrals along the elements of the group (7) (or even (7′) if $\deg P \leq d-2$) arise, and in the exceptional case $n = 2$, when (3) is logarithmic, we remember that we are interested not in the potential, but in its first partial derivatives with respect to the parameter x (i.e. in the components of the attraction force vector). Therefore we integrate not the form (3) but its partial derivatives

$$\frac{x_i - y_i}{(x_1 - y_1)^2 + (x_2 - y_2)^2}\omega_F, \; i = 1, 2; \tag{9}$$

these forms are already single-valued and there is no problem in integrating them along the elements of the group (7) (or (7′) if $\deg P \leq d-2$).

3.2. Homology bundles and ramification of potential functions.

It follows easily from the Thom isotopy lemma (see Theorem I.8.5) that for almost all $x \in \mathbf{C}^n$ the groups $\mathcal{H}(x)$ are naturally isomorphic to each other. The set of exceptional x (for which the pair $(W_F, W_F \cap S(x))$ is not homeomorphic to these for all neighbouring x') belongs to a proper algebraic subvariety in $\mathbf{C}^n$ consisting of three components: W_F itself, the set of those x such that $S(x)$ and W_F are tangent outside x in $\mathbf{C}^n$, and the set of those x such that the closure of $S(x)$ in $\mathbf{C}P^n$ is "more nontransversal" to the closure of W_F at their infinitely distant points. For a generic F the last component is empty, and the second is irreducible provided additionally that $n \geq 3$.

Denote this algebraic set of all exceptional $x \in \mathbf{C}^n$ by $\Sigma(F)$.

Consider two fibre bundles over $\mathbf{C}^n \setminus \Sigma(F)$ whose fibres over a point x are the spaces $W_F \setminus S(x)$, $\bar{W}_F \setminus \bar{S}(x)$, and associate with them the homological bundles whose fibres over the same point are the groups $\mathcal{H}(x)$ and $\mathcal{PH}(x)$. As usual, the Gauss–Manin connection in these bundles defines the monodromy representations

$$\pi_1(\mathbf{C}^n \setminus \Sigma(F)) \to \text{Aut } \mathcal{H}(x), \tag{10}$$

$$\pi_1(\mathbf{C}^n \setminus \Sigma(F)) \to \text{Aut } \mathcal{PH}(x). \tag{10'}$$

Let $u(x)$ be the potential function of the polynomial charge $P \cdot \omega_F$, i.e. the function defined for any x by the integral of the form (3) along the cycle $A(x)$.

Proposition 1. *If $n \neq 2$, then the analytic continuation of the potential function u of a compact hyperbolic layer from a point $x \in \mathbf{R}^n \setminus \Sigma(F)$ to the point x' along a path in $\mathbf{C}^n \setminus \Sigma(F)$ is equal to the integral of the differential form $G(x'-y)P(y)\omega_F(y)$, obtained from the similar form (3) by analytic continuation of the parameter x' along the same path, over the cycle in the group $\mathcal{H}(x')$ obtained from $A(x)$ by the Gauss–Manin connection in our homological bundle $[\mathcal{H}(\cdot) \to \cdot\,]$ over this path.*

If $\deg P \leq d-2$, then the same is true for arbitrary (maybe, noncompact) hyperbolic layers and the Gauss–Manin connection in the bundle $[\mathcal{PH}(\cdot) \to \cdot\,]$.

Indeed, by definition this is true for x' from a small neighbourhood of x in $\mathbf{R}^n$.

The two-dimensional version of this proposition is obvious; see Proposition 2 below.

Thus the ramification of (the analytic continuation of) the potential function depends on the monodromy action (10) (respectively, (10′)) on the Arnold element in $\mathcal{H}(x)$ (respectively, in $\mathcal{PH}(x)$).

Namely, consider the linear form $N : \mathcal{H}(x) \to \mathbf{C}$ whose value on the element γ is equal to the integral of the form (3) along the cycle γ (if $n \neq 2$, and if $n = 2$ then N is the linear map $\mathcal{H}(x) \to \mathbf{C}^2$ defined by the integrals of two forms (9)). Tautologically, the analytic function u (or, in the case $n = 2$, the vector function grad u) is finite-valued at x if and only if the linear function N takes finitely many values on the orbit of the cycle $A(x)$ under the action of the monodromy group.

More generally, for any multiindex $\nu \in \mathbf{Z}^n_+$ consider the linear forms

$$N^{(\nu)} : \mathcal{H}(x) \to \mathbf{C}, \tag{11}$$

$$PN^{(\nu)} : \mathcal{PH}(x) \to \mathbf{C}, \tag{11'}$$

whose values on the cycle γ are equal to the integral along γ of the ν-th partial derivative of the form (3) with respect to the parameter x.

Proposition 2. *For any ν ($\neq 0$ if $n = 2$) and $x \in \mathbf{R}^n \setminus \Sigma(F)$, the ν-th partial derivative of the potential function of the standard charge of the compact hyperbolic* *surface M_F with density P is finite-valued at x if and only if the linear form $N^{(\nu)}$ takes finitely many values on the orbit of the cycle $A(x)$ under the action of the monodromy group (10). If P is a polynomial of degree $\leq d-2-|\nu|$, then the same is true for noncompact hyperbolic surfaces if we replace $A(x)$ by $PA(x)$, $N^{(\nu)}$ by $PN^{(\nu)}$, and the action (10) by (10′).*

This is again a tautology.

3.3. Reduced Arnold cycle.

For an arbitrary element γ of the group $\mathcal{H}(x)$ or $\mathcal{PH}(x)$, the corresponding potential function $u_\gamma(x)$ is defined as the integral of the form (3) along the cycle γ (if this integral exists), in particular the usual potential $u(x)$ coincides with $u_{A(x)}(x)$ (or $u_{PA(x)}(x)$ if M_F is noncompact).

According to the Newton–Ivory–Arnold–Givental theorem, if the class γ in $\mathcal{H}(x)$ (respectively, in $\mathcal{PH}(x)$) is obtained by the Gauss–Manin connection from the Arnold cycle $A(\mathbf{x})$, where $\mathbf{x}$ is a point in the hyperbolicity domain of a compact hyperbolic surface (respectively, from the projective cycle $PA(\mathbf{x})$ provided that the degree of the density polynomial is $\leq d-2$), then the potential function $u_\gamma(x)$ is a single-valued holomorphic function; therefore the ramification of our integrals depends only on the class of γ in the quotient group of $\mathcal{H}(x)$ (respectively, $\mathcal{PH}(x)$) modulo the line spanned by an arbitrary homology class obtained by the Gauss–Manin connection from the Arnold class (respectively, projective Arnold class) of a hyperbolic point $\mathbf{x}$.

For any point $x \in \mathbf{R}^n \setminus M_F$ (where M_F is compact) we choose canonically some class thus obtained. Namely, we connect x with an arbitrary $\mathbf{x}$ from the hyperbolicity domain by a complex line and take the path in this line that goes from $\mathbf{x}$ to x along the real axis and misses any point of W_F along a small arc in the *lower* half-plane with respect to this direction (i.e. the half-plane into which the vector $i \cdot (\mathbf{x} - x)$ is directed).

For any $x \in \mathbf{R}^n \setminus M_F$, denote by $A_{\mathsf{hyp}}(x)$ the class in $\mathcal{H}(x)$ thus obtained. We are interested in the monodromy of classes $A(x) - A_{\mathsf{hyp}}(x)$, which will be called the *reduced Arnold classes* and denoted by $\tilde{A}(x)$.

In almost the same way, the *reduced projective Arnold classes*

$$P\tilde{A}(x) \equiv PA(x) - PA_{\mathsf{hyp}}(x)$$

are also defined in the case of noncompact M_F. The only extra ambiguity in this case arises from the fact that the hyperbolicity domain can consist of two components. In this situation we agree to choose the hyperbolic point $\mathbf{x}$ participating in the construction of the class $P\tilde{A}(x)$ in such a component that x lies in the $\leq [d/2]$-th zone with respect to it.

3.4. Groups $\mathcal{H}(x)$ and the vanishing homology of complete intersections.

In what follows we shall consider particularly carefully the case when the attracting surface M_F satisfies certain genericity conditions at its infinitely distant points, namely, the following ones.

Definition 2. The polynomial F (and the corresponding complex surface W_F) is called:

∞-*nondegenerate* if the closure $\bar{W}_F$ of W_F in the projective compactification $\mathbf{C}P^n$ of $\mathbf{C}^n$ is smooth and transversal to the improper hyperplane $\mathbf{C}P^n \setminus \mathbf{C}^n$;

G-compatible if additionally the "infinite part" $\bar{W}_F \setminus \mathbf{C}^n$ is transversal in the improper hyperplane to the standard quadric $\{z_1^2 + \cdots + z_n^2 = 0\}$, i.e. to the boundary of any cone $S(x)$.

In this subsection we consider only the case of G-compatible F.

Theorem 1. *Suppose that the algebraic surface $W_F = \{F = 0\}$ in $\mathbf{C}^n$ is G-compatible,* deg $F = d$. *Then for a generic x the ranks of both groups (7), (8) (in particular, of the group $\mathcal{H}(x)$) are equal to*

$$(d-1)^n + \frac{1}{d-2}(2(d-1)^n - d)$$

if $d > 2$, and $2n$ if $d = 2$.

Indeed, let $\bar{F}$ be the principal (of degree d) homogeneous part of F, and $r^2 \equiv z_1^2 + \cdots + z_n^2$. Then the pair of functions $(f_1, f_2) \equiv (\bar{F}, r^2)$ is a homogeneous complete intersection in $\mathbf{C}^n$ with isolated singularity at 0, and the pair $(F, r^2(\cdot - x))$ is a perturbation of it. Since this perturbation changes only terms of lower degree in the polynomials $\bar{F}$ and r^2, the pair $(W_F, W_F \cap S(x))$ for smooth W_F and nondiscriminant x is homeomorphic to the pair (X', V_λ) from formula (55) of Chapter I for this complete intersection. Thus, for the group (8) the assertion of the theorem follows from Theorems I.13.3 and I.13.2 and from the fact that the Milnor number of the isolated homogeneous function singularity $\bar{F}$ of degree d in $\mathbf{C}^n$ is equal to $(d-1)^n$. Let us prove the assertion about the group (7).

Denote by ∂W_F the "infinite part" $\bar{W}_F \setminus \mathbf{C}^n$ of $\bar{W}_F$. Then the group (7) is Poincaré–Lefschetz dual to the group $\tilde{H}_{n-1}(\bar{W}_F, \partial W_F \cup (\bar{W}_F \cap S(x)))$. Consider the homological exact sequence of the triple $(\bar{W}_F, \partial W_F \cup (\bar{W}_F \cap S(x)), \partial W_F)$. By Theorem I.13.1, the only nontrivial fragment in this sequence is

$$0 \to \tilde{H}_{n-1}(\bar{W}_F, \partial W_F) \to \tilde{H}_{n-1}(\bar{W}_F, \partial W_F \cup (\bar{W}_F \cap S(x))) \to$$

$$\to \tilde{H}_{n-2}(\bar{W}_F \cap S(x), \partial W_F \cap S(x)) \to 0,$$

and the assertion of our theorem about the group (7) follows from Theorem I.13.2 and Poincaré–Lefschetz duality in the manifolds W_F and $W_F \cap S(x)$.

So we have identified the pair $(W_F, W_F \cap S(x))$ with a standard object of the theory of singularities of complete intersections. In particular, we get *two* monodromy actions on the group $\mathcal{H}(x)$: the first one is the representation (10) of the group $\pi_1(\mathbf{C}^n \setminus \Sigma(F))$, and the second is the similar action defined by the loops in the base of any versal deformation of our complete intersection $(\bar{F}, r^2)$ that contains the perturbation $(F, r^2(\cdot \; - x))$. Recall that the latter monodromy action is generated by reflections in the vanishing cycles, corresponding to the critical values of the restriction of a generic perturbation of the function r^2 on the level set W_F of a generic perturbation F of $\bar{F}$; see Chapter I, § 13.

By the definition of versality (and since the set $\mathbf{C}^n$ of all functions of the form $(F, r^2(\cdot \; - x))$ can be embedded in a deformation of $(\bar{F}, r^2)$) the first monodromy group is a subgroup of the second. In § 4 we shall see that this subgroup is proper. We shall call these groups the *small* and the *big* monodromy groups, respectively. To describe the small monodromy group we need several more reductions and notions.

The subgroup $\mathcal{J}(x) \subset \mathcal{H}(x)$ for even n is defined as the image of the tube monomorphism

$$\tilde{H}_{n-2}(W_F \cap S(x)) \to \tilde{H}_{n-1}(W_F \setminus S(x)); \tag{12}$$

for odd n we set $\mathcal{J}(x) \equiv \mathcal{H}(x)$. In other words, for all n the group $\mathcal{J}(x)$ is generated by the vanishing cycles in $M_F \setminus S(x)$ (defined as in Chapter I, § 9 if n is even and as in Chapter I, § 12.4 if n is odd).

On this subgroup there is a symmetric bilinear form $\langle \cdot, \cdot \rangle$: in the case of odd n this is the usual intersection index (which is well defined because the local system $@(x)$ is selfdual), and in the case of even n it is induced by the tube monomorphism (12) from the intersection index on the first group in (12). By Propositions I.2.2 and I.12.3, the square $< \alpha, \alpha >$ of any vanishing cycle $\alpha \in \mathcal{H}(x)$ is equal to 2 if $[\frac{n+1}{2}]$ is odd and -2 if $[\frac{n+1}{2}]$ is even.

Lemma 1. *For any n, the action of the big monodromy group on $\mathcal{H}(x)$ preserves the subgroup $\mathcal{J}(x)$ and the natural bilinear form on it.*

This follows immediately from the Picard–Lefschetz formulae.

Theorem 2. *a) If M_F is compact, then for any point x from the k-th zone of $\mathbf{R}^n \setminus \Sigma(F)$ the reduced Arnold class $\tilde{A}(x) = A(x) - A_{\mathsf{hyp}}(x)$ is equal to the sum of k pairwise orthogonal vanishing cycles. In particular, $\tilde{A}(x)$ belongs to the subgroup $\mathcal{J}(x)$, and its square $\langle \tilde{A}(x), \tilde{A}(x) \rangle$ is equal to $2k$ if $[\frac{n+1}{2}]$ is odd, and $-2k$ if $[\frac{n+1}{2}]$ is even.*

b) For arbitrary M_F and $k \leq [d/2]$, the projective reduced Arnold class $P\tilde{A}(x)$ is equal to the image of such a sum under the obvious inclusion map $\mathcal{H}(x) \to \mathcal{PH}(x)$.

This theorem will be proved by induction over the number of the zone. First, we study the local problem, i.e. the "jump" of Arnold cycles when x traverses a component of M_F.

Since the problem is topological and local, we can reduce it to the study of a certain variation operator for a model topological representation of the picture. Namely, we fix an ε-disc B in $\mathbf{C}^n$ centred at the origin, and suppose that W_F is given by the local equation $x_1 = 0$ and x goes from the point $x_- = (-\varepsilon^2, 0, \dots, 0)$ to $x_+ = (\varepsilon^2, 0, \dots, 0)$. The *localized Arnold classes* $A_{\text{loc}}(x_.)$, $x_. = x_+$ or x_-, are considered as elements of the groups

$$\tilde{H}_{n-1}(W_F \cap B \setminus S(x_.), W_F \cap \partial B) \tag{13}$$

if n is even, and of the groups

$$\tilde{H}_{n-1}(W_F \cap B \setminus S(x_.), W_F \cap \partial B; @(x_.)) \tag{14}$$

if n is odd, presented by the cycle $M_F \equiv \operatorname{Re} W_F$ taken with the appropriate orientation.

This cycle generates the groups (13), (14): for even n this assertion is Corollary I.2.1, and in the case of odd n it is verified in exactly the same way.

Consider the bundle over the punctured disc of radius ε^2 in $\mathbf{C}^1$, whose fibre over the point ξ is $W_F \cap B \setminus S((\xi, 0, \dots, 0))$. Close to the boundary of B this bundle can be trivialized (since it extends there to a bundle over the nonpunctured disk), hence, as in the construction of the operator Var, the transportation of relative cycles over any path in the punctured disc can be chosen so that it respects this trivialization close to ∂B. We can define this trivialization in such a way that it respects the complex conjugation in our bundle, i.e. if $y, \bar{y}$ are two complex conjugate points in the fibre over a real value $\xi \in \mathbf{C}^1$, and $\xi', \bar{\xi}'$ are two complex conjugate points in $\mathbf{C}^1$, then the point of the fibre over ξ' that corresponds to y via our trivialization is complex conjugate to the point over $\bar{\xi}'$ corresponding to $\bar{y}'$. Then after the transportation over a path connecting $-\varepsilon^2$ with ε^2 in the lower half-disc the cycle $A_{\text{loc}}(x_-)$ becomes a cycle $A'_{\text{loc}}(x_-)$ that coincides with $A_{\text{loc}}(x_+)$ close to ∂B. Thus the difference $A_{\text{loc}}(x_+) - A'_{\text{loc}}(x_-)$ of these cycles is an absolute cycle in $W_F \cap B \setminus S(x_+)$. This cycle (more exactly, its class in $\mathcal{H}(x_+)$) is exactly the jump to be calculated. See Figure 5.

Lemma 2. *The difference $A_{\text{loc}}(x_+) - A'_{\text{loc}}(x_-)$ is equal to a vanishing cycle generating the group $\tilde{H}_{n-1}(W_F \cap B \setminus S(x_.))$ (if n is even) or $\tilde{H}_{n-1}(W_F \cap B \setminus S(x_.), @(x_.))$ (if n is odd).*

Proof. The spaces $W_F \cap B \setminus S(x_-)$ and $W_F \cap B \setminus S(x_+)$ coincide, and the family of spaces $W_F \cap B \setminus S((\xi, 0, \dots, 0))$, where ξ runs over the arc connecting x_- and x_+ in the upper half-plane, coincides with the *closed* one-parameter family which participated in the definition of the variation operator for a Morse singularity; see Chapter I, §§ 2, 12. Thus our lemma follows from the Picard–Lefschetz formulae; see Proposition I.2.4 and Theorem I.12.2. The epithet "vanishing" for the cycle generating the group $\tilde{H}_{n-1}(W_F \cap B \setminus S(x_.))$ or $\tilde{H}_{n-1}(W_F \cap B \setminus S(x_.), @(x_.))$ is justified by the fact that it actually vanishes in the standard way in the one-parameter family of functions $r^2(\cdot - x_-) - t$ (or $r^2(\cdot - x_+) - t$) restricted to W_F at the instant $t = \varepsilon^4$.

Now let us calculate the result of such jumps iterated k times, i.e. when we go into the k-th zone from the hyperbolicity domain. Consider the model hyperbolic surface that is the union of $[d/2]$ concentric close spheres of radii $1, 1 + \varepsilon, 1 + ([d/2] - 1)\varepsilon$ (which do not intersect each other even in the complex domain) and, if d is odd and we prove statement b) of Theorem 2, one plane distant from these spheres.

By the previous calculation, executed inductively, the class $A(x)$ (respectively, $PA(x)$) of the point x in the k-th zone is equal to the class $A_{\mathsf{hyp}}(x)$ (respectively, $PA_{\mathsf{hyp}}(x)$) plus the sum of k vanishing cycles, each of which lies on the complexification of the corresponding sphere. Thus, all of them do not intersect each other, and our assertion is proved for the (very degenerate) model hyperbolic surface. We can change this surface arbitrarily weakly so that the complex surface $\tilde{W}_F$ becomes topologically generic, i.e. nonsingular and irreducible in the complex domain. This weak change does not move the topological picture in an arbitrarily prescribed compact subset in $\mathbf{C}^n$, e.g. in the disc containing all these k vanishing spheres. Therefore the answer is the same also for a certain generic hyperbolic polynomial. Finally, the set of nongeneric real hyperbolic polynomials, all the "nongenericity" of which lies in the complex domain, has codimension at least 2 in the space of all hyperbolic polynomials. Hence the statements a) and b) of Theorem 2 follow from the last and first statements of Lemma 2.2, respectively.

Remark 1. In the previous proof, for any point x in the k-th zone, $1 \leq k \leq [d/2]$, we have constructed explicitly a class in $\mathcal{J}(x)$ which becomes the reduced projective Arnold class $P\tilde{A}(x)$ after the obvious map $\mathcal{H}(x) \to \mathcal{PH}(x)$ (and coincides with the usual reduced Arnold class $\tilde{A}(x)$ if M_F is compact). We shall denote this class by $\tilde{A}(x)$ and call it the *reduced Arnold class* even if M_F is noncompact.

3.5. Example: the two-dimensional case.

Let $n = 2$. Then the surface $S(x)$ consists of two complex lines through x, collinear to the lines $\{x_1 = \pm i \cdot x_2\}$. The reduced Arnold class $\tilde{A}(x)$ corresponding to a point x from the k-th zone is represented by $2k$ small circles in $W_F \setminus S(x)$ around the

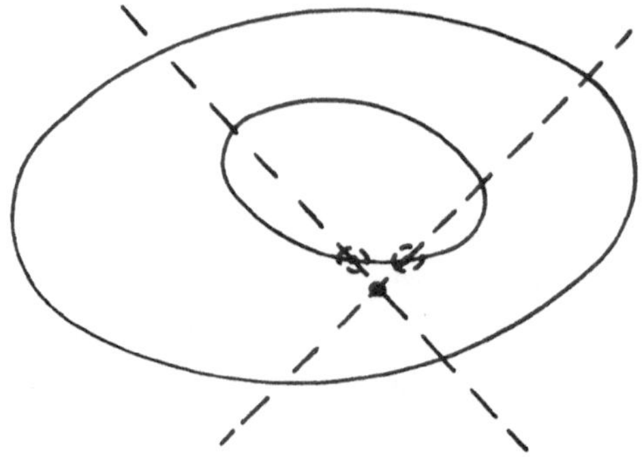

Fig. 33. Integration cycle for the attraction force of a hyperbolic plane curve

intersection points of these two lines with W_F: k circles around the points of any line, see Figure 33. It follows from the construction of Arnold cycles that all these circles close to one line are oriented in accordance with the complex structure of the normal bundle of this line, while close to all points of the other they are oriented clockwise. The total number of such intersection points in the finite domain for any line is equal to $d-s$, where s is the number of factors $x_1^2+x_2^2$ in the principal part of F. Moving the crucial point x in $\mathbf{C}^2 \setminus \Sigma(F)$ we can only permute these $d-s$ circles (and, if W_F is smooth, all permutations can be realized). Therefore the orbit of the monodromy group consists of $\binom{d-s}{k}^2$ elements; this implies Theorem 1 and Proposition 1 of the Introduction to this chapter.

Remark about Ivory's second theorem. Given a hyperbolic surface, do there exist other surfaces defining the same attraction force in some exterior zone? If yes, these surfaces should define at least identical ramification loci of the analytic continuations of these forces. In the case of irreducible plane curves this locus consists of $d(d-1)$ lines tangent to W_F and parallel to the line $x_1 = i \cdot x_2$ plus $d(d-1)$ lines parallel to the line $x_1 = -i \cdot x_2$. If $d = 2$, the set of curves for which these ramification loci coincide consists of all conics inscribed in a given rectangle whose sides are parallel to these two directions. It is easy to see that this set is one-parameter and coincides with the family of confocal conics. For larger d, such copotential families do not exist or at least are exceptional, because the number $2d(d-1)$ of conditions that the curves of such a family should satisfy becomes much more than the dimension of the space of curves.

3.6. Proof of Main Theorem 2.

Denote by Ker$\mathcal{J}(x)$ the kernel of our symmetric bilinear form $\langle\cdot,\cdot\rangle$ on the group $\mathcal{J}(x)$, i.e. the set of all $\gamma \in \mathcal{J}(x)$ such that $\langle\gamma,\alpha\rangle = 0$ for any α. By the Picard–Lefschetz formula, this subspace is invariant under the monodromy action, and hence this action on the quotient lattice

$$\tilde{\mathcal{J}}(x) \equiv \mathcal{J}(x)/\mathrm{Ker}\mathcal{J}(x)$$

is well defined.

Theorem 3. *If F is G-compatible, $x_0 \in \mathbf{C}^n \setminus \Sigma(F)$, and P is a polynomial of degree p, then for any $\gamma \in$ Ker $\mathcal{J}(x_0)$ the corresponding potential function $u_\gamma(x)$ of the standard charge with density P (i.e. the integral of the corresponding forms (3) along the cycles $\gamma(x)$ obtained by the Gauss–Manin connection over the paths connecting x_0 and x) coincides with a polynomial function of degree $\leq p+1-d$ in the case $n>2$; in the two-dimensional case the attraction force vector defined by such charge and density is polynomial of degree $\leq p-d$. In particular, any form $N^{(\nu)}$ with $|\nu| \geq p+2-d$ takes the zero value on* Ker$\mathcal{J}(x)$.

Proof. Let n be even. By Poincaré duality in $W_F \cap S(x_0)$, the condition $\gamma \in$ Ker $\mathcal{J}(x_0)$ implies that the cycle $t^{-1}(\gamma) \in \tilde{H}_{n-2}(W_F \cap S(x_0))$ is homologous in the projective closure $\bar{W}_F \cap \bar{S}(x_0) \subset \mathbf{C}P^n$ of $W_F \cap S(x_0)$ to a cycle which lies in the improper subspace $\bar{W}_F \cap \bar{S}(x_0) \cap (\mathbf{C}P^n \setminus \mathbf{C}^n)$. The tube of this homology provides the homology of γ to some cycle belonging to $\partial W_F \setminus \bar{S}(x_0) \equiv (\bar{W}_F \setminus \bar{S}(x_0)) \cap (\mathbf{C}P^n \setminus \mathbf{C}^n)$. The last space is an $(n-2)$-dimensional Stein manifold, thus γ is homological to zero in $\bar{W}_F \setminus \bar{S}(x_0)$. On the other hand, the forms

$$D_x^{(\nu)}|_{x=x_0} G(x-y)P(y)\omega_F(y)$$

with $|\nu| \geq 2+p-d$ extend to holomorphic forms on $\bar{W}_F \setminus \bar{S}(x_0)$, and hence their integrals along γ are equal to zero.

In the case of odd n, the condition $\gamma \in Ker\ \mathcal{J}(x_0)$ also implies that γ is homological in $\bar{W}_F \setminus \bar{S}(x_0)$ (as a cycle with coefficients in $@(x_0) \otimes \mathbf{C}$) to a cycle in the improper plane: indeed, by Poincaré duality this condition implies that γ defines a trivial element of the group $\tilde{H}^{lf}_{n-1}(\bar{W}_F \setminus S(x_0), \partial W_F \setminus S(x_0); @(x_0))$, and hence, by the relative version of Theorem I.13.3, also of the group $\tilde{H}_{n-1}(W_F \setminus S(x_0), \partial W_F \setminus S(x_0); @(x_0) \otimes \mathbf{C}))$. The rest of the proof is the same as for even n.

As a corollary, we get a proof of Theorem 2 of the Introduction to this chapter. Indeed, if F is a *generic* quadric, then the form $\langle\cdot,\cdot\rangle$ is parabolic (see Theorem I.13.4), and hence the induced form on the quotient space $\tilde{\mathcal{J}}(x)$ is elliptic. Since the group $\tilde{\mathcal{J}}(x)$ is an integer lattice, the orbit of any class in it (in particular of the coset of the reduced Arnold class) under the reduced monodromy action in $\tilde{\mathcal{J}}(x)$ is finite, and any linear form takes finitely many values on it.

Finally, the *nongeneric* quadric F can be approximated by a one-parameter family F_τ, $\tau \in (0, \epsilon]$, of generic quadrics. The analytic continuation of the potential function $u = u(F)$ is equal to the limit of similar continuations of potentials $u(F_\tau)$. Hence the number of leaves of $u(F)$ is majorized by the (common) number of leaves of any of the $u(F_\tau)$.

If n is even, $n > 2$, then the pair consisting of the lattice $\mathcal{J}(x)$ for a generic quadric F in $\mathbf{C}^n$ and the intersection form $\langle\cdot,\cdot\rangle$ on it coincides with that defined by the extended root system $\tilde{D}_{n+1}$ (for the definition of the latter see [Bourbaki 68]); for odd n this pair is a direct sum of the lattice $\tilde{D}_{n+1}$ and the $(n-1)$-dimensional lattice with zero form on it. This fact is essentially proved in [Ebeling 87], although the result is formulated there only for the case of even n and nontwisted homology.

Hence also the pair consisting of the lattice $\tilde{\mathcal{J}}(x)$ and the (lowered) form $\langle\cdot,\cdot\rangle$ on it coincides with the lattice D_{n+1} with the natural bilinear form on it. This allows us to estimate the number of leaves of potential functions of quadrics by the numbers of elements of length $\sqrt{-2}$ in the latter quotient lattice. As we shall see in the next section, this majorization is not realistic: a more precise upper bound is the number of integer points in the intersection of the sphere of radius $\sqrt{-2}$ and a certain affine sublattice of corank 1 of this quotient lattice that does not pass through the origin.

§ 4. Description of the small monodromy group

In the rest of this chapter we suppose that the surface W_F is G-compatible.

Recall that the subgroup $\mathcal{J}(x) \subset \mathcal{H}(x)$ was defined as the group spanned by all vanishing cycles; the "big" monodromy group studied in Chapter I, § 13 acts on it by reflections in these cycles with respect to the natural bilinear form on $\mathcal{J}(x)$.

In this section we describe the subgroup in it that is the image of the representation (10).

Namely, $\mathcal{J}(x)$ contains a subspace $\bar{\mathcal{J}}(x)$ of codimension 1 that is also spanned by several vanishing cycles (and *does not* contain the reduced Arnold cycles $\tilde{A}(x)$ for x not in the hyperbolicity domain). The subgroup in question is generated by all the reflections in the vanishing cycles from $\bar{\mathcal{J}}(x)$, in particular any orbit of it lies entirely in a subspace parallel to $\bar{\mathcal{J}}(x)$. Also, this subspace $\bar{\mathcal{J}}(x)$ does not belong to the kernel of the linear function N defined by integrals of the form (2); see § 3.2.

In § 5 below we prove that if $d \geq 3, n \geq 3, n + d \geq 8$, and F is a generic polynomial of degree n in $\mathbf{C}^n$, then the monodromy group (10) is rich enough to ensure that this function N takes infinitely many values on the elements in $\bar{\mathcal{J}}(x)$

that are the differences between different elements of the orbit of $\tilde{A}(x)$ for any x not in the hyperbolicity domain. This will imply Main Theorem 3 of this chapter.

4.1. The subspace $\bar{\mathcal{J}}(x)$.

Lemma 1. *For arbitrary $x \in \mathbf{C}^n$ the group*

$$\tilde{H}_{n-1}(\mathbf{C}P^n \setminus \bar{S}(x)) \tag{15}$$

(if n is even) or

$$H_{n-1}(\mathbf{C}P^n \setminus \bar{S}(x), @(x)) \tag{16}$$

(if n is odd) is one-dimensional.

Here is an equivalent statement. Let PS be the projectivization in $\mathbf{C}P^{n-1}$ of an arbitrary standard cone $S(x)$, $x \in \mathbf{C}^n$, and @ the natural lowering on $\mathbf{C}P^{n-1} \setminus PS$ of the local system $@(x)$. Obviously the space $\mathbf{C}P^n \setminus \bar{S}(x)$ is the space of a (trivializable) bundle with fibre $\mathbf{C}^1$ over $\mathbf{C}P^{n-1} \setminus PS$.

Lemma 1′. *If n is even, then the group $\tilde{H}_{n-1}(\mathbf{C}P^{n-1} \setminus PS)$ is one-dimensional. If n is odd, then the group $\tilde{H}_{n-1}(\mathbf{C}P^{n-1} \setminus PS, @)$ is one-dimensional. In both cases, these groups are generated by the vanishing cycles of the simplest degeneration of the quadric PS.*

The proof is elementary.

The identical embeddings

$$W_F \setminus S(x) \to \mathbf{C}P^n \setminus \bar{S}(x), \tag{17}$$

$$\bar{W}_F \setminus \bar{S}(x) \to \mathbf{C}P^n \setminus \bar{S}(x) \tag{17'}$$

induce homomorphisms of the groups $\mathcal{H}(x)$, $\mathcal{PH}(x)$ into the group (15) or (16). Define the subgroup $\bar{\mathcal{J}}(x) \subset \mathcal{J}(x)$ as the kernel of the restriction to $\mathcal{J}(x)$ of the homomorphism of $\mathcal{H}(x)$ induced by (17).

Theorem 1. *The subgroup* Ker $\mathcal{J}(x) \subset \mathcal{J}(x)$ *belongs to* $\bar{\mathcal{J}}(x)$.

The proof follows from that of Theorem 3.3.

Theorem 2. *Let x be a point in the k-th zone of $\mathbf{R}^n \setminus M_F$, $1 \leq k \leq [d/2]$. Then the embedding (17) (respectively, (17′)) maps the reduced Arnold cycle $\tilde{A}(x)$ (respectively, the reduced projective Arnold cycle $P\tilde{A}(x)$) into k times a generator of the group (15) or (16), where k is the number of the zone containing x.*

Indeed, any induction step from the proof of Theorem 3.2 obviously increases the image of $\tilde{A}(x)$ or $P\tilde{A}(x)$ under the map (17) (respectively, (17′)) by a generator of the target homology group; all such k steps are locally topologically equivalent, and hence add a fixed generator of this target group with the same sign.

Theorem 3. *Suppose that* $n > 2$. *Then*

a) the basis of vanishing cycles generating the group $\mathcal{J}(x)$ *can be chosen so that the subgroup* $\bar{\mathcal{J}}(x)$ *is spanned by some* $\dim\mathcal{J}(x) - 1$ *elements of it;*

b) the small monodromy group (10) in $\mathcal{J}(x)$ *is generated by reflections in the vanishing cycles belonging to* $\bar{\mathcal{J}}(x)$*, in particular, the action of this group preserves all planes parallel to* $\bar{\mathcal{J}}(x)$*;*

c) any basis vanishing cycle in $\bar{\mathcal{J}}(x)$ *can be transposed into any other by the action of the small monodromy group (10).*

For any $k = 1, 2, \ldots, [d/2]$, and any point x in the k-th zone, consider the difference of the projective Arnold class $PA(x)$ and the element in $\mathcal{PH}(x)$ obtained as in the definition of the reduced Arnold cycles (i.e. by transportation along an arc in the lower half-plane) from a similar class $PA(x')$, x' in the $(k-1)$-st zone. In Lemma 3.2 we have shown that if x and x' are sufficiently close to each other and to the k-th component of M_F separating them, then this class can be realized by a cycle contained in a small common neighbourhood B of these points x, x'. Denote by $a(x)$ the class of this cycle in the group $\mathcal{H}(x)$; by continuity this class $a(x)$ is well defined also for arbitrary x from the same zone (not necessarily close to M_F). By Theorem 2, for all x not in the hyperbolicity domain the corresponding inclusions (17) and the subsequent projections

$$\mathbf{C}P^n \setminus \bar{S}(x) \to \mathbf{C}P^{n-1} \setminus PS$$

map the elements $a(x)$ into the same element of the homology group participating in Lemma 1′.

Theorem 4. *If* $n > 2$, *then*

a) all classes $a(x)$*, corresponding to all points* $x \in \mathbf{R}^n \setminus \Sigma(F)$*,* x *not in the hyperbolicity domain or in the* $([d/2]+1)$*-th zone, can be obtained from one another by the Gauss–Manin connection in the homology bundle* $\{\mathcal{H}(x) \to x\}$ *over some path in* $\mathbf{C}^n \setminus \Sigma(F)$*. These classes* $a(x)$ *do not belong to* $\bar{\mathcal{J}}(x)$*, and any of them, being added to the set of* $\dim\mathcal{J}(x) - 1$ *elements mentioned in statement a) of Theorem 3, completes this set to a basis in* $\mathcal{J}(x)$*;*

b) for arbitrary x *in the* k*-th zone,* $1 \le k \le [d/2]$*, the linear form* $\langle a(x), \cdot\rangle$ *on the group* $\bar{\mathcal{J}}(x)$*, defined by our bilinear form, is nontrivial.*

A little later we shall prove also that any similar form $\langle \tilde{A}(x), \cdot\rangle$ or $\langle P\tilde{A}(x), \cdot\rangle$ for x from such zones also is nontrivial; see Theorem 5.1.

4.2. Proof of Theorems 3 and 4.

First we compare the fundamental groups of $\mathbf{C}^n \setminus \Sigma(F)$ and of the complement of the discriminant variety of any versal deformation of the complete intersection $(\bar{F}, r^2)$. Since F is G-compatible, the set $\Sigma(F)$ consists of only two components, W_F and the set of $x \notin W_F$ such that $S(x)$ is tangent to W_F; if $n > 2$, then the latter component is irreducible.

Consider the $(n+1)$-parameter family of functions

$$\Phi_{x_1,\ldots,x_n,t}(z_1,\ldots,z_n) \equiv (z_1 - x_1)^2 + \cdots + (z_n - x_n)^2 - t$$

in $\mathbf{C}^n$, and the family of divisors $\mathcal{S}(x;t)$ distinguished by these equations, so that $S(x) \equiv \mathcal{S}(x;0)$. Let $T \simeq \mathbf{C}^{n+1}$ be the space of parameters $(x_1,\ldots,x_n,t)$ of these families; the space $\mathbf{C}^n$ in which W_F lies is naturally identified with the subspace $\{t=0\}$ of T. Let Σ_T be the set of $(x;t) \in T$ such that $\mathcal{S}(x;t)$ is nontransversal to W_F, in particular, $\Sigma(F) \equiv \Sigma_T \cap \mathbf{C}^n$. Suppose that the distinguished point $\mathbf{x}$ of the space $T \setminus \Sigma_T$ lies in the subspace $\mathbf{R}^n \setminus \Sigma(F)$. The group $\pi_1(T \setminus \Sigma_T)$ acts in the usual way on the group $\mathcal{H}(\mathbf{x})$. The image of this action in $\mathrm{Aut}(\mathcal{H}(\mathbf{x}))$ coincides with that of the similar action of the "big" monodromy group: indeed, by Zariski's theorem the fundamental group of the complement of the appropriate versal deformation of $(\bar{F}, r^2)$ is generated by the loops that lie in the complement of Σ_T in the line $\{(x_0,t), t \in \mathbf{C}^1\} \subset T$ for a generic x_0.

Let Λ be a generic 2-plane in T, and $L = \Lambda \cap \mathbf{C}^n$; $\bar{U}$ a small neighbourhood of L in the projective compactification of T, and $U = \bar{U} \cap T$ the affine part of $\bar{U}$. Let L' be a generic line in Λ through $\mathbf{x}$ sufficiently close to L, so that $L' \subset U$ and L' intersects Σ_T transversally.

Lemma 2. *The obvious maps* $\pi_1(L \setminus \Sigma(F)) \to \pi_1(\mathbf{C}^n \setminus \Sigma(F))$ *and* $\pi_1(L' \setminus \Sigma_T) \to \pi_1(U \setminus \Sigma_T) \to \pi_1(T \setminus \Sigma_T)$ *are epimorphic.*

The proof follows immediately from the generalized Lefschetz theorem (see [Goresky & MacPherson 86]).

Thus the small and big monodromy groups are generated by simple loops lying in $L \setminus \Sigma_T$ and $L' \setminus \Sigma_T$, respectively. Let us compare these collections of loops.

Lemma 3. *The group* $\mathcal{J}(\mathbf{x})$ *is generated by the cycles vanishing along the paths of an arbitrary distinguished system in* L' *connecting the distinguished point* $\mathbf{x}$ *with all points of* $L' \cap \Sigma_T$.

Indeed, the group $\pi_1(L' \setminus \Sigma_T)$ acts on the group $\mathcal{J}(\mathbf{x})$; this monodromy action is described by the Picard–Lefschetz formulae, see Chapter I, §§ 13.3, 13.4. Lemma 3 follows from these formulae, from Lemma 2, and from the fact that the group $\mathcal{J}(\mathbf{x})$ coincides with the linear hull of the orbit of any vanishing cycle under the action

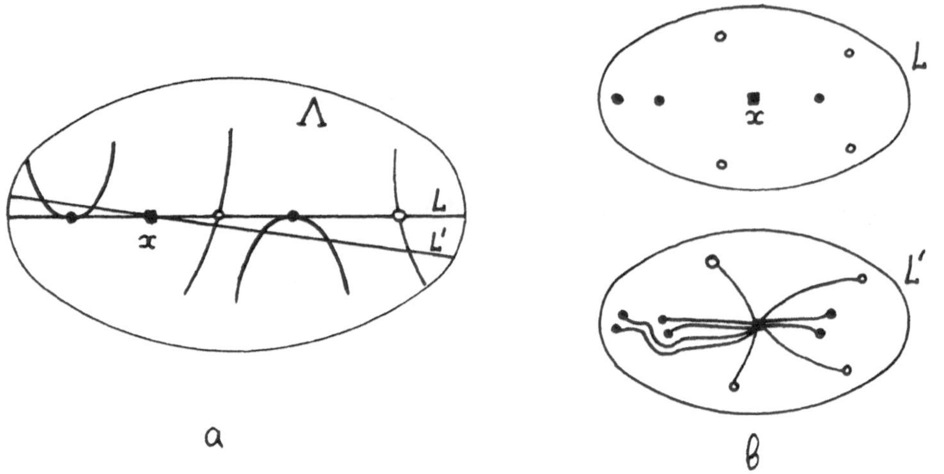

Fig. 34. Lines L and L' and discriminant points in them

of the big monodromy group (which follows, in turn, from the irreducibility of the discriminant of a versal deformation, see [Gabrielov 74], [Ebeling 87]).

The set $L \cap \Sigma(F)$ consists of several points of two different kinds: the points of transversal intersection of L and W_F and points $x \notin W_F$ such that $S(x)$ is tangent to W_F.

Lemma 4. *a) Close to a generic point y of the submanifold $W_F \subset \mathbf{C}^n \subset T$ (i.e. to a point at which the generating lines of the cone $S(y)$ are transversal to W_F) the variety Σ_T is smooth and has simple tangency with $\mathbf{C}^n$ along W_F. In particular, the intersection of Σ_T with any 2-plane Λ transversal to W_F coincides close to the points of $\Lambda \cap W_F$ with a smooth curve having simple tangency with the line $\Lambda \cap \mathbf{C}^n \equiv L$;*

b) if F is generic, then close to a generic point of the variety $(\Sigma(F) \setminus W_F) \subset \mathbf{C}^n \subset T$ the variety Σ_T is smooth and intersects $\mathbf{C}^n$ transversally along the variety $(\Sigma(F) \setminus W_F)$.

The proof is immediate.

Thus the cardinality of $L' \cap \Sigma_T$ is equal to the cardinality of $L \cap \Sigma(F)$ plus $\deg F$: to any point of $L \cap (\Sigma(F) \setminus W_F)$ there corresponds one close point of $L' \cap \Sigma_T$, while to any point of $L \cap W_F$ there correspond two such points; see Figure 34a.

We can (and shall) assume that the distinguished point $\mathbf{x}$ of $T \setminus \Sigma_T$ lies in the hyperbolicity domain in $\mathbf{R}^n$, hence all points of $L \cap W_F$ are real. For any such point y belonging to the k-th component of M_F, let $y_+ \in \mathbf{R}^n \setminus \Sigma(F)$ be a close point in the

k-th zone. For such a point y_+, the class $a(y_+)$ was defined before the formulation of Theorem 4.

Let us agree to choose the distinguished system of paths in L' in such a way that the paths connecting $\mathbf{x}$ with any two points of $L' \cap \Sigma_T$ arising from the same point y of $L \cap W_F$ go together up to a small common neighbourhood of these two points and are close to the real segment in L connecting $\mathbf{x}$ and y, while the paths in L' connecting $\mathbf{x}$ with any other points of $L' \cap \Sigma_T$ do not touch this small neighbourhood; see Figure 34b.

Definition 1. A point of $L' \cap \Sigma_T$ is of the first kind (respectively, of the second kind) if it arises from a close point of W_F (respectively, of $\Sigma(F) \setminus W_F$) in L after the move $L \to L'$. A cycle in $\mathcal{J}(\mathbf{x})$ vanishing over a path of our distinguished system in L' that connects $\mathbf{x}$ with a point $y \in \Sigma_T$ is called a cycle of the first kind (respectively, of the second kind) if this point y is of the first (respectively, the second) kind.

In Figure 34b the points of $L \cap W_F$ and of the first kind in L' are shown by small black circles, while the points of $L \cap (\Sigma(F) \setminus W_F)$ and the points of the second kind in L' are shown by white circles.

Lemma 5. *a) Two cycles of the first kind in $\mathcal{H}(\mathbf{x})$, vanishing over two distinguished paths connecting $\mathbf{x}$ with two points of $L' \cap \Sigma_T$ arising from the same close point y of $L \cap W_F$, coincide (maybe up to sign);*

b) this cycle coincides (maybe up to sign) with the cycle $a(y_+)$ transported from the point y_+ to $\mathbf{x}$ along the path described in the definition of the reduced Arnold class, see § 3.3. In particular, the embedding (17) maps any such cycle into a generator of the corresponding group (15) or (16);

c) the monodromy action in the group $\mathcal{H}(\mathbf{x})$, defined by any simple loop in $L \setminus \Sigma(F)$ going around some point of $L \cap W_F$, is trivial;

d) any cycle in $\mathcal{H}(\mathbf{x})$ vanishing over a path in $L \setminus \Sigma(F)$ connecting $\mathbf{x}$ with a point of $\Sigma(F) \setminus W_F$ belongs to the subspace $\bar{\mathcal{J}}(\mathbf{x})$. In particular, the same is true for any cycle of the second kind defined by a path of our distinguished system in $L' \setminus \Sigma_T$ connecting $\mathbf{x}$ with a point (of the second kind) of Σ_T.

Proof. Consider the space of complex lines through $(\mathbf{x}, 0)$ transversal to Σ_T in the plane Λ. Obviously this space is a projective line with several points removed, one of which is the point $\{L\}$. Consider a small loop in this space, which starts and finishes at the point $\{L'\}$ and goes once around the point $\{L\}$. This loop takes one of the two distinguished paths from statement a) of lemma into the other, thus this statement follows.

Statement c) is a direct consequence of a). Indeed, the loop considered there is homotopic in $\Lambda \setminus \Sigma_T$ to a loop in $L' \setminus \Sigma_T$ which turns around two discriminant points defining the same vanishing cycle, thus its monodromy action is equal to the square of the reflection in the hyperplane orthogonal to this vanishing cycle.

Statement b) follows from Lemma 3.2 and the local shape of the pair (W_F, $\mathcal{S}(x,t)$) for the discriminant point (x,t) of the first kind. The way in which the pairs of distinguished paths connecting x with different pairs of points of the first kind miss each other is not important, because by the proof of Theorem 3.2 all the cycles of the first kind that vanish over the paths going from x to the points arising from different points of $L \cap W_F$ on the same side of x in $\operatorname{Re} L$ are pairwise orthogonal.

Statement d) of the lemma follows immediately from the constructions.

Thus, the vanishing cycles of the first (respectively, second) kind are exactly those that are mapped by the embedding (17) into a generator of the group (15) or (16) (respectively, into a zero class).

Lemma 6. *Any vanishing cycle of the first kind in $\mathcal{J}(\mathbf{x})$ can be transposed into any other by a sequence of reflections in the hyperplanes orthogonal to cycles of the second kind and to this cycle itself.*

By the Picard–Lefschetz formula, this lemma follows from the next one.

Lemma 6′. *There exists a distinguished system of paths in $L' \setminus \Sigma_T$ connecting $\mathbf{x}$ with all points of $L' \cap \Sigma_T$, such that all vanishing cycles of the first kind defined by this system are equal to each other.*

Proof. (This proof simulates that of the well-known fact that the fundamental group of the complement of a smooth irreducible algebraic hypersurface in $\mathbf{C}^n$, $n \geq 2$ is isomorphic to $\mathbf{Z}$.)

Let y_1 be any point of $L \cap W_F$. Let us fix an arbitrary path γ_1 in $L' \setminus \Sigma_T$ connecting $\mathbf{x}$ with y_1. Denote by A^n the space of complex lines in $\mathbf{C}^n$, and by $\operatorname{Reg}(\Sigma(F))$ the subset of A^n consisting of lines transversal to $\Sigma(F)$. Consider a path $\chi_1 : [0,1] \to A^n$ such that $\chi_1(0) = L$, $\chi_1([0,1)) \subset \operatorname{Reg}(\Sigma(F))$, the last point $\chi_1(1)$ is a line transversal to $\Sigma(F)$ everywhere except for one point of simple tangency with W_F, and one of the two points of $\chi_1(\tau) \cap W_F$, $\tau = 1 - \varepsilon$ that coalesce at this tangency point is obtained from the point y_1 of the similar set corresponding to the value $\tau = 0$ during the deformation of the set $\chi_1(\tau) \cap W_F$, $\tau \in [0, 1-\varepsilon]$.

Consider the continuous deformation $\gamma_1[\tau]$, $\tau \in [0,1]$, of the path γ_1 such that $\gamma_1[0] = \gamma_1$, $\gamma_1[\tau] \subset \chi_1(\tau)$, and for any τ the path $\gamma_1[\tau]$ connects in $\chi_1(\tau) \setminus \Sigma(F)$ a point of $\chi_1(\tau) \cap W_F$ with some distinguished point $\mathbf{x}(\tau) \in \gamma_1[\tau] \setminus \Sigma(F)$, $\mathbf{x}(0) = \mathbf{x}$. At almost the final instant $\tau = 1 - \varepsilon$, the endpoint $\gamma_1[1-\varepsilon](1)$ of the path $\gamma_1[1-\varepsilon]$ lies very close to some other point of $\chi_1(1-\varepsilon) \cap W_F$ (with which it coalesces at the instant $\tau = 1$). Connect this new point with $\mathbf{x}(1-\varepsilon)$ by a path $\gamma_2[1-\varepsilon]$ in $\chi_1(1-\varepsilon) \setminus \Sigma(F)$ that goes very close to $\gamma_1[1-\varepsilon]$ but does not intersect it except for the initial point.

Then construct a continuous family of paths $\gamma_2[\tau] \subset \chi_1(\tau)$, $\tau \in [0, 1-\varepsilon]$, such that for any τ the corresponding path $\gamma_2[\tau]$ connects a point of $\chi_1[\tau] \cap W_F$ with $\mathbf{x}(\tau)$ and does not intersect other points of $\chi_1(\tau) \cap \Sigma(F)$ or of the path $\gamma_1[\tau]$. At the instant $\tau = 0$ we get a path $\gamma_2 \equiv \gamma_2[0] \subset L$ connecting $\mathbf{x}$ with some point y_2 of W_F.

Then consider a new path $\chi_2 : [0,1] \to A^n$, $\chi_2([0,1)) \subset \mathrm{Reg}\,(\Sigma(F))$, connecting L with some new simple tangent line to W_F and having no extra nontransversalities with $\Sigma(F)$, in such a way that at the last instant $\tau = 1$ one of the two points of $\chi_2(\tau) \cap W_F$ that coalesce at the tangency point is obtained by deformation along our path χ_2 from one of the points y_1 or y_2, and the other two points of these two pairs do not coincide. Arguing as before, we construct a third path in $L \setminus \Sigma(F)$, connecting $\mathbf{x}$ with some third point of $L \cap W_F$, and so on.

After the $(d-1)$-th step we get a system of d nonintersecting paths in $L \setminus \Sigma(F)$, connecting $\mathbf{x}$ with all points of $L \cap W_F$. Complete this family to any distinguished collection of paths conecting $\mathbf{x}$ with all points of $L \cap \Sigma(F)$. For the close perturbation $L' \subset T$ of L, take a close distinguished system of paths in L', connecting the point $\mathbf{x}$ with all points of $L' \cap \Sigma_T$ in such a way that to any path in L connecting $\mathbf{x}$ with W_F there correspond two paths connecting $\mathbf{x}$ with two close points of the first kind. This system of paths is the desired one. For instance, the cycles vanishing along the (perturbed) paths γ_1 and γ_2 define the same vanishing homology class in $\mathcal{J}(\mathbf{x})$: indeed, a similar assertion for the cycles in the group $\mathcal{H}(\mathbf{x}(1-\varepsilon)) \equiv \tilde{H}_{n-1}(W_F \setminus S(\mathbf{x}(1-\varepsilon)), \mathbf{Z})$ or $\tilde{H}_{n-1}(W_F \setminus S(\mathbf{x}(1-\varepsilon)), @(\mathbf{x}(1-\varepsilon)))$ is proved just as the statement a) of Lemma 5, and for other values of $\tau \in [0, 1-\varepsilon]$ it follows by continuity. Lemmas 6′ and 6 are thus proved.

Now we are ready to prove statement a) of Theorem 3. Indeed, by Lemma 3 the group $\mathcal{J}(\mathbf{x})$ is generated by the vanishing cycles of the first and second kind. By Lemma 6 and the Picard–Lefschetz formula, all vanishing cycles of the first kind lie in the linear span of an arbitrary one of them (for which we can take the class obtained by the Gauss–Manin connection from $a(x)$, x from the $\mathbf{k}$-th zone, $1 \leq k \leq [d/2]$, see statement b) of Lemma 5) and the cycles of the second kind (which lie in $\bar{\mathcal{J}}(\mathbf{x})$, see statement d) of Lemma 5).

Statement b) of Theorem 3 follows immediately from statement c) of Lemma 5, and statement c) of Theorem 3 follows from the fact that the variety $\Sigma(F) \setminus W_F$ is irreducible, cf. Theorem I.3.8 and the Corollary after it.

Statement a) of Theorem 4 follows from Lemma 3.2. Indeed, we can assume that the points y_1 and y_2, whose classes $a(y_1)$ and $a(y_2)$ we want to transfer to each other, lie very close to the "interior" (i.e. closest to the hyperbolicity domain) components of M_F bounding corresponding zones. For such y_i the class $a(y_i)$ is realized by a cycle generating the group $\tilde{H}_{n-1}(W_F \cap B \setminus S(y_i))$ or $\tilde{H}_{n-1}(W_F \cap B \setminus S(y_i), @(y_i))$, where B is a small neighbourhood of y_i; see Lemma 3.2. Thus, for the desired path connecting y_1 and y_2 we can take the path that goes very close to the set of generic points of W_F (i.e. of such points y close to which all the generating lines of the cones $S(y)$ are transversal to W_F and hence the pairs $(W_F, S(y))$ have locally the same topological structure).

Finally, statement b) of Theorem 4 follows from the connectedness of the Dynkin diagram of an isolated singularity of the complete intersection; see for example [Ebeling 87].

Remark 1. The hyperbolicity (and even reality) of the polynomial F is unnecessary for the construction of the groups $\mathcal{H}(x)$, $\mathcal{J}(x)$ and $\bar{\mathcal{J}}(x)$. For instance, the group $\bar{\mathcal{J}}(x)$ can be defined for an arbitrary complex polynomial F and any point x of $\mathbf{C}^n \setminus \Sigma(F)$ as the subgroup in $\mathcal{H}(x)$ spanned by all cycles vanishing over the paths connecting x with generic points of $\Sigma(F) \setminus W_F$. For all generic F of given degree these groups are isomorphic, and some such isomorphism for different F and x is canonically defined by any path in the space of generic pairs (F, x) connecting them.

§ 5. Proof of Main Theorem 3

Throughout this section we assume that $n \geq 3$.

5.1. About the Main Conjecture.

Definition 1. A triple

$$(A; \langle \cdot, \cdot \rangle; g) \tag{18}$$

consisting of an integer lattice A, an integer-valued symmetric bilinear form $\langle \cdot, \cdot \rangle$ on it such that $\langle a, a \rangle \in 2\mathbf{Z}$ for all $a \in A$, and a group $g \subset \mathrm{Aut}(A)$ generated by the reflections in hyperplanes orthogonal to several elements a_i of length -2 in A, is called *completely infinite* if for any element $a \in A$ of nonzero length, any nonzero linear form $A \otimes \mathbf{C} \to \mathbf{C}$ takes infinitely many values on the orbit of a under the action of the group g.

Conjecture 1. *For any generic polynomial F of degree $d \geq 3$ in $\mathbf{C}^n$, $n \geq 3$, the triple consisting of the group $\bar{\mathcal{J}}$, the bilinear form equal (up to sign if $[\frac{n+1}{2}]$ is odd) to the form $\langle \cdot, \cdot \rangle$ defined before Lemma 3.1, and the "small" monodromy group (10) on $\bar{\mathcal{J}}(x)$, is completely infinite.*

A little later we reduce the Main Conjecture of this chapter to this Conjecture 1.

Very strong methods of proving (some properties ensuring) the complete infiniteness of certain triples (18) were developed in [Ebeling 87]; see Theorem 4.1.2 there. Unfortunately I cannot prove that the triples described in Conjecture 1 always satisfy the requirements that are essential for these results of Ebeling. Nevertheless I am sure that this conjecture is true and will be proved in the near future.

Recall that for any nondiscriminant point x from the k-th zone, $1 \leq k \leq [d/2]$, the *reduced Arnold class* $\tilde{A}(x) \in \mathcal{J}(x)$ was constructed in the proof of Theorem 3.2, see Remark 3.1. The ramification of the potential function is controlled by the monodromy action (10) on it.

Theorem 1. *For any G-compatible hyperbolic polynomial F and any nondiscriminant point x in the k-th zone, $1 \leq k \leq [d/2]$, the pairing $\langle \tilde{A}(x), \cdot \rangle$ with the reduced Arnold class $\tilde{A}(x)$ defines a nonzero linear form on the group $\bar{\mathcal{J}}(x)$.*

For G-compatible F the bilinear form $\langle \cdot, \cdot \rangle$ has obvious extension to the group $\mathcal{PJ}(x)$ ($\equiv \mathcal{PH}(x)$ for odd n and $t(\tilde{H}_{n-2}(\bar{W}_F \cap \bar{S}(x)))$ for even n); our bilinear form on $\mathcal{J}(x)$ is induced from this form by the obvious inclusion $W_F \setminus S(x) \to \bar{W}_F \setminus \bar{S}(x)$. Therefore Theorem 1 is equivalent to the following.

Theorem 1′. *For any G-compatible hyperbolic polynomial F and any nondiscriminant point x in the k-th zone, $1 \leq k \leq [d/2]$, the pairing $\langle \tilde{P}A(x), \cdot \rangle$ with the reduced projective Arnold class $\tilde{P}A(x)$ defines a nonzero linear form on the image of the group $\bar{\mathcal{J}}(x)$ in $\mathcal{PJ}(x)$.*

Proof. First of all, this is true in the case when M_F is an ellipsoid with different eigenvalues. Indeed, by Theorem 3.2 in this case $\tilde{A}(x)$ is a vanishing cycle, and the assertion follows from the connectedness of the Dynkin diagram and the fact that the group $\bar{\mathcal{J}}(x)$ is nontrivial for such F; see § 3.6 and [Ebeling 87].

For arbitrary d, consider the model (not G-compatible) hyperbolic surface $M_{F'}$ consisting of $[d/2]$ ellipsoids

$$\alpha_1 x_1^2 + \cdots + \alpha_n x_n^2 = j, \quad j = 1, 2, \ldots, [d/2], \tag{19}$$

where all α_i are positive and distinct,

plus, if d is odd, a sufficiently distant hyperplane.

The class $\tilde{A}(x)$ for x from the k-th zone, $1 \leq k \leq [d/2]$, is then equal to the sum of k vanishing cycles, each of which lies in the complexification of its own ellipsoid; see the proof of Theorem 3.2. By the previous special case of a single ellipsoid, in each of these k complexified ellipsoids $\mathcal{E}_i$ there is a compact cycle Γ defining an element of the group $\tilde{H}_{n-1}(\mathcal{E}_i \setminus S(x))$ if n is even, or in $\tilde{H}_{n-1}(\mathcal{E}_i \setminus S(x), @(x))$ if n is odd, such that $\langle \tilde{A}(x), \Gamma \rangle \neq 0$ and the embedding (17) maps Γ into the zero homology class.

Let us move our model hyperbolic polynomial slightly in such a way that it becomes generic (in particular G-compatible) by a perturbation so small that it does not move the topology of the variety $W_{F'} \cap S(x)$ inside a sufficiently large disc, in which the cycles Γ and $\tilde{A}(x)$ lie. The cycle $\tilde{\Gamma}$ neighbouring to Γ in the moved manifold W_F satisfies all the above conditions, and Theorem 1′ is proved for some G-compatible hyperbolic polynomial. For an arbitrary such polynomial Theorem 1′ follows from the connectedness of the space of all strictly hyperbolic G-compatible surfaces in $\mathbf{R}P^n$; see Lemma 2.2.

Theorem 2. *For a generic F and a generic choice of the distinguished point* $\mathbf{x}$, *the integral of the differential form (2) along any vanishing cycle of the second kind in the space $\bar{\mathcal{J}}(\mathbf{x})$ is nonzero. Moreover, there are arbitrarily high partial derivatives of this form on the parameter x, whose integrals along the vanishing cycles are also not equal to zero.*

Proof. It follows from statement a) of Theorem 4.3 that it is sufficient to prove this theorem for *one* vanishing cycle; let us do this. If this assertion does not hold, then by the Picard–Lefschetz formula the restriction of the potential function to any component of $\mathbf{R}^n \setminus M_F$ coincides there with a single-valued holomorphic function. This is surely impossible if M_F is an ellipsoid with different eigenvalues; see for example [Ivory 1809]. For arbitrary *even* d, consider again the model hyperbolic surface of degree d: $d/2$ ellipsoids given by the equations (19). Since the polynomial that distinguishes any of these ellipsoids is constant on any other, the corresponding potential function is a linear combination with nonzero coefficients of the potential functions of all these ellipsoids taken separately. Since the ramification sets of these $d/2$ potential functions are all different, the resulting sum also has a nontrivial ramification.

By continuity the same is also true for a close generic hyperbolic surface, and hence, by the considerations of analytic continuation, for all generic hyperbolic surfaces; cf. the proof of Theorem 3.2.

Finally, in the case of odd d a similar model hyperbolic surface is obtained by adding one hyperplane, namely the "plane of infinity" $\mathbf{R}P^n \setminus \mathbf{R}^n$: it is easy to see that the arguments of continuity for going to a "close" generic surface stay valid.

Reduction of the Main Conjecture to Conjecture 1. Let x be a nondiscriminant point in the k-th zone, $1 \leq k \leq [d/2]$. Choose an arbitrary path in $\mathbf{C}^n \setminus \Sigma(F)$ connecting x and $\mathbf{x}$ and transport the class $\tilde{A}(x)$ by the Gauss–Manin connection in the bundle $\{\mathcal{H}(\cdot) \to \cdot\}$ over this path to the point $\mathbf{x}$. Denote by $\tilde{\mathbf{A}}$ the resulting class in $\mathcal{J}(\mathbf{x})$. By Theorem 1 there exists a vanishing cycle $\Gamma \in \bar{\mathcal{J}}(\mathbf{x})$ of the second kind such that $\langle \tilde{\mathbf{A}}, \Gamma \rangle \neq 0$. Let $\tilde{\mathbf{A}}_1$ be the image of $\tilde{\mathbf{A}}$ under the monodromy along the simple loop in $\mathbf{C}^n \setminus \Sigma(F)$ corresponding to the path along which Γ vanishes. By the Picard–Lefschetz formula, the difference $\tilde{\mathbf{A}}_1 - \tilde{\mathbf{A}}$ is a nonzero multiple of Γ. By Conjecture 1 and Theorem 2, the form N (as well as forms $N^{(\nu)}$ with arbitrarily large $|\nu|$) takes infinitely many values on the orbit of this difference under the action of the small monodromy group (10). On the other hand, this infinite number obviously does not exceed the square of the number of values of N (respectively, $N^{(\nu)}$) on the orbit of $\tilde{\mathbf{A}}$ under the same action, hence the last number is also infinite.

Finally, for the points x from the $([d/2]+1)$-th zone (if it exists) the assertion of the Main Conjecture follows from the fact that the potential function defined by the charge (2) (or (3) if $\deg P \leq d-2$) obviously extends to an analytic function on $\mathbf{R}P^n \setminus M_F$, hence its algebraicity in the $([d/2]+1)$-th zone is equivalent to that in the zone separated from it by a piece of the improper plane in $\mathbf{R}P^n$; the number of the latter zone is surely less than $[d/2]+1$.

5.2. Embedding the monodromy groups of function singularities into the small monodromy groups.

Definition 2. A symmetric bilinear form is called *hyperbolic* if one of its definite inertia indices μ_+, μ_- is equal to 1; compare with Definition I.13.5.

Proposition 1 (see for example [Ebeling 87]). *Let $f : (\mathbf{C}^m, 0) \to (\mathbf{C}, 0)$ be an isolated function singularity, and suppose that the intersection form in the corresponding vanishing homology group $\tilde{H}_{m-1}(V_\lambda)$ (if m is odd) or in the similar homology group $\tilde{H}_m(V_\lambda^1)$ of its stabilization $f^{(1)} \equiv f(x_1, \dots, x_n) + x_{m+1}^2$ (if m is even) is neither elliptic nor parabolic or hyperbolic. Then the triple (18) in which A is this group $\tilde{H}_{m-1}(V_\lambda)$ (respectively, $\tilde{H}_m(V_\lambda^1)$) and g is the monodromy group of f (respectively, of $f^{(1)}$) defined in Chapter I, § 3, is completely infinite.*

Recall that by Givental's theorem the intersection form in the stabilized group $\tilde{H}_m(V_\lambda^1)$ (and the monodromy action defined by it) is equal to that for the twisted vanishing homology group $\tilde{H}_m(B \setminus V_\lambda^1; \pm\mathbf{Z})$; see Chapter I, § 12.5.

Example 1 (see for example [AVG 82]). A generic homogeneous polynomial $\mathbf{C}^m \to \mathbf{C}$ of degree d defines an isolated singularity at zero. If $m \geq 2, d \geq 3$, and $m + d \geq 7$, then the intersection form in the vanishing homology of the q-fold stabilization of this singularity (where $m + q$ is odd) is neither elliptic nor parabolic or hyperbolic. For such a generic polynomial we can always take the function $x_1^d + \cdots + x_m^d$.

Our next aim is to include such a triple (18) in a similar triple ($\bar{\mathcal{J}}(x); \langle \cdot, \cdot \rangle$; the image of the homomorphism (10)) for any generic surface F of degree d in $\mathbf{C}^n$, where n and d satisfy all the conditions of Main Theorem 3.

Proposition 2. *For any natural numbers n and d such that $n \geq 3$, $d \geq 3$ and $n+d \geq 8$, there exists an open subset U in the space of complex polynomials of degree d in $\mathbf{C}^n$ and an open subset V in $\mathbf{C}^n$ such that for any polynomial $F \in U$ and any $x \in V$ there is a* **completely infinite** *triple (18) consisting of a sublattice in $\bar{\mathcal{J}}(x) \equiv \bar{\mathcal{J}}(x, F)$ spanned by several vanishing cycles, the restriction of our standard bilinear form in $\bar{\mathcal{J}}(x)$ to this sublattice, and a subgroup of the monodromy group (10) preserving this sublattice.*

(Namely, this triple of subobjects will be isomorphic to the triple considered in Proposition 1 for some function singularity of Example 1.)

In this Proposition we do not require that F is hyperbolic or at least real, see the concluding remark of § 4.

Since all discrete topological properties of generic surfaces W_F of fixed degree are the same, Proposition 2 implies that there also exists a generic *hyperbolic* polynomial F satisfying all the conditions of this proposition.

By Theorem 2, the restriction of the form N (or forms $N^{(\nu)}$ with appropriate arbitrarily large $|\nu|$) to this subspace of $\bar{\mathcal{J}}(x)$ for this hyperbolic polynomial is nonzero, and Main Theorem 3 will follow from the definition of complete infiniteness.

The rest of this section is devoted to the proof of Proposition 2.

Let $\phi : (\mathbf{C}^{n-1}, 0) \to (\mathbf{C}, 0)$ be a function singularity.

Definition 3. The hypersurfaces W_F and $S(x)$ *have tangency of type ϕ* at a point $y \neq x$, if W_F is smooth at y, and close to y there exist local holomorphic coordinates $v_1, \ldots, v_n$ such that W_F is locally distinguished by the equation $v_n = 0$, and $S(x)$ by the equation

$$v_n = \phi(v_1, \ldots, v_{n-1}).$$

Given a pair of varieties W_F, $S(x)$ having tangency of type ϕ, there is a deformation of the singularity ϕ parametrized by the points $x' \in \mathbf{C}^n$ neighbouring to x. Namely, for any such x' the corresponding function $\phi_{x'}$ is defined by the following condition: the variety $S(x')$ is distinguished by the equality $v_n = \phi_{x'}(v_1, \ldots, v_{n-1})$ in the same local coordinates at y. Denote by Σ^ϕ the discriminant variety of this deformation. Obviously $\Sigma^\phi \subset \Sigma(F)$.

By the definition of versality, this deformation $\{\phi_{x'}\}$ is equivalent to one induced from some miniversal deformation of ϕ by a map of parameters

$$(\mathbf{C}^n, x) \to (\mathbf{C}^{\mu(\phi)}, 0). \tag{20}$$

Definition 4. The tangency of W_F and $S(x)$ at a point y is *perfect* if the map (20) induces an epimorphism of fundamental groups of complements of (local) discriminant varieties of both deformations.

Lemma 1. *If $n \geq 3, d \geq 3$ and $n + d \geq 8$, then for an arbitrary point $y \neq 0$ of the variety $S(0)$ there exists a polynomial $F' : \mathbf{C} \to \mathbf{C}$ of degree d such that the corresponding variety $W_{F'} = \{F' = 0\}$ and $S(0)$ have perfect tangency of type $\phi = v_1^d + \cdots + v_{n-1}^d$ at y.*

Indeed, for such a polynomial we can take $F' \equiv G + v_1^d + \cdots + v_{n-1}^d$, where G is the equation of $S(0)$ and $v_1, \ldots, v_{n-1}$ are a generic set of linear (nonhomogeneous) functions whose restrictions to the tangent plane to $S(0)$ at y define an affine coordinate system in this plane, cf. Lemmas I.3.1 and II.4.1.

Let us fix a point y and a polynomial F' satisfying this lemma. Let B be a small disc in $\mathbf{C}^n$ around y, in which the function ϕ and the deformation $\{\phi_{x'}\}$ are defined, and D a very small neighbourhood of the point x in $\mathbf{C}^n$. Let $x_0 \in D$ be a nondiscriminant value of the parameter x', so that $W_{F'} \cap S(x_0)$ is smooth on B. Denote by $h'(x_0)$ the group $\tilde{H}_{n-1}(W_{F'} \cap B \setminus S(x_0))$ (if n is even), or the group $\tilde{H}_{n-1}(W_{F'} \cap B \setminus S(x_0); @(x_0))$ (if n is odd). In the case of even n this group is naturally isomorphic to the standard $(n-2)$-dimensional vanishing homology group of the perturbation ϕ_{x_0} of the singularity ϕ (and the isomorphism is realized by the tube operation). In the case of odd n it is isomorphic to the similar $(n-1)$-dimensional homology group of the stabilization $\phi^{(1)}$ of ϕ; see Chapter I, § 12.5. In both cases the group $h'(x_0)$ is generated by the cycles vanishing over some paths joining x_0 to close points of $\Sigma(F')$.

The fundamental group of the set of nondiscriminant points in a neighbourhood of x acts in an obvious way on the group $h'(x_0)$. By the condition of perfectness of tangency at the point y, the image of this action is isomorphic to the local monodromy group of the singularity ϕ (if n is even) or of its stabilization $\phi + w^2$ (if n is odd), in particular by Proposition 1 it is completely infinite.

Let F be a small perturbation of F' such that the manifold W_F is generic (in particular, G-compatible). If this perturbation is sufficiently close to F' in B, then the corresponding group $h(x_0)$ (defined just as $h'(x_0)$ with F instead of F') is isomorphic to $h'(x_0)$ and is again generated by vanishing cycles; also the monodromy group in $\operatorname{Aut}(h(x_0))$ defined by the obvious action

$$\pi_1(D \setminus \Sigma(F)) \to \operatorname{Aut}(h(x_0)) \tag{21}$$

is isomorphic to that defined by the above representation of $\pi_1(D \setminus \Sigma(F'))$ in $\operatorname{Aut} h'(x_0)$. In particular, the triple (18) consisting of a) the image of the group $h(x_0)$ under the obvious homomorphism of this group into the group $\mathcal{H}(x_0)$, b) the restriction of the standard bilinear form $\langle \cdot, \cdot \rangle$ on this image, and c) this action (21), is completely infinite. Proposition 2 is thus completely proved.

Chapter IV. LACUNAS AND THE LOCAL PETROVSKIĬ CONDITION FOR HYPERBOLIC DIFFERENTIAL OPERATORS WITH CONSTANT COEFFICIENTS

The principal fundamental solution of a hyperbolic operator with constant coefficients is a distribution in $\mathbf{R}^N_+$ that has singularities on a surface (called the *wave front* of the operator) and coincides with analytic functions close to all points of its complement. The lacuna problem studies the asymptotic behaviour of these functions when the argument tends to the wave front: the front is holomorphically (respectively, $C^\infty-$) sharp at some point of it from the side of some local component of its complement if the restriction of the fundamental solution to this component can be extended to a real analytic (respectively, C^∞-smooth) function on the whole neighbourhood of our point (respectively, on the closure of this local component); in this case the component is called a *local lacuna* of our operator. The main problem arizing there is to give geometrical or topological criteria recognizing the sharpness of fronts in terms of their local shape.

The methods of this theory are very similar to those used in Newton's integrability problem. Indeed, the fundamental solution of a hyperbolic equation (or at least sufficiently high partial derivatives of it) outside the wave front is given by an explicit integral formula (Herglotz–Petrovskii–Leray integral) along a cycle (Petrovskii cycle) in the space of momenta, which depends smoothly on the point at which we calculate this fundamental solution. In particular, the analytic properties of it depend strongly on the ramification of Petrovskii cycles in the same way as the behaviour of the volume function depends on the monodromy of the "cap" cycle. Here is a dictionary of parallel concepts in these two theories.

Integrability problem $\leftrightarrow$ *lacuna problem.*

The spaces P and P_C of affine hyperplanes in $\mathbf{R}P^n$ and $\mathbf{C}P^n$ $\leftrightarrow$ *the projectivizations $\mathbf{R}P^{N-1}$ and $\mathbf{C}P^{N-1}$ of the real and complex "physical spaces".*

The spaces $\mathbf{R}^n$ and $\mathbf{C}^n$ $\leftrightarrow$ *the projectivizations $\check{\mathbf{R}}P^{N-1}$ and $\check{\mathbf{C}}P^{N-1}$ of the real and complex spaces of momenta.*

The surface $A_f \subset \mathbf{C}^n$ $\leftrightarrow$ *the projectivized set of zeros of the principal symbol, A^*.*

The secant hyperplane X $\leftrightarrow$ *the projectivized plane X^* dual to the point at which we calculate the fundamental solution.*

The set tang (A) $\leftrightarrow$ *the wave front.*

The "cap" element in $\tilde{H}_n(\mathbf{C}^n, X \cup A_f)$ $\leftrightarrow$ *the element in $\tilde{H}_{N-1}(\check{\mathbf{C}}^{N-1} \setminus (X^* \cup A^*))$ equal to the tube around the Petrovskii cycle.*

The volume form $\leftrightarrow$ *the Herglotz–Petrovskii–Leray form.*

The localization of the "cap" element $\leftrightarrow$ *the local Petrovskii cycle.*

However, the construction of Petrovskii cycles and the Herglotz–Petrovskii–Leray formula is more complicated than their analogues from Chapter I. One more difference is that in the theory of hyperbolic equations one traditionally studies not the algebraicity of integrals ("algebraic integrability") but their local regularity ("sharpness") or even triviality. However, the obstructions to these nice properties are of the same nature: they lie in the monodromy group of cycles of integration.

It turns out that the sharpness problem for generic hyperbolic equations is a problem in pure singularity theory.

Indeed, for any real isolated singularity $f : (\mathbf{C}^n, \mathbf{R}^n, 0) \to (\mathbf{C}, \mathbf{R}, 0)$ and any nondiscriminant real perturbation f_λ of it, the two cycles in the relative homology group $\tilde{H}_{n-1}(V_\lambda, \partial V_\lambda)$ of the corresponding local level variety $V_\lambda \equiv f_\lambda^{-1}(0) \cap B$ are well defined: the even and odd local Petrovskii cycles $\Pi_{\mathrm{ev}}(\lambda)$ and $\Pi_{\mathrm{odd}}(\lambda)$; see Chapter V, § 1.

Let $F = F(x, \lambda)$ be a real deformation of f, and $\Sigma(F)$ its discriminant. If for certain $\lambda \notin \Sigma(F)$ the even (respectively, odd) local Petrovskii cycle is homologous to zero, then the component of the complement of $\Sigma(F)$ that contains λ is called an even (respectively, odd) formal lacuna. The problem of counting all formal lacunas in the spaces of deformations of real singularities will be studied in Chapter V.

This problem is related to the "physical" lacuna problem by means of the generating functions and generating families of partial differential equations; see [Hörmander 71], [Zakalyukin 76], and Chapter II, § 4.

Namely, to any point of the wave front of a strictly hyperbolic operator there corresponds a germ of a function with singularity at 0 (generating function) and a deformation of this singularity (generating family). The singular points of wave fronts are naturally classified in acorrdance with the classification of the corresponding generating functions up to $\mathcal{K}$-equivalence, and the local connected components of the complement of a wave front close to a singular point are in one-to-one correspondence with the components of the complement of the discriminant of their generating family.

For almost all hyperbolic operators (for all but a proper subset in the space of operators of degree d in $\mathbf{R}^N$ whose codimension grows rapidly with d and n) all singularities of the corresponding generating functions are isolated, and hence for the nondiscriminant domains in the bases of their deformations the notion of a formal lacuna is well defined.

Theorem 1. *Suppose that the generating function of the wave front at a point has an isolated singularity. Then for even (respectively, odd) n the local component of the complement of a wave front is a holomorphic local lacuna if and only if the corresponding component of the set of nondiscriminant parameters of its generating family is an even (respectively, odd) formal lacuna.*

"If" in this theorem was essentially proved in [ABG 73], and "only if" was conjectured in [ABG 73] and proved in [Vassiliev 86]. The proof is based on the study of the action of local monodromy group on Petrovskii cycles; this approach to the proof was also suggested in [ABG 73].

Remark. In fact, in [ABG 73] a stronger result was proved: that the so-called *local Petrovskii condition* implies the sharpness close to arbitrary (not necessarily with isolated singularities of generating functions) points of wave fronts. In the case of isolated singularities this condition coincides with the condition from Theorem 1 that the corresponding component is a formal lacuna, and hence in this case "if" of Theorem 1 follows.

Also the "only if" part was conjectured in [ABG 73] in this general setting; in § 7 we construct a counterexample to this general conjecture.

Close to nonsingular (of type A_1) points of wave fronts the lacuna problem was solved completely in [Davydova 45] and [Borovikiv 59]; see also [Leray 62] and [ABG 70 + 73]. The answer is expressed in terms of inertia indices of the second fundamental form of the wave front (or, equivalently, of the corresponding (Morse) generating function; see § 3.

Close to the simplest singular points (of types A_2 and A_3) all local lacunas were found in [Gårding 77]; see § 3 below. In particular, it turns out that close to the cuspidal edge (=singularity A_2) of the wave front the "bigger" part of the complement of the front is never a local lacuna.

In § 6 of Chapter V we find all formal lacunas in the spaces of parameters of versal deformations of all simple singularities of functions (and hence all local lacunas close to simple singularities of generic wave fronts); a description of these lacunas is given in § 2 of Chapter V; see Table 12. In particular this calculation solves completely the problem about local lacunas for all generic hyperbolic operators with constant coefficients in the spaces $\mathbf{R}^n$, $n \leq 7$. These lacunas have the following geometrical characterization.

Theorem 2. *1. A local component of the complement of a wave front close to any stable simple singular point of it is a holomorphic local lacuna if and only if the Davydova–Borovikov signature condition is satisfied at all nonsingular points of its boundary, and this boundary does not contain cuspidal edges with respect to which this component is the "bigger one" (i.e., situated as component 1 in Figure 3a).*

2. Close to stable simple singularities, all local C^∞-lacunas are also local holomorphic lacunas.

(The "stability" condition demanded in these statements means the following: a singularity of a wave front is stable if its generating family is a $\mathcal{K}$-versal deformation of the corresponding generating function. A generic front satisfies this condition close to all its simple singularities.)

Conjecture. *The assertions of Theorem 2 are true close to all singularities of wave fronts.*

One half of part 1 of this conjecture is obvious: it follows from the results of Davydova and Gårding, that both conditions should be satisfied on the boundary of any lacuna. The problem is whether these conditions are also sufficient.

We also present many results about the existence (and quantity) of local lacunas close to nonsimple singularities, including all singularities of corank 2 whose Milnor numbers do not exceed 11; for the summary of these results see § 2 of Chapter V. Almost all these lacunas were discovered by a FORTRAN–algorithm which counts all topologically different morsifications of any given real isolated singularity. Of course, this algorithm can be used not only for the lacuna problem.

The idea of this algorithm is as follows.

Any two morsifications of a real singularity can be joined by a one-parameter family along which these morsifications undergo only finitely many standard metamorphoses. Knowing a sufficiently complete set of topological characteristics of a morsification at the beginning of the path (such as the Dynkin diagram, homology classes of local Petrovskii cycles, quantity and Morse indices of real critical points, the order and signs of their critical values, etc.), we can determine their values after any admissible sequence of these surgeries. This allows us to model any path in the parameter space of a deformation of a singularity as a sequence of arithmetical transformations over a set of discrete characteristics; the problem of studying morsifications thus reduces to a combinatorial algorithm. We describe this algorithm in §§ 8 and 9 of Chapter V and present the corresponding program in the Appendix to this book.

§ 0. Hyperbolic polynomials

Let $P : (\mathbf{C}^N, \mathbf{R}^N, 0) \to (\mathbf{C}, \mathbf{R}, 0)$ be a real homogeneous polynomial of degree d, and ϑ a nonzero vector in $\mathbf{R}^N$. Denote by $A(P)$ the cone in $\mathbf{C}^N$ defined by the equation $P = 0$, and by Re $A(P)$ the cone $A(P) \cap \mathbf{R}^N$.

Definition 1. The polynomial P is *hyperbolic* with respect to ϑ if $P(\vartheta) \neq 0$ and any line in $\mathbf{R}^N$ collinear to ϑ intersects the cone Re $A(P)$ in exactly d points (counted with multiplicities). The polynomial P is *strictly hyperbolic* with respect to ϑ if any line in $\mathbf{R}^N$ collinear to ϑ but not passing through the origin intersects Re $A(P)$ in exactly d geometrically distinct points.

In other words, a polynomial is strictly hyperbolic if and only if it is hyperbolic and has no singular points in $\mathbf{R}^N$ apart from the origin.

Example 1. Any hyperbolic quadratic form in $\mathbf{R}^N$ (i.e. a quadratic form that can be written as $x_1^2 - x_2^2 - \cdots - x_N^2$ in an appropriate coordinate system) is strictly hyperbolic with respect to any line directed to the cone where this quadratic form is positive.

Example 2. If two operators are hyperbolic with respect to a vector ϑ, then so is their product.

Example 3. A nondegenerate cubic curve in $\check{\mathbf{R}}P^2$ can have two or one connected components (and looks like either the two curves in Figure 35a or only the right-hand one of them). The polynomial of degree 3 in $\mathbf{R}^3$ that defines this curve is never hyperbolic in the second case, and in the first case it is hyperbolic with respect to all vectors directed inside the left oval.

In general, the set of real zeros of a strictly hyperbolic polynomial of degree d always consists of $[(d+1)/2]$ components: if d is even, then all these components are isotopic to the standard cone given by the equation of Example 1, and if d is odd, then there is one more component homeomorphic to a hyperplane.

The space of all polynomials of degree d hyperbolic with respect to a fixed vector ϑ is denoted by $\mathrm{Hyp}(\vartheta, d)$.

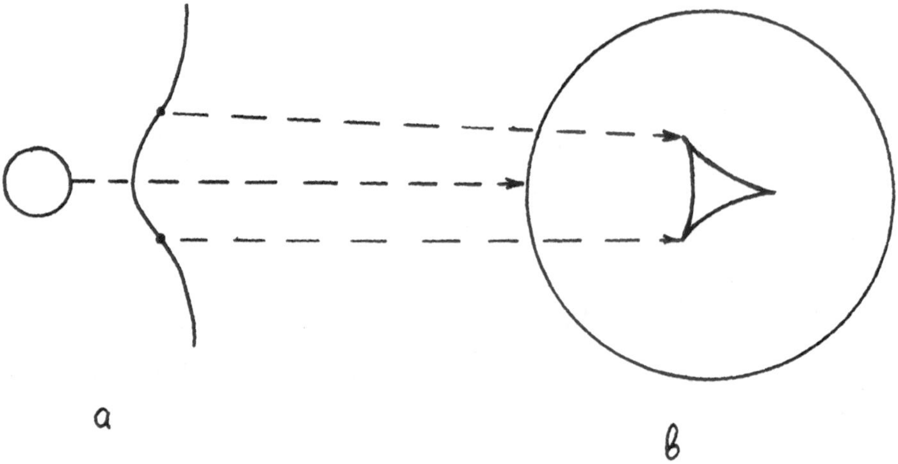

Fig. 35. Standard cubic curve and its projective dual

Theorem 1 (see [Nuij 68]). *For any nonzero vector* $\vartheta \in \mathbf{R}^N$ *the space* $\operatorname{Hyp}(\vartheta, d)$ *consists of exactly two connected components (which go into one another under the map* $\{P \to -P\}$*). Both these components are contractible. The set of all strictly hyperbolic polynomials is dense in the space* $\operatorname{Hyp}(\vartheta, d)$*, and the nonstrictly hyperbolic polynomials are exactly the boundary points of this set.*

Theorem 2 (see [ABG 70]). *If a polynomial* P *is hyperbolic with respect to a vector* ϑ*, then it is also hyperbolic with respect to any other vector from the same component of the set* $\{x | P(x) \neq 0\}$.

Definition 2. For any real homogeneous polynomial P, a component of the set $\{x | P(x) \neq 0\}$ is called the *hyperbolicity domain* if it consists of vectors ϑ such that P is hyperbolic with respect to all of them.

Theorem 3 (see [ABG 70]). *The hyperbolicity domain is always a convex cone in* $\mathbf{R}^N$.

Here is an important example of a nonstrict hyperbolic polynomial. Let K_m be the space of all real symmetric $m \times m$ matrices (or, which is the same, of all real quadratic forms $\mathbf{R}^m \to \mathbf{R}$).

Proposition 1 (see [Arnold 88′]). *The determinant of symmetric matrices is a hyperbolic polynomial on* K_m *with respect to an arbitrary positive quadratic form.*

Indeed, using the canonical basis we can assume that our positive form is given by the unit matrix. In this case our assertion follows from the fact that all eigenvalues of a real symmetric matrix are real.

§ 1. Hyperbolic operators and hyperbolic polynomials. Sharpness, diffusion and lacunas

1.1. Main definitions.

Let $P \equiv P(D)$ be a linear partial differential operator of order d with constant coefficients in $\mathbf{R}^N$, i.e. a polynomial in the symbols $D_j \equiv \partial/i\partial x_j$, $j = 1, \ldots, N$.

The *Cauchy problem* in the half-space $\mathbf{R}^N_+ \equiv \{x | x_1 \geq 0\}$ for this operator is stated as follows: for any given function ϕ in this half-space and any d functions $\chi_0, \ldots, \chi_{d-1}$ on the hyperplane $\mathbf{R}^{N-1} \equiv \{x | x_1 = 0\}$, to find a function (or at least distribution) u with support in this half-space such that

$$P(u) = \phi \quad \text{in } \mathbf{R}^N_+,$$

$$\frac{\partial^k u}{\partial x_1^k} = \chi_k \quad \text{on } \mathbf{R}^{N-1}, \quad k = 0, \ldots, d-1.$$

A *fundamental solution* of the operator P is a distribution Ψ satisfying the condition $P\Psi \equiv \delta_0$ in $\mathbf{R}^N$, where δ_0 is the Dirac delta-function located at the origin.

The operator P is said to be *hyperbolic* in the half-space $\mathbf{R}^N_+$ if either of the following conditions holds:

a) P possesses a fundamental solution whose support is a proper cone in the half-space $\mathbf{R}^N_+$;

b) any Cauchy problem in $\mathbf{R}^N_+$ for this operator has an unique solution, and its solution at any point a depends only on the values of the initial data χ_k and the right-hand side ϕ of the problem, lying in some proper cone with vertex a in the half-space $\{x | x_1 < x_1(a)\}$.

(We say that a closed cone in a half-space is *proper* if its intersection with the boundary of this half-space consists of only one point, its vertex.)

Proposition 1 (see for example [ABG 70]). *The above conditions a) and b) are equivalent. The fundamental solution satisfyung condition a) is unique (if it exists).*

Such a fundamental solution of a hyperbolic operator P is called its *principal fundamental solution* and is denoted by E_P.

Example 1. The wave equation

$$\frac{\partial^2 u}{\partial t^2} = c^2 \sum_{j=2}^{N} \frac{\partial^2 u}{\partial x_j^2}$$

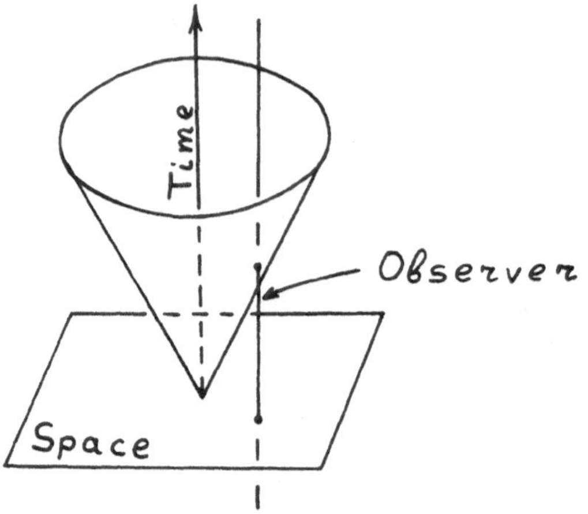

Fig. 36. Wave front of the wave equation

describing the propagation of waves with speed c is hyperbolic in the half-space $\{t \geq 0\}$. Its principal fundamental solution can be considered as an elementary wave arising from a pointwise instantaneous perturbation. It has singularities on the cone defined by the conditions

$$c^2t^2 = \sum_{j=2}^{N} x_j^2, \quad x_1 \geq 0. \tag{1}$$

The asymptotic behaviour of this fundamental solution close to this cone depends on the parity of the number of variables. For example, for $N = 2, 3, 4$ these fundamental solutions E_N are given by following formulae (in which θ is the Heaviside function): $E_2 = \theta(ct - |x|)/2c$, $E_3 = \theta(ct - |x|)\sqrt{c^2t^2 - |x|^2}/2\pi c$, $E_4 = \theta(t)\delta(c^2t^2 - |x|^2)/2\pi c$. In higher even dimensions the qualitative picture is the same as for $N = 4$, i.e. the signal is accepted at only one instant, when it passes by a fixed observer; see Figure 36. In the odd-dimensional case the signal continues to sound the whole time after the instant of the first meeting. The first circumstance allows us to communicate via sound, while by the second the "acoustical layer" in the ocean, which is an excellent conductor of individual signals, cannot serve for rapid transmission of complicated information. Both variants of the expansion of sound waves have analogues for arbitrary hyperbolic equations. In the language of general theory (applied to the wave equation) we say that in the even-dimensional case the interior component of the complement of the cone (1) of singularities (= *wave front*) is a *lacuna*, and in the odd-dimensional case there is *diffusion* of waves from the side of this component; the exterior component is a lacuna for arbitrary dimensions (and arbitrary hyperbolic equations).

1.2. Hyperbolic operators and hyperbolic polynomials.

We identify the partial derivatives $\partial/i\partial_j$ in "physical space" with the coordinates ξ_j in the dual space $\check{\mathbf{R}}^N$ (called the *space of momenta*). Thus the differential operator P of order d becomes a polynomial P_ξ of degree d in the variables $\xi_1, \ldots, \xi_N$.

Denote by a bar over P and P_ξ the action of taking the principal (of highest order or highest degree) homogeneous part $\bar{P}$ or $\bar{P}_\xi$ of an operator or polynomial.

Theorem 1 (see [ABG 70 + 73]). *1. If the operator P is hyperbolic in the half-space $\mathbf{R}^N_+$, then the principal (of highest degree) homogeneous part of the corresponding polynomial P_ξ is a hyperbolic polynomial with respect to the vector $\partial/\partial\xi_1$, in particular it is proportional to a polynomial with real coefficients.*

2. Conversely, if this principal part $\bar{P}_\xi$ is strictly hyperbolic, then the operator P is hyperbolic.

3. If the operator P is homogeneous, $P = \bar{P}$, then it is hyperbolic if and only if the polynomial P_ξ is hyperbolic.

If the hyperbolic homogeneous polynomial $\bar{P}_\xi$ has singularities in $\check{\mathbf{R}}^N \setminus 0$, i.e. it is not strictly hyperbolic, then for the operator P with principal part equal to $\bar{P}$ to be hyperbolic it is necessary that some additional conditions on the lower terms of P_ξ hold; see [ABG 70], [Svensson 70] (and, by item 3 of the previous theorem, it is sufficient that these lower terms are trivial).

1.3. Wave front.

The singularities of the principal fundamental solution of a hyperbolic operator P are located on some conical semialgebraic surface in $\mathbf{R}^N_+$, its wave front $W(P)$. Here is a description of this surface in the case of strictly hyperbolic operators: roughly speaking, this is the dual cone to the cone of zeros of the principal symbol of the operator.

More precisely, denote by $A(P)$ the conical hypersurface in $\check{\mathbf{C}}^N$ defined by the vanishing of the principal part $\bar{P}_\xi$ of the polynomial P_ξ. Let $\operatorname{Re} A(P)$ be the set of its real points, $A(P) \cap \check{\mathbf{R}}^N$.

The spaces $\check{\mathbf{R}}^N$ and $\check{\mathbf{C}}^N$ will be considered as the spaces dual to $\mathbf{R}^N$ and $\mathbf{C}^N$ via the pairing

$$\langle x, \xi \rangle = \sum x_j \xi_j. \tag{2}$$

Then any point $x \in \mathbf{C}^N \setminus 0$ defines its orthogonal hyperplane $X(x) \subset \check{\mathbf{C}}^N$. The *wave front* $W(P)$ of a strictly hyperbolic operator P is defined as the cone consisting of all points in $\mathbf{R}^N_+$ such that the plane $X(x) \cap \check{\mathbf{R}}^N$ orthogonal to x is tangent to the cone Re $A(P)$ along some line.

For an arbitrary (not strictly) hyperbolic operator the wave front is defined in a more complicated manner (see [ABG 70]); in any case it belongs to the set of points x such that the plane $X(x)$ is not in general position with the cone Re $A(P)$.

It follows directly from the definition that the wave front of a strictly hyperbolic operator of degree d consists of $[(d+1)/2]$ components, each of which is dual to a component of the cone Re $A(P)$. The component dual to the boundary of the hyperbolicity domain is the "exterior" one (i.e. it separates all other components from the half-space $\{x | x_1 < 0\}$), and is convex by Theorem 0.3.

For instance, the projectivization of the wave front of the operator from Example 0.3 (see Figure 35a) is as shown on Figure 35b; the cuspidal points of the front correspond to the inflection points of the curve $P = 0$. (The third inflection point of this curve is placed at infinity in Figure 35a).

Theorem 2 (see [ABG 70]). *The principal fundamental solution of a hyperbolic operator is an analytic function everywhere in* $\mathbf{R}^N$ *outside its wave front and is identically zero outside the convex hull of the wave front.*

1.4. Sharpness, diffusion, local and global lacunas.

Let P be a hyperbolic operator, W and E_P the wave front and principal fundamental solution of P; let y be a point of W. The front W divides any small neighbourhood of y into several connected components; the restriction of E_P to any of them is a real analytic function by Theorem 2.

Definition 1. The principal fundamental solution E_P is *holomorphically sharp at the point* $y \in W$ from the side of a local (close to y) component l of the complement of W if its restriction to l coincides with the restriction of a holomorphic function defined in some neighbourhood of the point y. Similarly, E_p is C^∞*-sharp* from the side of l if it has a C^∞-smooth extension from l to the closure $\bar{l}$ of this component close to the point y. In these cases the component l is called a (holomorphic or C^∞-) *local lacuna* of the operator P close to y. If E_P is not sharp from the side of the component l, then one says that there is *diffusion of waves* at l.

Definition 2. A (global) component L of the complement of a wave front is called a *holomorphic lacuna* (respectively, C^∞*-lacuna*) if the solution E_P is holomorphically (respectively, C^∞-) sharp from the side of L at any point of the closure of L (or, equivalently, at the single point 0). If the solution E_P is identically zero at L, then L is called a *strong lacuna*.

Conjecture (see [ABG 70 + 73]). *The notions of holomorphic and C^∞- lacunas (both local and global) are equivalent.*

Example 2. The "exterior" component of the complement of a wave front is always a strong lacuna. In the case of the wave equation in $\mathbf{R}^N$, the unique interior component is a strong lacuna if N is even and greater than 2, is a holomorphic but not strong lacuna for $N = 2$, and is diffuse for all odd N. The exclusion of the case $N = 2$ depends on the fact that in this case N does not exceed the order of the operator; see § 4.

Example 3. Let P be a strictly hyperbolic operator of order 3 in $\mathbf{R}^3$; see Example 0.3. The most interior component of the complement of the front is in this case a holomorphic, but not a strong lacuna, while the intermediate component (into which there are directed the wedges of the interior part of the front) is diffuse close to any point of its boundary.

§ 2. Generating functions and generating families of wave fronts for hyperbolic operators with constant coefficients. Classification of the singular points of wave fronts

The asymptotic properties of the principal fundamental solution E_P close to a point of a wave front are determined by the local geometry of the front; a convenient way to describe this geometry is the (projective) generating function and generating family of the front. They were essentially determined in Chapter II, § 4; now we repeat this definition in accordance with our new situation.

Since a wave front is a cone, it is sufficient to consider its projectivization $W^* \equiv W^*(P) \in \mathbf{R}P^{N-1}$.

In general, the projectivizations of all conical (i.e. invariant under dilations) surfaces in $\mathbf{C}^N$, $\mathbf{R}^N$, $\check{\mathbf{C}}^N$ or $\check{\mathbf{R}}^N$ will be denoted by a star. For instance, $A^*(P)$ is the projectivized set of zeros of the principal symbol of the operator P in $\check{\mathbf{C}}P^{N-1}$; this operator is strictly hyperbolic if and only if the real part Re $A^*(P)$ of this set is smooth.

Let a^* be a nonsingular point of the surface Re $A^*(P)$, and Re Y the plane tangent to Re $A(P)$ at the line a^*. Let y^* be the point in $\check{\mathbf{R}}P^{N-1}$ such that any covector $y \in \check{\mathbf{R}} \setminus 0$ representing it vanishes on *Re* Y. By the definition, this point y^* belongs to the projectivized wave front $W^*(P)$. Choose in $\check{\mathbf{R}}P^{N-1}$ a local affine coordinate system $\{\zeta_1, \dots, \zeta_{N-2}, \eta\}$ with origin at a^* such that the plane Re Y^* is distinguished by the equation $\eta = 0$. Then the divisor $A^*(P)$ is locally defined by the condition

$$\eta = f(\zeta_1, \ldots, \zeta_{N-2}),$$

where f is a smooth algebraic function with a critical point at the origin.

Definition 1. The function f thus defined is called a *projective generating function* of the projective wave front at the point y^*.

Remark 1. The wave front can have several locally irreducible components at the point y^*: if the corresponding plane $\operatorname{Re} Y^* \subset \check{\mathbf{R}}P^{N-1}$ is tangent to $\operatorname{Re} A^*$ at several real points a_i^*, then to any such point a_i^* there corresponds a local component consisting of planes tangent to $\operatorname{Re} A^*$ close to this point. By the above rule any such component determines a germ of a generating function; the collection of generating functions thus obtained defines the germ of the (projectivized) wave front at the point y^*.

Convention. In what follows we study the *local* properties of wave fronts. We define a germ of wave front at a point y^* as the germ of a locally irreducible component of it at this point.

Of course, the generating function depends on the choice of appropriate coordinates ζ_i, η; however, the following assertion holds.

Proposition 1. *Any two germs of projective generating functions, defined as above in any two appropriate affine coordinate systems by the same germ of the wave front of a strictly hyperbolic operator, are $\mathcal{K}$-equivalent to each other; see Chapter I, § 13.*

The proof is obvious.

Therefore the $\mathcal{K}$-classification of singular points of functions defines a classification of singular points of wave fronts.

Example 1. If the singularity of the function f at the point a^* is Morse, then the corresponding piece of the wave front is nonsingular. If this Morse function is a minimum or maximum, then the corresponding piece of the wave front is also locally convex.

Moreover, it is easy to verify that the signatures of the second fundamental forms of the mutually dual smooth pieces of surfaces $W^*(P)$ and $\operatorname{Re} A^*(P)$ coincide with one another. (Since the signatures of the second fundamental forms of real hypersurfaces depend on the choice of transversal directions to these hypersurfaces, for the above assertion to be correct we should specify a concordance of these directions for the dual surfaces. This concordance is defined by the following rule. If such a direction for one surface, say $\operatorname{Re} A^*$, at its point a^* is chosen, v is positively directed vector transversal to $\operatorname{Re} A^*$, and $\operatorname{Re} Y^*$ is the tangent plane of $\operatorname{Re} A^*$ at a^*, then a positively directed transversal vector to W^* consists of points orthogonal to planes in $\check{\mathbf{R}}P^{N-1}$ obtained from $\operatorname{Re} Y^*$ by parallel shift (in any local affine coordinates) towards the direction v.)

Example 2. If the germ of f at the origin is of class A_2 (see the Classification Table 1), then the wave front has a semicubical cuspidal edge (see Figs. 3, 1).

Example 3. Suppose that the germ of f at 0 is of class A_3. Then, in suitable (generally speaking, curvilinear) local coordinates, f can be written in the form

$$\pm\zeta_1^4 \pm \zeta_2^2 \pm \cdots \pm \zeta_{N-2}^2. \tag{3}$$

If such a canonical coordinate system is generic with respect to the affine structure, then the corresponding piece of the projectivized wave front is locally diffeomorphic to the direct product of an $(N-5)$-dimensional linear space and the *swallow-tail* sketched in Figure 3b. The genericity condition that we use here is formulated as follows. For any choice of coordinate system in which f is written in the form (3), the 2-jet of the coordinate curve given by the equation $\zeta_2 = \cdots = \zeta_{N-2} = 0$ is the same: all such curves have second-order tangency at the origin. We shall say that a singularity of type A_3 is *stable* if the first and second derivatives of this curve are linearly independent. This is exactly the desired genericity condition; its analogue for arbitrary singularities is formulated in terms of the (projective) generating family of the wave front.

Indeed, any hyperplane $\operatorname{Re} X^* \subset \check{\mathbf{R}}P^{N-1}$ sufficiently close to $\operatorname{Re} Y^*$ is distinguished by the linear equation

$$\eta = \lambda_1\zeta_1 + \cdots + \lambda_{N-2}\zeta_{N-2} + \lambda_0, \tag{4}$$

where λ_i are certain coefficients. It is easy to see that these coefficients define a local affine coordinate system in the space $\mathbf{R}P^{N-1}$ with origin at y^*. Consider the deformation $F = F(\zeta, \lambda)$ of the generating function f that is given by the condition

$$F(\zeta,\lambda) = f(\zeta) - \lambda_1\zeta_1 - \cdots - \lambda_{N-2}\zeta_{N-2} - \lambda_0. \tag{5}$$

Definition 2. The family of functions $f_\lambda \equiv F(\cdot, \lambda)$ is called the *projective generating family* of (a local component of) the wave front at the point y^*.

By the definition, for λ close to 0 the plane $\operatorname{Re} X^*(\lambda)$ given by equation (4) is tangent to the surface $\operatorname{Re} A^*(P)$ if and only if the corresponding point λ belongs to the wave front. It is easy to see that this holds if and only if the function f_λ has near the origin a real critical point with zero critical value. Thus we have a one-to-one correspondence between the germ of a wave front and the *real discriminant* of the corresponding generating family; see Chapter I, § 7.

Definition 3. A germ of a projective wave front is *stable* if the corresponding generating family is a $\mathcal{K}$-versal deformation of the corresponding projective generating function of this germ.

It is easy to see that this definition does not depend on the choice of local affine coordinates ζ_i that define the generating function and family, and in the case of a singularity A_3 this definition coincides with the one from Example 3 above. Moreover, for an arbitrary singularity of class A_k this condition has the following explicit reformulation.

The generating function of such a germ has the form

$$\pm\zeta_1^{k+1} \pm \zeta_2^2 \pm \cdots \pm \zeta_{N-2}^2 \tag{6}$$

in a suitable curvilinear local coordinate system. The $(k-1)$-germ of the coordinate curve $\{\zeta | \zeta_2 = \cdots = \zeta_{N-2} = 0\}$ is well defined: all similar curves, defined by all coordinate systems in which f has the canonical form (6), have tangency of order $k-1$ at the origin.

Proposition 2. *The germ of a singularity of a wave front of class A_k is stable if and only if the first $k-1$ derivatives of the curve in $\mathbf{R}^{N-2}$ defined by the equation $\{\zeta | \zeta_2 = \cdots = \zeta_{N-2} = 0\}$ in any canonical coordinate system are linearly independent.*

The proof is elementary.

§ 3. Local lacunas close to nonsingular points of fronts and to singularities A_2, A_3 (after Davydova, Borovikov and Gårding)

The sharpness from the side of a given component of the complement of the front is satisfied or not simultaneously for all points of the front that lie on the same ray through the origin in $\mathbf{R}^N$. Therefore the notions of local lacunas, sharpness and diffusion are well defined for points of the projectivized front $W^*(P) \subset \mathbf{R}^{N-1}$: we have sharpness at a point $y^* \in W^*(P)$ from the side of a certain local component L^* of the complement of W^* if there is sharpness in the sense of Definition 1.1 at any nonzero point $y \in W(P)$ of the line y^* from the side of the corresponding component L; the component L^* in this case is called a (projectivized) local lacuna.

At the nonsingular points of the front the condition of sharpness can be recognized in the following terms.

3.1. The Davydova–Borovikov condition.

Let $W^*(P)$ be the projectivized wave front of a strictly hyperbolic operator, and y^* a nonsingular point of the hypersurface $W^*(P)$; in particular, this hypersurface divides its complement close to y^* into two local components. Let L^* be one of them. The second fundamental form of W^* is nondegenerate at y^*: otherwise the dual surface Re $A^*(P)$ is singular. Fix some local affine coordinates $z_0, z_1, \dots, z_{N-2}$ in $\mathbf{R}P^{N-1}$ with origin at y^* in such a way that the plane $\{z_0 = 0\}$ is tangent to the front at this point, and the basis vector $\partial/\partial z_0$ is directed "inside" the component L^*. Then the front can be expressed close to y^* by a condition of the form $z_0 = g(z_1, \dots, z_{N-2})$, where g is a smooth function, $g(0) = 0, dg(0) = 0$. Obviously the signature of the quadratic form $d^2g|_0$ does not depend on the choice of affine coordinates satisfying the above conditions.

Definition 1. The local component L^* of the complememt of the projectivized discriminant satisfies the *Davydova-Borovikov condition* at the point y^* if the negative inertia index of the quadratic form $d^2g|_0$ thus defined is even.

Theorem 1 (see [Davydova 45], [Borovikov 59], [Leray 62], [ABG 70 + 73]). *Close to a nonsingular point of the projectivized wave front, the local component of the complement of the front is a local C^∞-lacuna if and only if it satisfies the Davydova–Borovikov condition at this point. If this condition is satisfied, it is also a holomorphic lacuna.*

3.2. The same in terms of generating functions.

The Davydova–Borovikov condition can also be reformulated in terms of the quadratic part of the corresponding generating function (or, which is the same, of the second fundamental form of the dual hypersurface Re $A^*(P)$).

Indeed, consider the point a^* of this hypersurface such that y^* is orthogonal to the tangent hyperplane Y^* of Re $A^*(P)$ at a^*. (In what follows we shall identify the points in $\mathbf{R}P^{N-1}$ with the hyperplanes orthogonal to them in $\check{\mathbf{R}}^{N-1}$). Let f be a generating function of our front defined as in § 2 by some local affine coordinates at the point a^*. Since y^* is a nondegenerate point of the front, the second quadratic part of this function is nondegenerate. Its signature depends on the choice of these coordinates in the following way only: if we change the sign of the coordinate η, then the negative inertia index becomes the positive one, and vice versa. On the other hand, for an arbitrarily small positive constant ε, only one of the two (points orthogonal to) hyperplanes $\{\eta = \varepsilon\}$ and $\{\eta = -\varepsilon\}$ belongs to the distinguished component L^*. We choose the coordinate η in such a way that this will be the hyperplane $\{\eta = \varepsilon\}$. Let us call such a coordinate system and the corresponding generating function *adapted.*

Definition 2. The projectivized wave front satisfies the *Davydova–Borovikov condition* close to its nonsingular point y^* if the negative inertia index of the adapted generating function of the front close to the corresponding point of $\operatorname{Re} A^*(P)$ is even.

Proposition 1. *This definition is equivalent to Definition 1.*

This is a trivial exercise in projective duality: the best known version of it asserts that the Legendre transform of a convex function is again convex. See also Example 2.1.

3.3. The local lacunas close to singularities A_2 and A_3 (after Lars Gårding).

The simplest singular points of wave fronts (besides the self-intersections) are the points of types A_2 and A_3. The lacuna problem close to these singularities was investigated by Lars Gårding; see [Gårding 77]. His results, reformulated in terms of projective generating functions, are as follows.

Recall that the singularity A_2 is of corank 1 (see Table 1), in particular, the rank of any generating function of such a singularity of a front in $\mathbf{R}^N$ is equal to $N-3$.

Theorem 2. *If the dimension N is even, or N is odd and both inertia indices of the quadratic part of the generating function of a singular point of type A_2 of a projectivized wave front in $\mathbf{R}P^{N-1}$ are odd, then close to this singular point there are no local lacunas. If N is odd and these inertia indices are even, then the "smaller" (i.e. situated like the component 2 in Figure 3a) component of the complement of the front is a local lacuna, and the "bigger" component is not.*

Corollary 1: *the "only if" part of statement 1 of Theorem 2 of the introduction to this chapter.*

Corollary 2. *Statement 2 of Theorem 2 of the introduction to this chapter follows from statement 1.*

Indeed, if a local component of the complement of the discriminant is not a holomorphic lacuna, then by statement 1 its boundary has forbidden nonsingular or cuspidal points. The behaviour of the principal fundamental solution close to such points was explicitly studied in [Davydova 45], [Borovikov 59], [Gårding 77], [Varchenko 87], etc.; in particular, it is known that close to them the C^∞-sharpness fails.

Now consider a singularity of type A_3 and suppose that the sign of the coordinate η from the definition of the generating function of this singularity is chosen in such a way that this generating function is of type $+A_3$ (see Table 1), i.e. the sign before ζ_1^4 in the corresponding D_0-normal form is $+$. This assumption fixes the signature of the quadratic part of this generating function, in particular, its positive inertia index i_+.

Theorem 3. *Close to the singularities of type A_3, the local lacunas are only the following components of the complement of the front (see Figure 3b):*
component 3 for odd i_+ and any N;
component 2 for even i_+ and odd N.

The fact that the component 1 is never a local lacuna follows from the similar fact for component 1 in Figure 3a for the singularity A_2. Indeed, the semicubical edges in Figure 3b are the points of type A_2; by Theorem 2 we always have diffusion close to these points from the side of component 1.

§ 4. Petrovskii and Leray cycles. The Herglotz–Petrovskii–Leray formula and the Petrovskii condition for global lacunas

All the main theorems in this chapter depend on the fact that the principal fundamental solution of a hyperbolic operator, as well as all partial derivatives of this solution, are given by explicit integral formulae: the following *Herglotz–Petrovskii–Leray integrals* (7), (7′) in the case of homogeneous operators, and similar formulae for arbitrary hyperbolic operators; see Subsection 4.3 below. Now we define all the ingredients of these formulae.

$$D^\nu E_P(x) = c \cdot \int_{\alpha(x)} \Lambda^*_{x,\nu,P} \quad \text{for } q \geq 0, \tag{7}$$

$$D^\nu E_P(x) = c \cdot \int_{\gamma(x)} \Lambda^*_{x,\nu,P} \quad \text{for } q < 0. \tag{7'}$$

4.1. The case of homogeneous operators.

In formulae (7), (7′), P is a homogeneous hyperbolic operator of degree d in $\mathbf{R}^N$, $\nu = (\nu_1, \dots, \nu_N) \in \mathbf{Z}_+^N$ is a multiindex, and $D^\nu E_P(x)$ is the ν-th partial derivative of the principal fundamental solution of P, evaluated at the point $x \in \mathbf{R}^N \setminus W(P)$.

The constant c is nonzero, and $q = q(\nu) \equiv d-(\nu_1+\ldots+\nu_N)-N$. The integration cycles $\alpha(x)$ and $\gamma(x)$ lie in the dual projective space $\check{\mathbf{C}}P^{N-1}$ and will be defined in the next subsection 4.2; the differential form $\Lambda^*_{x,\nu,P}$ in $\check{\mathbf{C}}P^{N-1}$ is defined as follows. First, define the universal $N-1$-form ω in $\check{\mathbf{C}}^N$ by

$$\omega(\xi) = \sum_{k=1}^{N} (-1)^k \xi_k \cdot d\xi_1 \wedge \ldots \wedge \widehat{d\xi_k} \wedge \ldots \wedge d\xi_N. \tag{8}$$

Then for every multiindex ν define the form

$$\Lambda_{x,\nu,P}(\xi) = \langle x, \xi \rangle^q \xi^\nu P(\xi)^{-1} \omega(\xi), \tag{9}$$

where $\langle \cdot, \cdot \rangle$ is the scalar multiplication of elements of dual spaces, see (2). This form is homogeneous of degree 0 with respect to dilations in $\check{\mathbf{C}}^N$, and its convolution with any ray through the origin in $\check{\mathbf{C}}^N$ is trivial; hence it defines a form on the quotient space $\check{\mathbf{C}}P^{N-1}$ of $\check{\mathbf{C}}^N$ by these dilations: this form is exactly the desired form $\Lambda^*_{x,\nu,P}$.

Obviously, this form has singular points on the set $\operatorname{Re} A^*(P)$ and, if $q < 0$, also on the hyperplane X^* orthogonal to the vector x; hence the integration cycles must lie outside these singular sets.

4.2. The Leray and Petrovskii cycles.

The cycle $\alpha(x)$ from formula (7) is a relative cycle in $\check{\mathbf{C}}P^{N-1} \setminus A^*(P) \bmod X^*$; it is constructed as follows. Let $P : (\check{\mathbf{C}}^N, \check{\mathbf{R}}^N) \to (\mathbf{C}, \mathbf{R})$ be a homogeneous polynomial strictly hyperbolic with respect to the vector ϑ, $W(P)$ its wave front, x an arbitrary point in $\mathbf{R}^N_+ \setminus W(P)$, and X the corresponding hyperplane in $\check{\mathbf{R}}^N$. The unit sphere S^{N-1} in $\check{\mathbf{R}}^N$ is divided into two hemispheres by the plane X. We orient these two hemispheres in a discordant way via the differential $(N-1)$-form $\langle x, \xi \rangle \omega(\xi)$, where $\omega(\xi)$ is the form (8): i.e. this form takes positive values on positively oriented frames in these hemispheres.

Lemma 1 (see [ABG 70]). *There exists a continuous family v of arbitrarily small vectors in $\check{\mathbf{R}}^N$, applied at the points of S^{N-1}, such that*

a) at any point $a \in S^{N-1}$ the corresponding vector $v(a)$ is parallel to the hyperplane $\operatorname{Re} X$ *or is equal to zero;*

b) at any point a of the set $S^{N-1} \cap A(P)$, $v(a)$ is transversal to $\operatorname{Re} A(P)$ *and is directed into the same component of the open cone $T_a\mathbf{R}^N \setminus T_a(\operatorname{Re} A(P))$ as the vector parallel to ϑ.*

Moreover, such a vector-field v can be chosen to be zero everywhere outside an arbitrarily small neighbourhood of the set $\operatorname{Re} A(P)$.

This lemma follows immediately from the definitions. (In [ABG 70] a similar family v was constructed for arbitrary (maybe, nonstrictly) hyperbolic polynomials.)

Fix such a vector-field and assume that it is sufficiently small. Move the sphere S^{N-1} in $\check{\mathbf{C}}^N \setminus 0$ slightly by sending each point a of it to $a - i \cdot v(a)$. Since P is

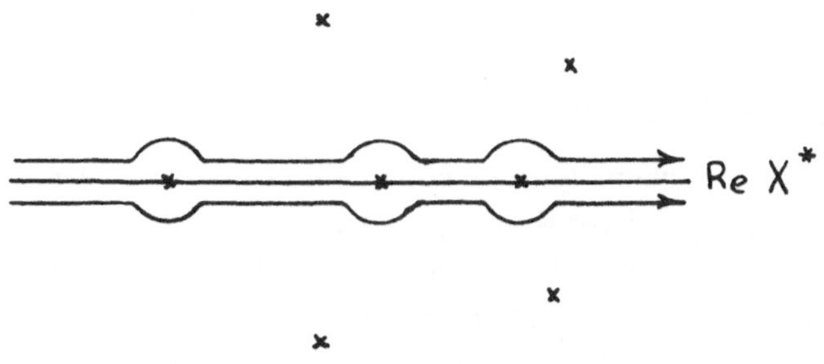

Fig. 37. Petrovskii cycle for $N = 3$

hyperbolic, the resulting submanifold does not intersect the surface $A(P)$. The two components of this manifold coming from the two hemispheres inherit the above-defined orientations. Their sum is a relative $(N-1)$-dimensional cycle in $\check{\mathbf{C}}^N \setminus A(P)$ mod X, and the obvious projection of this cycle into $\check{\mathbf{C}}P^{N-1}$ is a relative $(N-1)$-dimensional cycle in $\check{\mathbf{C}}P^{N-1} \setminus A^*(P)$ mod X^*. This cycle is called the *Leray cycle* and is denoted by $\alpha(x)$. The class of this cycle in the group $\tilde{H}_{N-1}(\check{\mathbf{C}}P^{N-1} \setminus A^*(P), X^*)$ is well defined and is called the *Leray class.*

The boundary of the Leray cycle, $\partial\alpha(x)$, is an absolute cycle in $X^* \setminus A^*(P)$. It is called the *Petrovskii cycle* corresponding to the point x and is denoted by $\beta(x)$.

The cycle $\gamma(x) \in \tilde{H}_{N-1}(\check{\mathbf{C}}P^{N-1} \setminus (A^*(P) \cup X^*))$ participating in formula (7′) is defined as the Leray tube around the Petrovskii cycle: $\gamma(x) \equiv t(\partial\alpha(x))$, see Chapter I, § 2.

Remark. Suppose that outside some small neighbourhood of the set $S^{N-1} \cap A(P)$ the vectors of the family v were chosen to be trivial. Then for even N the pieces of the cycle $\alpha(x)$ arriving from diametrically opposite points of the sphere annihilate each other, and this cycle turns out to be concentrated close to the set Re $A^*(P)$. Of course, the same is also true for the boundary of this cycle in $X^* \setminus A^*(P)$, moreover, this class has the following explicit description.

Proposition 1. *In the case of even N the Petrovskii cycle $\beta(x)$ is homologous to the Leray tube in $X^* \setminus A^*(P)$ around the (suitably oriented) manifold Re $A^*(P) \cap X^* \subset A^*(P) \cap X^*$.*

Conversely, in the case of odd N the cycle $\alpha(x)$ coincides far from $A^*(P)$ with the manifold $\check{\mathbf{R}}P^{N-1}$ taken twice. In a similar way, the cycle $\beta(x)$ coincides far from $A^*(P)$ with the manifold X^* taken twice; see Figure 37 for the case $N = 3$.

Theorem 1 (see [Petrovskii 45], [ABG 70]). *For any x not in the wave front of a homogeneous hyperbolic operator P, and any multiindex ν, the ν-th partial derivative* *of the principal fundamental solution E_P, considered as a function of x, is given by the corresponding formula (7) or (7′), where c is a constant not depending on x.*

4.3. The nonhomogeneous case.

Let $P \equiv P(D)$ be a nonhomogeneous hyperbolic operator (= a nonhomogeneous polynomial in the partial derivatives $D_j \equiv \partial/i\partial x_j$), and $P = \bar{P} + P_<$ its decomposition into the sum of the homogeneous operator of highest degree (which is, by Theorem 1.1, again hyperbolic) and an operator of strictly lower degree. By item 3 of Theorem 1.1, all degrees $\bar{P}^k$, $k \geq 1$, of the operator $\bar{P}$ are hyperbolic with respect to the same vector ϑ, in particular, their principal fundamental solutions $E_{\bar{P}^k}$ are well defined.

Theorem 2 (see [ABG 70]). *The principal fundamental solution of the nonhomogeneous hyperbolic operator P with constant coefficients is given by the following series:*

$$E_P(x) = \sum_{k=0}^{\infty}(-1)^k P_<(D)^k E_{\bar{P}^{k+1}}(x), \tag{10}$$

which converges in the sense of distributions.

In principle this theorem reduces the calculation of the fundamental solution to the "homogeneous" Theorem 1.

4.4. Global Petrovskii condition.

Definition 1. The point $x \in \mathbf{R}^N \setminus W(P)$ satisfies the *Petrovskii condition* if the corresponding Petrovskii cycle $\beta(x) \in \tilde{H}_{N-2}(X^* \setminus A^*)$ is equal to zero.

Theorem 3 (see [Petrovskii 45], [ABG 70]). *1. If the Petrovskii condition $\beta(x) = 0$ is satisfied, then the component L of the complement of the wave front $W(P)$ that contains x is a holomorphic lacuna for the hyperbolic operator P and for all other hyperbolic operators with the same principal part $\bar{P}$. If, moreover, deg $P < N$ and $P = \bar{P}$, then L is a strong lacuna.*

2. If the set A^* is a smooth manifold or at least a divisor with normal crossings in $\mathbf{C}P^{N-1}$, then the converse is also true: if L is a C^∞-lacuna, then $\beta(x) = 0$ for any $x \in L$.

The first assertion of this theorem in the case of a homogeneous operator follows immediately from formula (7′): indeed, by this formula all the partial derivatives of degree > deg $P - N$ of the principal fundamental solution vanish at x, and hence this solution close to x coincides with a polynomial of degree $\leq$ deg $P - N$. The case of nonhomogeneous P reduces to the previous one by formula (10) and the fact that the Petrovskii cycles of all degrees of P coincide.

Assertion 2 follows immediately from the fact that, conversely, if L is a lacuna, then $E_P|_L$ is a holomorphic homogeneous function in $\mathbf{C}^N$, hence a polynomial, and from the following two lemmas.

Lemma 2 (see [ABG 73], [Gabrielov 86]). *Under the hypotheses of Theorem 3, the tube operator*

$$\tilde{H}_{N-2}(X^* \setminus A^*(P)) \to \tilde{H}_{N-1}(\check{\mathbf{C}}P^{N-1} \setminus (A^*(P) \cup X^*))$$

is a monomorphism.

Lemma 3 ("Grothendieck's lemma", see [Grothendieck 66], adaptation to the lacuna theory from [ABG 73]). *If $A^* \equiv A^*(P)$ is a divisor with normal crossings, then for a sufficiently large natural number M the group $\tilde{H}^{N-1}(\check{\mathbf{C}}P^{N-1} \setminus (A^* \cup X^*), \mathbf{C})$ is generated by the differential forms $\Lambda^*_{x,\nu,P}$ with $\nu_1 + \cdots + \nu_N = M$.*

Example 1. For $N = 3$ the Petrovskii class $\beta(x) \in \tilde{H}_1(X^* \setminus A^*)$ is represented by two identically oriented copies of the circle Re X^*, missing the real points of the set $A^* \cap X^*$ on different sides, see Figure 37. It is easy to see that $\beta(x) = 0$ if and only if all the points of the set $A^* \cap X^*$ are real, or, which is equivalent by the definition of a hyperbolic polynomial, the plane Re X intersects the hyperbolicity domain of P. Therefore for $N = 3$ there is only one lacuna, the complement of the convex hull of the wave front.

Similar methods prove the following sharpening of Theorem 1.2.

Theorem 4 (see [ABG 1973]). *If the operator P is hyperbolic and the divisor A^* has only normal crossings, then any point of the wave front $W(P)$ is a singular point of the principal fundamental solution E_P. For almost any hyperbolic operator P with given principal part $\bar{P}$, the support of the solution E_P coincides with the convex hull of the wave front. If $N = 3$, then the last assertion is true for any (and not only for "almost any") hyperbolic operator.*

For example, the (homogeneous) wave operator for even $N \geq 4$ is not "almost any". For other examples of such exceptional lower parts, which, being added to the wave operator, do not make it "almost any", see [Berest & Veselov 94].

Another important problem is as follows: for which homogeneous P in the second assertion of this theorem can "almost any" be replaced by "any". The reasons for the fact that this is not true for the wave operator are very similar to those for the fact that the set Re A^* for the even-dimensional wave operator is exactly the same ellipsoid in an odd-dimensional space as the exceptional surface for the algebraic integrability problem from Chapter II.

§ 5. Local Petrovskii condition and local Petrovskii cycle. The local Petrovskii condition implies sharpness (after Atiyah, Bott and Gårding)

The same component of the complement of a wave front can be a local lacuna close to certain points of its boundary and not be close to the others. The main obstruction to sharpness at an arbitrary point of a front is the local ramification of the corresponding integral (7′), which is defined by the monodromy of Petrovskii classes. This ramification depends only on some localization of the Petrovskii class close to the set of nontangency of A^* and discriminant hyperplanes in $\check{C}P^{N-1}$.

5.1. Local Petrovskii condition.

Atiyah, Bott and Gårding have introduced a local analogue of the Petrovskii condition, which ensures the triviality of this localization and, hence, of the ramification of fundamental solutions.

Namely, let $y \neq 0$ be a point of the wave front W, l a local component of the complement of W close to y, and $Y^* \subset \check{C}P^{N-1}$ the projectivization of the plane orthogonal to y. By the definition of the wave front, Y^* is nontransversal to A^*. If the point $x \in l$ is sufficiently close to y, then the orthogonal projection $p_x : Y \to X$ (defined by an arbitrary Hermitian structure in $\check{C}^N$) induces a homomorphism

$$(p_x)_* : \tilde{H}_{N-2}(Y^* \setminus A^*) \to \tilde{H}_{N-2}(X^* \setminus A^*). \tag{11}$$

Indeed, the group $\tilde{H}_{N-2}(Y^* \setminus A^*)$ can be generated by several compact cycles. If x is sufficiently close to y, then the projections of these cycles into X^* do not meet the set $A^* \cap X^*$, and hence define a certain homomorphism (11). Another choice of these generators, of the cycles representing them, and of the projection can lead only to a strengthening of the requirements on the closeness of x and y; for those x for which these requirements will also be satisfied, the new homomorphism (11) will coincide with the former one.

Definition 1 (see [ABG 73]). The *local Petrovskii condition* is the condition

$$\beta(x) \in (p_x)_*(\tilde{H}_{N-2}(Y^* \setminus A^*)). \tag{12}$$

5.2. The local Petrovskii condition implies sharpness.

Theorem 1 (see [ABG 73]). *If the local Petrovskii condition is satisfied for all points of a local (close to y) component l of the complement of a wave front, then* *the principal fundamental solution E_P is holomorphically sharp at the point y from the side of l.*

Proof. First, let P be homogeneous. Suppose that there exists a cycle $\beta(y) \in \tilde{H}_{N-2}(Y^* \setminus A^*)$ such that $\beta(x) = (p_x)_*(\beta(y))$. Let $b = b(y)$ be a compact cycle in $Y^* \setminus A^*$ realizing $\beta(y)$. Let G be a Leray tube in $\check{\mathbf{C}}P^{N-1} \setminus (A^* \cup Y^*)$ around the cycle $\beta(y)$. Then G can be considered also as a tube around all cycles $b(x) \subset X^* \setminus A^*$ obtained from $b(y)$ by orthogonal projections p_x of Y^* onto all sufficiently close planes X^*. For any point $x \in \mathbf{C}^N$ close to y and any multiindex ν with $\nu_1 + \cdots + \nu_N > \deg P - N$, define the number $I_\nu(x)$ as the integral of the form $\Lambda^*_{x,\nu,P}$ over the cycle G. By construction, $I_\nu(x)$ is a holomorphic function of x. By Theorem 4.1, in l it coincides with $D^\nu E_p$, and hence E_p is holomorphically sharp from the side of l in y.

Finally, in the case of nonhomogeneous P formula (10) reduces this theorem to its homogeneous version.

5.3. Local Petrovskii cycle.

If the operator $\bar{P}$ is not very degenerate, then the local Petrovskii condition (12) is equivalent to the triviality of a relative homology class, the *local Petrovskii class*. To define it, fix an arbitrary semialgebraic Whitney stratification of the variety $A^* = \{z|\bar{P}(z) = 0\}^*$; let y be a point of the wave front $W(P)$.

Definition 2. The point $y \in W(P) \setminus \{0\}$ (and the corresponding point $y^* \in W^*(P)$) is *of discrete type* with respect to the fixed stratification if there is a neighbourhood Ξ of $\check{\mathbf{R}}P^{N-1}$ in $\check{\mathbf{C}}P^{N-1}$ such that the hyperplane $Y^* \equiv Y^*(y)$ is nontransversal to this stratification only at finitely many points in Ξ.

For almost any operator $\bar{P}$ and the appropriate stratification of the corresponding surface $A^*(\bar{P})$, all the planes Y^* satisfy this condition everywhere in $\check{\mathbf{C}}P^{N-1}$; the codimension of the set of operators such that the set $A^*(\bar{P})$ does not admit a stratification satisfying this condition increases rapidly together with N and $\deg P$. It is also clear that if y is of discrete type, then the neighbourhood Ξ from Definition 2 can be chosen so narrow that all points of nontransversality in Ξ are real.

Let $y \in W(P)$ be a point of discrete type, let $a_1, \ldots, a_t \in Y^*(y) \cap \operatorname{Re} A^*$ be all real points of nontransversality of A^* and $Y^*(y)$, and let B_j be small discs in $\check{\mathbf{C}}P^{N-1}$ with centres at these points. Let x be a real point very close to y and such that the orthogonal plane $X \equiv X^*(x)$ is transversal to A^*. Then the Petrovskii class $\beta(x) \in \tilde{H}_{N-2}(X^* \setminus A^*)$ is defined. Define the *localized homology group* $h(x)$ by

$$h(x) = \oplus_j \tilde{H}_{N-2}(X^* \cap B_j \setminus A^*, X^* \cap \partial B_j \setminus A^*). \quad (13)$$

Definition 3. The *local* (close to y) *Petrovskii class* is the image of the class $\beta(x)$ under the obvious mapping of the group $\tilde{H}_{N-2}(X^* \setminus A^*)$ into the group (13) (defined by the reduction modulo the complement of the union of the discs B_j). This image is denoted by $\beta(x,y)$.

Theorem 2. 1. *Let $y \in W(P)$ be a point of discrete type, and x a point close to y such that $X^*(x)$ is transversal to A^*, so that the local Petrovskii class $\beta(x,y)$ is welldefined. Then the local Petrovskii condition for this pair of points is equivalent to the triviality of this class $\beta(x,y)$.*

2. If moreover the set $X^* \cap A^*$ is nonsingular in the union of the discs B_j, then the Leray tube operator

$$t : \oplus \tilde{H}_{N-3}(X^* \cap A^* \cap B_j, \partial B_j) \to \oplus \tilde{H}_{N-2}(X^* \cap B_j \setminus A^*, \partial B_j \setminus A^*) \tag{14}$$

is an isomorphism.

Definition 4. If all the hypotheses of Theorem 2 are satisfied, then the inverse image of the class $\beta(x,y)$ under the isomorphism (14) is also called the *local Petrovskii class* and is denoted by $\Pi(x,y)$.

Proof of Theorem 2. Denote by Ξ a tubular neighbourhood of the set $\check{\mathbf{R}}P^{N-1}$ in $\check{\mathbf{C}}P^{N-1}$ that is so narrow that

1) all points of nontransversality of $Y^* \equiv Y^*(y)$ and A^* inside Ξ are real,

2) the plane Y^* is transversal to the boundary of Ξ.

The discs B_j will be chosen so small that they lie in this neighbourhood Ξ; denote the union of them by B. We shall assume also that X^* is so close to Y^* that A^* and X^* are transversal in $\Xi \setminus B$, and the set $A^* \cap X^*$ is transversal in X^* to the manifolds ∂B and $\partial \Xi$.

Denote the set $\Xi \setminus B$ by $\check{\Xi}$.

The Petrovskii cycle can be chosen to be so close to $\check{\mathbf{R}}P^{N-1}$ that it lies in Ξ and thus defines a class $\tilde{\beta}(x) \in \tilde{H}_{N-2}(X^* \cap \Xi \setminus A^*)$; the Petrovskii class $\beta(x)$ is its image under the inclusion

$$X^* \cap \Xi \setminus A^* \to X^* \setminus A^*, \tag{15}$$

and the local Petrovskii class $\beta(x,y)$ is the image of $\tilde{\beta}(x)$ under the homomorphism

$$j : \tilde{H}_{N-2}(X^* \cap \Xi \setminus A^*) \to \tilde{H}_{N-2}(X^* \cap \Xi \setminus A^*, \check{\Xi}) \simeq \tilde{H}_{N-2}(X^* \cap B \setminus A^*, \partial B).$$

The exact sequences of pairs $(Y^* \cap \Xi \setminus A^*, Y^* \cap \check{\Xi} \setminus A^*), (X^* \cap \Xi \setminus A^*, X^* \cap \check{\Xi} \setminus A^*)$ constitute the commutative diagram

$$\begin{array}{ccc}
\tilde{H}_{N-1}(Y^* \cap \Xi \setminus A^*, \breve{\Xi}) & & \\
\downarrow & & \\
\tilde{H}_{N-2}(Y^* \cap \breve{\Xi} \setminus A^*) & \longrightarrow & \tilde{H}_{N-2}(X^* \cap \breve{\Xi} \setminus A^*) \\
\downarrow & & \downarrow i \\
\tilde{H}_{N-2}(Y^* \cap \Xi \setminus A^*) & \xrightarrow{(p_x)_*} & \tilde{H}_{N-2}(X^* \cap \Xi \setminus A^*) \\
\downarrow & & \downarrow j \\
\tilde{H}_{N-2}(Y^* \cap \Xi \setminus A^*, \breve{\Xi}) & \longrightarrow & \tilde{H}_{N-2}(X^* \cap \Xi \setminus A^*, \breve{\Xi}).
\end{array}$$

The highest and lowest left terms of this diagram are trivial, since $(Y^* \cap \Xi \setminus A^*)/\breve{\Xi} \equiv (Y^* \cap B \setminus A^*)/\partial B$ and the pair $(Y^* \cap B, Y^* \cap A^* \cap B)$ is homeomorphic to the cone over the pair $(Y^* \cap \partial B, Y^* \cap A^* \cap \partial B)$; cf. Theorem I.8.13. Hence the middle vertical arrow in the left column is an isomorphism, as well as the highest horizontalal arrow (which is defined by an diffeomorphism). Therefore,

$$\tilde{H}_{N-2}(X^* \cap \Xi \setminus A^*)/(p_x)_*(\tilde{H}_{N-2}(Y^* \cap \Xi \setminus A^*)) =$$

$$= \tilde{H}_{N-2}(X^* \cap \Xi \setminus A^*)/i(\tilde{H}_{N-2}(X^* \cap \breve{\Xi} \setminus A^*)) =$$

$$= \tilde{H}_{N-2}(X^* \cap \Xi \setminus A^*)/\mathrm{Ker}\, j \simeq \mathrm{Im}\, j \subset \tilde{H}_{N-2}(X^* \cap \breve{\Xi} \setminus A^*, \breve{\Xi}) \simeq$$

$$\simeq \tilde{H}_{N-2}(X^* \cap B \setminus A^*, \partial B),$$

in particular, if the local Petrovskii class $\beta(x,y) \equiv j(\tilde{\beta}(x))$ is trivial, then $\tilde{\beta}(x) \in (p_x)_*(\tilde{H}_{N-2}(Y^* \cap \Xi \setminus A^*))$, and hence also the class $\beta(x)$ (which is the image of $\tilde{\beta}(x)$ under the inclusion homomorphism (15)) belongs to the subgroup $(p_x)_* \tilde{H}_{N-2}(Y^* \setminus A^*)$. We have proved that the triviality of the local Petrovskii class implies the local Petrovskii condition.

Conversely, suppose that $\beta(x) = p_x(\beta(y))$ for a certain cycle $\beta(y)$ in $Y^* \setminus A^*$. Again by Theorem I.8.13, the cycle $\beta(y)$ is homologous in $Y^* \setminus A^*$ to some cycle not intersecting B. If x is sufficiently close to y, then the projection p_x of this homology into X^* realizes a homology in $X^* \setminus A^*$ between $\beta(x)$ and a cycle lying outside B.

The isomorphism from assertion 2 of Theorem 2 is the composition of the boundary isomorphism $\tilde{H}_{N-2}(X^* \cap B \setminus A^*, \partial B \setminus A^*) \simeq \tilde{H}_{N-1}(X^* \cap B, (X^* \setminus A) \cup \partial B)$ from the exact sequence of the triple $(X^* \cap B, (X^* \cap B \setminus A^*) \cup \partial B, \partial B)$, and of the Thom isomorphism $\tilde{H}_{N-1}(X^* \cap B, (X^* \cap B \setminus A^*) \cup \partial B) \simeq \tilde{H}_{N-3}(X^* \cap A^* \cap B, \partial B)$ for the (trivial) normal bundle of the submanifold $X^* \cap A^*$ in $X^* \cap B$; it follows immediately from the construction of these two isomorphisms that their composition is given by the tube operation.

§ 6. Sharpness implies the local Petrovskii condition close to discrete-type points of wave fronts of strictly hyperbolic operators

Theorem 1. *Suppose that the operator P is hyperbolic, the point $y^* \in W^*(P)$ is of discrete type, and the variety $A^*(P)$ is smooth close to all points of its nontransversality with $Y^*(y)$. Then for any local component of the complement of $W^*(P)$ close to y^*, the holomorphic sharpness of E_P from the side of this component implies the local Petrovskii condition for all points x^* from this component for which this condition is defined (i.e. for points x^* such that the corresponding plane X^* is transversal to $A^*(P)$ not only at real, but also at close complex points).*

All the rest of this section is devoted to the proof of this theorem.

In fact, we shall prove that the sharpness implies the triviality of the local Petrovskii classes $\beta(x,y)$; by Theorem 5.2 this will imply Theorem 1.

Let U be a sufficiently small neighbourhood of y^* in $\mathbf{C}P^{N-1}$ such that for any $x^* \in U$ the corresponding plane X^* is transversal to ∂B and to $A^* \cap \partial B$. Let $\text{reg}(A^*,U)$ be the set of $x^* \in U$ such that X^* is transversal to A^*, in particular, $W^*(P) \cap U$ belongs to the complement $\text{tang}(A^*,U)$ of $\text{reg}(A^*,U)$ in U. Consider two fibre bundles over $\text{reg}(A^*,U)$ whose fibres over a point x^* are the pairs $(X^* \cap B \setminus A^*, X^* \cap \partial B \setminus A^*)$ and $(X^* \cap A^* \cap B, X^* \cap A^* \cap \partial B)$.

In the usual way, any loop C in $\text{reg}(A^*,U)$ defines the variation operators in these bundles, which are the horizontal maps in the following diagram:

$$\begin{array}{ccc} \tilde{H}_{N-2}(X^* \cap B \setminus A^*, X^* \cap \partial B \setminus A^*) & \to & \tilde{H}_{N-2}(X^* \cap B \setminus A^*) \\ \uparrow & & \uparrow \\ \tilde{H}_{N-3}(X^* \cap A^* \cap B, X^* \cap A^* \cap \partial B) & \to & \tilde{H}_{N-3}(X^* \cap A^* \cap B). \end{array}$$

The vertical arrows here are the tube operators, which are isomorphisms by assertion 2 of Theorem 5.2. It follows immediately from the construction of these operators that for any loop C this diagram is commutative. As in Chapter II, § 4, the generating family of the front close to the discriminant point y^* reduces the calculation of these variation operators to a standard problem in the local Picard–Lefschetz theory.

First we suppose for simplicity that Y^* and A^* are nontransversal at only one real point a, so that B consists of only one disc $B_1 \equiv B$.

Let $\varphi(\zeta_1,\ldots,\zeta_{N-2})$ and $F(\zeta_1,\ldots,\zeta_{N-2};\lambda_0,\ldots\lambda_{N-2})$ be the generating function and the generating family of the front $W^*(P)$ at y^* respectively; see § 2.

Let $x^{0*} \in \text{reg}(A^*,U) \cap \mathbf{R}P^{N-1}$ be the base point in $\text{reg}(A^*,U)$ that belongs to the investigated component of the complement of $W^*(P)$, and $\lambda^0 \equiv (\lambda_0^0,\ldots,\lambda_{N-2}^0)$

its coordinates in U, see (4). We can choose x^{0*} in such a way that the function

$$\varphi_{\lambda^0} \equiv \varphi - \lambda_1^0 \zeta_1 - \cdots - \lambda_{N-2}^0 \zeta_{N-2} - \lambda_0^0$$

takes $\mu(\varphi)$ different critical values at critical points close to a. Suppose that the local Petrovskii class $\Pi(x^0, y)$ is a nontrivial element in the vanishing homology group $\tilde{H}_{N-3}(V_{\lambda^0}, \partial V_{\lambda^0}) \equiv \tilde{H}_{N-3}(X^{0*} \cap A^* \cap B, X^{0*} \cap A^* \cap \partial B)$.

By Poincaré duality there exists a cycle $\Delta \in \tilde{H}_{N-3}(X^{0*} \cap A^* \cap B)$, vanishing over some path $\tilde{C}^*$ in $\mathbf{C}^1$ connecting 0 with some critical value of φ_{λ^0}, such that $\langle \Pi(x^0, y), \Delta \rangle \neq 0$. The coordinate λ_0 naturally identifies this line $\mathbf{C}^1$ with the complex line in U along which all coordinates $\lambda_i, i = 1, \ldots, N-2$, are equal to these for x^{0*}. After this identification, our path $\tilde{C}^*$ becomes a path in U that connects x^{0*} with some nonparabolic point $y!^* \in \operatorname{tang}(A^*, U)$. Let C^* be the simple loop around this path. Then the variation along C^* increases the Petrovskii cycle $\beta(x^0)$ by the cycle

$$(-1)^{(N-2)(N-1)/2} \langle \Pi(x^0, y), \Delta \rangle \cdot t(\Delta) \in \tilde{H}_{N-2}(X^{0*} \setminus A^*), \tag{16}$$

where t is the tube operation. Let $\nu = (\nu_1, \ldots, \nu_N)$ be any multiindex with $\nu_1 + \cdots + \nu_N > \deg P - N$. Let x^0 be any point in $\mathbf{R}_+^N \setminus W(P)$ representing the line x^{0*}. By formula (7′) the analytic continuation of the function $D^{(\nu)} E_P$ from x^0 along any closed path in $\mathbf{C}^N$ whose projection into $\mathbf{C}P^{N-1}$ is C^* increases this function by the integral of the form $\Lambda_{x^0,\nu,P}$ along the cycle $t(t(\Delta)) \in \tilde{H}_{N-1}(\mathbf{C}P^{N-1} \setminus (X^{0*} \cup A^*))$ taken with a nonzero coefficient. Thus we need only prove that for some ν this integral

$$\int_{t(t(\Delta))} \Lambda_{x^0,\nu,P}\,, \tag{17}$$

considered as a function of x^0, is not identically zero. To do this, consider some path $\tilde{C}$ in $\mathbf{C}^N$ whose projection into $\mathbf{C}P^{N-1}$ is the path $\tilde{C}^*$ along which Δ vanishes, and consider the analytic continuation of the integral (17) along this path to points very close to the discriminant. It is sufficient to prove the nontriviality of this continuation, i.e. of the integral of the form (17) in which the point x^0 is replaced by points x of this path $\tilde{C}$ that are arbitrarily close to the endpoint $y!$ of this path. (For such x the integration cycle $t(t(\Delta))$ lies in a small neighbourhood of the point of tangency of A^* and the plane $Y!^*$ orthogonal to $y!$; denote this point by $a!$)

Lemma 1. *If a local component of the complement of $W(P)$ close to the point y is a local lacuna for the operator P, then it is also a local lacuna for its principal part $\bar{P}$.*

This lemma follows immediately from formula (10): the fundamental solution $E_{\bar{P}}$ coincides with the term of lowest degree in the decomposition of E_P into homogeneous terms.

Thus it is sufficient to consider the case when P is homogeneous.

Choose the local affine chart with coordinates $s_1, \dots, s_{N-1}$ in $\check{C}P^{N-1}$ centred at the point $a!$ and such that $Y!^* = \{s_{N-1} = 0\}$. Then choose the linear coordinates with the same names $s_1, \dots, s_N$ in $\check{C}^N$ so that this affine chart in $\check{C}P^{N-1}$ is defined by the affine hyperplane $\{s_N = 1\}$ and by the restrictions of the remaining coordinates s_i on it.

We shall suppose that some last segment of the path $\tilde{C}$ lies in this hyperplane. An appropriate choice of this last segment and of the affine coordinates $s_1, \dots, s_{N-2}$ reduces our proof to the following lemma.

Lemma 2. *Let in an affine chart with coordinates $s_1, \dots, s_{N-1}$ the hypersurface A^* be given by the equation $s_{N-1} = G(z_1, \dots, z_{N-2})$, where the function G is equal to $s_1^2 + \dots + s_{N-2}^2 + O((|s_1| + \dots + |s_{N-2}|)^3)$; the plane $X^*(\tau)$ (for a small positive number τ) is given by the equation $s_{N-1} = \tau$. Let $\Delta(\tau) \in X^*(\tau) \cap A^*$ be the vanishing sphere generating the homology group of the intersection of $X^*(\tau) \cap A^*$ with the small boundary $\mathcal{B}$ of the origin, and let $t(t(\Delta(\tau))) \in \mathcal{B} \setminus (X^*(\tau) \cap A^*)$ be the image of this sphere under the iterated tube operation $X^*(\tau) \cap \mathcal{B} \cap A^* \to X^*(\tau) \cap \mathcal{B} \setminus A^* \to \mathcal{B} \setminus (X^*(\tau) \cap A^*)$. Let $q = \deg P - |\nu| - N$. Then there exists a multiindex $\nu = (\nu_1, \dots, \nu_N)$ with $|\nu| \equiv \nu_1 + \dots + \nu_N > \deg P - N$ such that the integral*

$$\int_{t(t(\Delta(\tau)))} s_1^{\nu_1} \cdot \dots \cdot s_{N-1}^{\nu_{N-1}} (s_{N-1} - \tau)^q (G - s_{N-1})^{-1} ds_1 \wedge \dots \wedge ds_{N-1} \tag{18}$$

is not equal to zero for all small $\tau > 0$.

Proof. Let us show that this assertion is satisfied by the multiindex $\nu = (0, \dots, 0, |\nu|)$. For such ν the Morse lemma reduces the integral (18) to the integral

$$\int_{t(t(\tilde{\Delta}(\tau)))} (s_{N-1} - \tau)^q (\tilde{G} - s_{N-1})^{-1} J(s_1, \dots, s_{N-2}) ds_1 \wedge \dots \wedge ds_{N-1}, \tag{19}$$

where J is the Jacobian determinant of the change of variables, $\tilde{G} \equiv s_1^2 + \dots + s_{N-2}^2$, and $t(t(\tilde{\Delta}(\tau)))$ is the double tube around the vanishing cycle in the manifold given by the two equations $\tilde{G} = s_{N-2} = \tau$. For such a cycle we can choose the sphere distinguished by the equations

$$s_{n-1} = \tau, \ \operatorname{Im} s_1 = \dots = \operatorname{Im} s_{N-2} = 0, \ s_1^2 + \dots + s_{N-2}^2 = \tau.$$

Let $J_0 + J_1 + \cdots$ be the Taylor expansion of the function J at 0, $\deg J_i = i$. Then $J_0 = 1$. Using the one-parameter group of quasihomogeneous dilations $(s_1, \cdots, s_{N-1}) \to (ts_1, \cdots, ts_{N-2}, t^2 s_{N-1})$, we see that the integral of the form (19) in which instead of J we have J_i is equal to $c_i \cdot \tau^{(i+2q+N-2)/2}$, where c_i is a constant depending only on J_i. It is easy to calculate that $c_0 \neq 0$. Therefore for small τ the integral (19) is not equal to zero.

Thus Theorem 1 is proved in the case when Y^* has only one point of nontransversality with A^*. Finally, let us prove it in the general case when B consists of several discs around several such points of nontransversality. In this case the proof is almost the same; we only need to make the following additions.

We choose the affine chart in $\check{C}P^{N-1}$ in such a way that all real points of nontransversality lie in it; the affine coordinate ζ_{N-1} is again the equation of the plane Y^*. The point $x^{0*} \in \operatorname{reg}(A^*, U) \cap \operatorname{Re} U$ should be chosen in such a way that any plane obtained from X^{0*} by a small parallel shift in the coordinates $\zeta_1, \ldots, \zeta_{N-1}$ is tangent to A^* at one point at most, and the surface A^* must be nonparabolic at these points. (The density of the set of such x^{0*} follows again from Sard's lemma; cf. Lemmas I.3.1 and II.4.1.) This property ensures that the line in U consisting of such planes with different τ intersects $\operatorname{tang}(A^*, U)$ at exactly $\sum \mu(\varphi_i)$ points, where φ_i is the generating function of the i-th point of tangency of Y^* and A^*.

§ 7. The local Petrovskii condition may be stronger than the sharpness close to singular points not of discrete type

Consider the polynomial $P = \xi_1(\xi_1^2 - \xi_2^2 - \xi_3^2)$. It is hyperbolic with respect to the vector $\partial/\partial\xi_1$. Its wave front in dual coordinate space consists of the cone $x_1^2 - x_2^2 - x_3^2 = 0$ and the line spanned by the vector $(1, 0, 0)$. By Hartogs' lemma about erasing singularities, the corresponding hyperbolic operator is sharp close to the point $y = (1, 0, 0)$. On the other hand, in this case $Y(y) \subset A(P)$, and hence the local Petrovskii condition for x close to y coincides with the global Petrovskii condition $\beta(x) = 0$. This condition is surely not satisfied, because the restriction of E_P to this component is, by the Davydova–Borovikov criterion, diffuse close to the exterior boundary of this component.

§ 8. Normal forms of nonsharpness close to singularities of wave fronts (after A.N. Varchenko)

Sharpness and diffusivity are the roughest characteristics of the behaviour of the fundamental solution close to a front. The question about the nature of the diffusion from the side of different components of the complement of the front is more refined. For example, close to nonsingular points of the front (corresponding to

Morse generating functions) the fundamental solution looks like a series in integer and half-integer powers of the distance to the front (see [Borovikov 59]); sharpness means that all the half-integer terms in this series occur with zero coefficients.

Analogously, the asymptotic behaviour of solutions close to singular points of fronts reduces to the behaviour of a finite number (determined by the type of the singularity) of standard special functions. This is connected with the fact that the integration chain occuring in the Herglotz–Petrovskii–Leray formula (7′) can be separated into two parts–one lying close to the point of nontransversality of Y^* and A^* (and equal to the tube around the local Petrovskii cycle) and the other lying far from it. The whole multitude of possible global behaviours of the set $\operatorname{Re} A^*(P) \cap X^*$ is reflected almost completely by the second part, but the integral over this part gives only the nonsingular portion of the asymptotics of E_P, and the nature of the diffusion is determined by the first, localized, part.

Varchenko found the normal forms of the diffusion close to semicubical edges, see [Varchenko 87]. For example, in the case of odd N they are described in the following terms.

The canonical form of a versal deformation of a real function of class A_2 of $n = N - 2$ variables is as follows:

$$\Psi(z, \lambda) = z_1^3 + \lambda_1 z_1 + \lambda_2 \pm z_2^2 \pm \cdots \pm z_n^2; \tag{20}$$

see Table 3. The discriminant $\Sigma(\Psi)$ of this deformation in the space $\mathbf{R}^2_\lambda$ is given by the condition $\mathcal{D} = 0$, where $\mathcal{D} = 4\lambda_1^3 + 27\lambda_2^2$; see Figure 3a. Three functions $\rho_i(\lambda)$, $i = 1, 2, 3$, are defined on $\mathbf{R}^2_\lambda \setminus \Sigma(\Psi)$, namely the roots of the polynomial $z^3 + \lambda_1 z + \lambda_2$. If λ belongs to the "smaller" domain 2 (see Figure 3a), then all these functions are real; we order them in increasing order. If λ belongs to domain 1, then one of these functions (say ρ_1) is real and the other two are complex conjugate. Define the functions $I_+(\lambda), I_-(\lambda)$ by the following conditions: in domain 1, $I_-(\lambda) = \rho_1(\lambda), I_+(\lambda) = \rho_2(\lambda) - \rho_3(\lambda)$, and in domain 2, $I_-(\lambda) = -2\rho_2(\lambda), I_+(\lambda) = 0$.

The generating function corresponding to a point y^* of the cuspidal edge of a front has a singularity of type A_2. By the definition of versal deformation there exists a mapping $G : (\mathbf{R}P^{N-1}, W^*, y^*) \to (\mathbf{R}^2_\lambda, \Sigma(\Psi), 0)$ that induces the generating family (5) of our front from the standard deformation (20). The composition of the mapping G and the projection $\pi : \mathbf{R}^N_+ \to \mathbf{R}P^{N-1}$ defines the functions $\tilde{I}_+ \equiv I_+ \circ G \circ \pi$, $\tilde{I}_- \equiv I_- \circ G \circ \pi$ and $\tilde{\mathcal{D}} \equiv \mathcal{D} \circ G \circ \pi$ on $\mathbf{R}^N_+$ close to the ray $\{y^*\}$.

Theorem 1 (see [Varchenko 87]). *Suppose that N is odd, and the index i_+ of the generating function of our front close to its point Y^* of type A_2 is even. Then for any natural number k there are complex-valued functions $\sigma_1, \sigma_2 \in C^\infty(\mathbf{R}^N_+)$ such that in any local (close to the point y) component of the complement of the front the* *difference*

$$E_P - (\sigma_1 \tilde{I}_+ + \sigma_2 \tilde{I}_+^2)/\tilde{\mathcal{D}}^r \tag{21}$$

is a smooth function which extends to a function of class C^k *on the closure of this component, and the number* r *in (21) depends only on* N *and the order of the operator.*

If N *is odd and* i_+ *is odd, the same result is true if* $\tilde{I}_+$ *is replaced in (21) by* $\tilde{I}_-$.

Similar results are obtained in [Varchenko 87] for the case of even N. These results are true also for hyperbolic operators with nonconstant coefficients.

§ 9. Several problems

1. All the examples known to me in which the local Petrovskii condition is stronger than the sharpness are related to the fact that the polynomial P_ξ is reducible and one of the irreducible components of A^* is flat (i.e. the set of its tangent planes has complex codimension ≥ 2 in $\mathbf{C}P^{N-1}$).

Is it true that the local Petrovskii condition is equivalent to the sharpness close to any singular point y *of a wave front, such that all strata of* A^* *nontransversal to* $Y^*(y)$ *are not flat?*

Perhaps, to solve this problem it would be convenient to answer the following question.

2. *What is the local Petrovskii cycle* $\beta(x, y)$ *for* y *not of discrete type?*

Such a cycle should be an object concentrated at the set of real points of non-transversality of A^* and Y^*: say, a class in the cohomology of some complex of sheaves with support at this set. It seems likely that the language of perverse sheaves would be essential to define this class, see [BBD 83].

3. *Perhaps it is possible to get some new obstructions to the existence of global lacunas in terms of the global monodromy action of* $\pi_1(\mathit{reg}\, A^*)$ *on the global Petrovskii class.*

Chapter V. CALCULATION OF LOCAL PETROVSKIĬ CYCLES AND ENUMERATION OF LOCAL LACUNAS CLOSE TO REAL FUNCTION SINGULARITIES

In this chapter we express the local Petrovskii cycles of strictly hyperbolic operators in the standard terms of singularity theory, and find the local lacunas close to many singularities of wave fronts. All the ingredients of the lacuna problem for such operators are reformulated in terms of the singularity theory of functions, so that this chapter can be read independently of the previous one.

§ 1. Main theorems

1.1. Definition of local Petrovskii cycles.

Let $f : (\mathbf{C}^n, \mathbf{R}^n, 0) \to (\mathbf{C}, \mathbf{R}, 0)$ be a real function with isolated singularity at 0.

Let f_λ be a small real nondiscriminant perturbation of the singularity f, i.e., $f_\lambda(\mathbf{R}^n) \subset \mathbf{R}$ and 0 is a noncritical value of f_λ. Recall the notation V_λ for the corresponding local level variety $f_\lambda^{-1}(0) \cap B$; see Chapter I, § 3. Suppose that an orientation of the space $\mathbf{R}^n$ is fixed once and for all.

The *even local Petrovskii cycle* $\Pi_{ev}(\lambda)$ is the set Re V_λ of real points of V_λ oriented in a such way that at any point of it the chosen orientation of $\mathbf{R}^n$ coincides with the one defined by the n-frame {grad f_λ; an $(n-1)$-frame positively oriented in the sense of this orientation of Re V_λ}. Its class in the group

$$\tilde{H}_{n-1}(V_\lambda, \partial V_\lambda) \qquad (1)$$

is called the *even Petrovskii class.*

To define the odd Petrovskii class we first define a cycle $b_\lambda \in \tilde{H}_n(B - V_\lambda, \partial B - V_\lambda)$. This cycle consists of two components, which coincide with the space $\mathbf{R}^n \cap B$ outside a very narrow neighbourhood of V_λ, and close to the set V_λ flow around this set in the complex domain in the sets $\{\mathrm{Im}\ f_\lambda(x) > 0\}$ and $\{\mathrm{Im}\ f_\lambda(x) < 0\}$ in such a way (see Figure 37) that geometrically the set of nonreal points of the cycle b_λ coincides with a tube around Re V_λ, but two components of this tube, which are separated by the set of real points, have discordant orientations. The orientation of both components of the cycle b_λ is chosen so that it coincides with that of $\mathbf{R}^n$ far from V_λ.

The groups $\tilde{H}_{n-1}(V_\lambda, \partial V_\lambda)$ and $\tilde{H}_n(B - V_\lambda, \partial B \setminus V_\lambda)$ are naturally isomorphic: this isomorphism is given by the tube operation (see Chapter I, § 3.6). The *odd local Petrovskii class* $\Pi_{\mathrm{odd}} \in \tilde{H}_{n-1}(V_\lambda, \partial V_\lambda)$ is defined as the inverse image of the class b_λ under this isomorphism.

Remark 1. The even Petrovskii class could be defined by the same scheme as the odd one, where the two components of the cycle b_λ from the above construction must be taken with opposite orientations at real points far from V_λ. Indeed, in this case these real parts of the two components annihilate each other, and the rest is just a tube in the complement of V_λ around the cycle $\Pi_{\mathrm{ev}}(\lambda)$.

Let $F \equiv F(x, \lambda)$, $\lambda \in \mathbf{C}^l$, be a *real* deformation of the function f, so that $F(\mathbf{R}^n \times \mathbf{R}^l) \subset \mathbf{R}$, and $\Sigma = \Sigma(F) \subset \mathbf{C}^l$ the (complex) discriminant of this deformation. For all real values of λ from the same component of the complement of Σ in $\mathbf{R}^l$, all corresponding even (respectively, odd) local Petrovskii classes are either simultaneously equal to 0 or not equal to 0.

Definition 1. A local (close to 0) component of the complement of Σ in $\mathbf{R}^l$ is called an *even* (respectively, *odd*) *formal lacuna* if for all λ from this component the even (respectively, odd) local Petrovskii class is a trivial element of the group (1).

Such a component is called a *formal lacuna* if it is an even (respectively, odd) formal lacuna and the number n of variables is even (respectively, odd).

Proposition 1. *Suppose that $f = f(\zeta)$ is a generating function of some projective wave front of a hyperbolic operator in $\mathbf{R}^{n+2}$ (see Chapter IV, § 2), and the deformation $F = F(\zeta; \lambda)$ of it is the projective generating family of the same front, i.e., F is given by formula (5) of Chapter IV in appropriate affine coordinates; in particular, for any λ close to 0 these coordinates $\zeta_1, \ldots, \zeta_n$ identify the level set $V_\lambda \equiv \{\zeta | f_\lambda(\zeta) = 0\} \cap B$ and the intersection of the surface $A^* \subset \mathbf{C}P^{n+1}$ with the hyperplane $X^*(\lambda)$ given by equation (4) of Chapter IV. Then the local Petrovskii class $\Pi(\lambda, 0) \in \tilde{H}_{n-1}(A^* \cap X^*(\lambda) \cap B, \partial B)$ defined in Chapter IV, § 4 is equal to the even local Petrovskii class if n is even and to the odd local Petrovskii class if n is odd.*

This follows immediately from the constructions.

Our next purpose is to calculate the local Petrovskii classes explicitly, and especially to recognize when they are trivial. Here are some of the simplest properties of them.

1.2. Complex conjugation.

For any $\lambda \in \mathbf{R}^l$ there is an involution of the group $\tilde{H}_{n-1}(V_\lambda, \partial V_\lambda)$, defined by complex conjugation.

Proposition 2. *For any n the even Petrovskii class is invariant under the involution of complex conjugation, and the odd Petrovskii class is antiinvariant.*

Indeed, it follows directly from the definitions that both the relative cycles $\operatorname{Re} V_\lambda \subset (V_\lambda, \partial V_\lambda)$ and $b_\lambda \subset (B - V_\lambda, \partial B - \partial V_\lambda)$ are invariant under complex conjugation, thus the assertion about the even Petrovskii class follows. Obviously,

$$\text{(tube operation)} \circ \text{(complex conjugation)} \equiv$$

$$-\text{(complex conjugation)} \circ \text{(tube operation)},$$

and the assertion about the odd Petrovskii class is also proved.

1.3. Boundary of the Petrovskii class.

The boundary of the Petrovskii cycle defines an element in the group

$$\tilde{H}_{n-2}(\partial V_\lambda). \tag{2}$$

The spaces ∂V_λ for different (even discriminant) λ form a locally trivial (and hence trivializable) fibre bundle over a small disc $D \subset \mathbf{C}^l$ around the origin in the space of parameters λ, therefore the groups (2) for all $\lambda \in D$ can naturally be identified with each other.

Proposition 3. *For all nondiscriminant values of $\lambda \in D$ (even from different components of the complement of the discriminant), the corresponding boundary elements $\Pi_{\text{ev}}(\lambda)$ (respectively, $\Pi_{\text{odd}}(\lambda)$) are the same with respect to the natural identification of the groups (2). In particular, if for some nondiscriminant λ this boundary is a nontrivial element of the group (2), then the same is also true for any other nondiscriminant λ, and hence the deformed singularity f does not have any even (respectively, odd) local lacunas in the space of any deformation of it.*

Indeed, the first assertion follows immediately from the construction of the Petrovskii classes; the last assertion follows from the first one and from the fact that any two deformations of a given singularity can be included as subdeformations of some common large deformation.

1.4. Reformulation of the Davydova–Borovikov condition in terms of singularity theory.

Let $f, F, \Sigma(F)$ be the same as above, while the deformation F is versal and contains together with any function f_λ also all functions of the form $f_\lambda + c$, $c \in \mathbf{C}^1$. Let L be a local component of the complement of $\Sigma(F)$ close to the origin in $\mathbf{R}^l$. The following two propositions follow immediately from the definitions.

Proposition 4. *The surface $\Sigma(F)$ is nonsingular close to a point $\lambda \in \mathbf{R}^l$ of it if and only if the corresponding function f_λ has in $\mathbf{R}^n$ only one critical point with critical value 0, and this point is Morse.*

Proposition 5. *Close to any nonsingular point λ of $\Sigma(F)$, the line in $\mathbf{R}^l$ consisting of all functions of the form $f_\lambda + c$ intersects the discriminant transversally.*

It follows from the last proposition that for any nonsingular point λ on the boundary of the component L we can say that this component lies above or below the discriminant close to this point: it lies above (respectively, below) if some interval of that line corresponding to small positive (respectively, negative) values of c belongs to this component.

Definition 2. A component L of the complement of $\Sigma(F)$ satisfies the Davydova–Borovikov condition close to a nonsingular point $\lambda_0 \in \Sigma(F) \cap \bar{L}$ if either this component lies above the discriminant close to λ_0 and the positive inertia index of the critical point of f_{λ_0} with zero critical value is even, or the component L lies below the discriminant close to λ_0 and the negative inertia index of such a point is even.

Proposition 6. *A local component of the complement of the wave front of a strictly hyperbolic operator satisfies the Davydova–Borovikov condition in the sense of Definition IV.3.1 if and only if the corresponding component of the complement of the discriminant of its generating family satisfies this condition in the sense of the previous definition.*

This is just another form of Proposition 1 from Chapter IV, § 3.

Now part 1 of Theorem 2 from the Introduction to Chapter IV is reduced to the following theorem.

Theorem 1. *A component of the complement in $\mathbf{R}^l$ of the discriminant variety of a real versal deformation of a simple real function singularity is a formal lacuna if and only if this component satisfies the Davydova–Borovikov condition close to all nonsingular points of its boundary, and its boundary does not contain cuspidal edges close to which this component is the "bigger" one.*

This theorem will be proved in § 7.

1.5. Local lacunas close to singularities that are stably equivalent to extrema.

Theorem 2. *Suppose that the function $f : (\mathbf{R}^n, 0) \to (\mathbf{R}, 0)$ has a minimum (respectively, a maximum) at 0. Then for small $\varepsilon > 0$ the perturbation $f + \varepsilon$ (respectively, $f - \varepsilon$) belongs to an even local lacuna, and the perturbation $f^{(0,1)} + \varepsilon$ (respectively, $f^{(1,0)} - \varepsilon$) of its stabilization $f^{(0,1)} : (\mathbf{R}^{n+1}, 0) \to (\mathbf{R}, 0)$ (respectively, $f^{(1,0)}$, see Chapter I, § 3.7) belongs to an odd lacuna.*

The assertion about the even lacunas for the function f follows immediately from the definition of the even Petrovskii cycle (which is empty if this condition is satisfied). The assertion about the odd lacunas for the stabilizations of f follows from it and Stabilization Theorems 4, 5 below.

1.6. Expression of the Petrovskii classes through the intersection matrices of singularities.

By the Poincaré duality theorem, the local Petrovskii classes can be considered as elements of the dual space to the group $\tilde{H}_{n-1}(V_\lambda)$, i.e., they are determined by their intersection indices with the basis vanishing cycles in this group. Now we describe a choice of these vanishing cycles which will be convenient for our calculations. Let f_λ be a nondiscriminant real strict morsification of f, in particular, f_λ has exactly $\mu(f)$ different critical values at the critical points close to the origin in $\mathbf{C}^n$, and none of these values is equal to 0. Since this morsification is real, the set of these values is invariant under complex conjugation.

A basis of vanishing cycles is defined by a system of nonintersecting paths in $\mathbf{C}^1$ joining the noncritical value 0 with all critical values, and by a choice of some orientation of the cycles in V_λ that vanish over these paths. We choose these paths and orientations as follows. (See Figure 38.)

(i) The paths joining 0 with the real critical values lie in the domain where Im z is nonnegative but is less than the absolute values of the imaginary parts of all nonreal critical values; the last small segment of such a path goes along the real axis in the domain of numbers smaller than the final critical value.

(ii) The paths joining 0 with complex conjugate critical values are also conjugate and have no intersections with the real axis (except for their common initial point 0).

(iii) Let α be a real critical value of f_λ. By the Morse lemma, in a neighbourhood of the corresponding critical point we can choose local coordinates such that in these

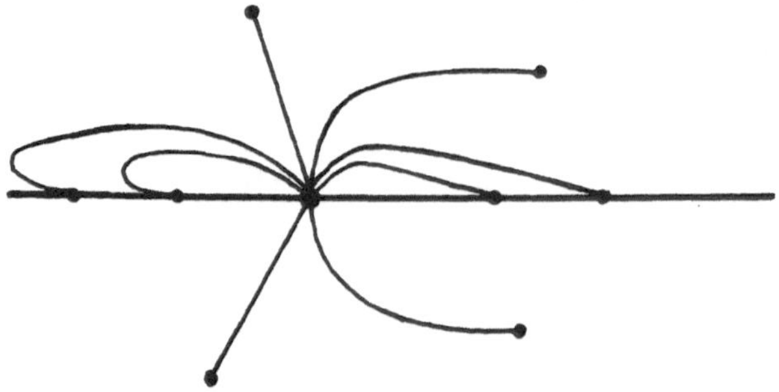

Fig. 38. Distinguished system of paths for a real singularity

coordinates

$$f_\lambda = z_1^2 + \cdots + z_r^2 - z_{r+1}^2 - \cdots - z_n^2 + \alpha. \tag{3}$$

Then for small $\varepsilon > 0$ for a cycle in the manifold $f^{-1}(\alpha - \varepsilon)$ vanishing at this point we can take the sphere Δ of radius $\sqrt{\varepsilon}$ in the plane which is the real linear span of the vectors

$$i\partial/\partial z_1, \ldots, i\partial/\partial z_r, \partial/\partial z_{r+1}, \ldots, \partial/\partial z_n. \tag{4}$$

We orient this plane by the condition that this sequence gives a positive orientation if and only if the sequence $\partial/\partial z_1, \ldots, \partial/\partial z_n$ gives the positive orientation of $\mathbf{R}^n$. Finally, we orient the sphere Δ by the condition that at any point of it the n-frame (grad f_λ, positive tangent frame of Δ) gives the positive (just defined) orientation of this plane.

Lemma 1. *This orientation does not depend on the choice of local coordinates z_i in which f_λ has the canonical form (3).*

The proof is elementary.

Transporting the resulting small oriented spheres over the paths indicated in item (i) above we obtain the correspondung vanishing cycles in V_λ.

(iv) The orientations of cycles corresponding to nonreal critical values should be chosen in such a way that the involution of complex conjugation takes any two cycles that vanish over the conjugate paths into one another preserving these orientations.

Remark 2. The basis of vanishing cycles defined by the paths of items (i), (ii) becomes a distinguished basis after some numbering of them. However, it will be convenient for us to use a different numbering; see part 2 of § 8.2 below.

In the next theorem we use the following standard notation: $\langle \cdot, \cdot \rangle$ is the intersection index in V_λ with respect to complex orientation, $\chi(\cdot)$ is the Euler characteristic, and S^k is the k-dimensional sphere. The Euler characteristic of a (-1)-dimensional sphere is assumed to be zero. A cycle that vanishes over the critical value α is denoted by Δ_α, and r is the positive Morse index of the corresponding real Morse critical point; see (iii).

Theorem 3. *If the critical value α of the morsification f_λ of the function f is positive, then*

$$\langle \Delta_\alpha, \Pi_{\text{ev}} \rangle = (-1)^{n(n-1)/2} \chi(S^{n-r-1}) + \sum \langle \Delta_\alpha, \Delta_{\alpha_i} \rangle, \tag{5}$$

$$\langle \Delta_\alpha, \Pi_{\text{odd}} \rangle = (-1)^{1+n(n-1)/2} \chi(S^{n-r}) - \sum \langle \Delta_\alpha, \Delta_{\alpha_i} \rangle, \tag{6}$$

where (as in the following formulae (7), (8)) the sum is taken over all vanishing cycles Δ_{α_i} corresponding to the real critical values α_i lying between α and 0; if α is negative, then

$$\langle \Delta_\alpha, \Pi_{\text{ev}} \rangle = (-1)^{n(n+1)/2} \chi(S^{r-1}) - \sum \langle \Delta_\alpha, \Delta_{\alpha_i} \rangle, \tag{7}$$

$$\langle \Delta_\alpha, \Pi_{\text{odd}} \rangle = (-1)^{1+n(n+1)/2} \chi(S^{r}) + \sum \langle \Delta_\alpha, \Delta_{\alpha_i} \rangle; \tag{8}$$

if α, $\bar{\alpha}$ are two complex conjugate critical values of f_λ, and the singularity f is simple, then

$$|\langle \Delta_\alpha, \Pi_{\text{ev}} \rangle| = |\langle \Delta_{\bar{\alpha}}, \Pi_{\text{ev}} \rangle| = |\langle \Delta_\alpha, \Delta_{\bar{\alpha}} \rangle|; \tag{9}$$

in the case of an arbitrary function f the same formula (9) is true, in which we need to replace the right-hand equality by the identity mod 2.

The proof will be given in §§ 3, 4.

Corollary. *A component of the complement of Σ cannot be simultaneously an even and an odd formal lacuna.*

Indeed, any such component in the base of a versal deformation contains a morsification having at least one real critical point (see for example the proof of Lemma 6.1 below). It follows immediately from the formulae (5)–(8) that the sum of the intersection indices of even and odd Petrovskii classes with the corresponding vanishing cycle is equal to 2 or -2.

Remark 3. If n is even, then complex conjugation changes the orientation of the manifold V_λ, so that the class in $(\tilde{H}_{n-1}(V_\lambda))^*$ corresponding to the even (respectively, odd) Petrovskii class by the Poincaré duality is antiinvariant (respectively, invariant), cf. subsection 1.2 above. In the case of odd n these classes are respectively invariant and antiinvariant.

1.7. The Petrovskii classes for stable equivalent singularities.

Let p, q be arbitrary nonnegative integers, and $f^{(p,q)}$, $F^{(p,q)}$ the corresponding (p,q)-stabilizations of the function f and its deformation $F = F(z, \lambda)$; see Chapter I, § 3.7. For any λ the functions f_λ and $f_\lambda^{(p,q)} \equiv F^{(p,q)}(\cdot, \cdot; \lambda)$ have the same critical values. In particular, there is a natural one-to-one correspondence between the components of complements of discriminants of the deformations F and $F^{(p,q)}$.

Let f_λ be a small real nondiscriminant strict morsification of f, and suppose that some basis of vanishing cycles in the group $\tilde{H}_{n-1}(V_\lambda)$ is fixed according to the rules (i)–(iv) of the previous subsection 1.6. Then the same system of paths defines a basis of vanishing cycles in the group $\tilde{H}_{n+p+q-1}(V_\lambda^{p,q})$, where $V_\lambda^{p,q}$ is the local zero level set for $f_\lambda^{(p,q)}$; let us orient the cycles of this basis in accordance with conditions (iii), (iv) above. Denote these cycles by $\Delta_\alpha^{p,q}$, and the local Petrovskii classes in $\tilde{H}_{n+p+q-1}(V_\lambda^{p,q}, \partial V_\lambda^{p,q})$ by $\Pi_{ev}^{p,q}$ and $\Pi_{odd}^{p,q}$.

Lemma 2. *For any ordering of our paths connecting* 0 *with the real critical values, the intersection indices of the corresponding vanishing cycles in the groups* $\tilde{H}_{n-1}(V_\lambda)$ *and* $\tilde{H}_{n+p+q-1}(V_\lambda^{p,q}, \partial V_\lambda^{p,q})$*, oriented according to item (iii) of the previous subsection 1.6, are expressed in terms of each other by the Gabrielov formula*

$$\langle \Delta_i^{p,q}, \Delta_j^{p,q} \rangle = [sign(i-j)]^{p+q}(-1)^{n(p+q)+(p+q)(p+q-1)/2}\langle \Delta_i, \Delta_j \rangle;$$

cf. formula (12) of Chapter I.

In other words, the choices of orientations of these vanishing cycles specified in the previous subsection for a singularity and its stabilization are concordant in the sense of Theorem I.3.4.

The proof of this lemma follows immediately from the construction of [Gabrielov 73].

Theorem 4. *If the critical value* α *of the real morsification* f_λ *is real, then the intersection indices of both local Petrovskii classes of the functions* $f_\lambda^{(1,0)}$, $f_\lambda^{(0,1)}$ *with cycles* $\Delta_\alpha^{1,0}$, $\Delta_\alpha^{0,1}$ *are related to similar indices for* f_λ *by the following formulae (in each of which all the letters* Π *should be replaced simultaneously either by* Π_{ev} *or by* Π_{odd}*):*

if $\alpha < 0$, *then*

$$\langle \Delta_\alpha^{0,1}, \Pi^{0,1} \rangle = (-1)^{n+1}\langle \Delta_\alpha, \Pi \rangle, \tag{10}$$

$$\langle \Delta_\alpha^{1,0}, \Pi^{1,0} \rangle = (-1)^{n+1}\langle \Delta_\alpha, \Pi \rangle - 2(-1)^{r(\alpha)+n(n-1)/2}; \tag{11}$$

if $\alpha > 0$, *then*

$$\langle \Delta_\alpha^{1,0}, \Pi^{1,0} \rangle = (-1)^{n}\langle \Delta_\alpha, \Pi \rangle, \tag{12}$$

$$\langle \Delta_\alpha^{0,1}, \Pi^{0,1} \rangle = (-1)^n \langle \Delta_\alpha, \Pi \rangle + 2(-1)^{r(\alpha)+n(n-1)/2}; \tag{13}$$

here $r(\alpha)$ is the positive inertia index of the Morse critical point of the morsification f_λ with critical value α.

This follows immediately from formulae (5)–(8) and Lemma 2.

Theorem 5. *If the critical value α of f_λ is not real, and $\bar{\alpha}$ is its complex conjugate critical value, then*

$$|\langle \Delta_\alpha^{1,0}, \Pi_{\mathrm{ev}}^{1,0} \rangle| = |\langle \Delta_\alpha, \Pi_{\mathrm{ev}} \rangle| = |\langle \Delta_\alpha^{0,1}, \Pi_{\mathrm{ev}}^{0,1} \rangle| =$$

$$= |\langle \Delta_\alpha^{1,0}, \Pi_{\mathrm{odd}}^{1,0} \rangle| = |\langle \Delta_\alpha, \Pi_{\mathrm{odd}} \rangle| = |\langle \Delta_\alpha^{0,1}, \Pi_{\mathrm{odd}}^{0,1} \rangle| =$$

$$= |\langle \Delta_{\bar{\alpha}}^{1,0}, \Pi_{\mathrm{ev}}^{1,0} \rangle| = |\langle \Delta_{\bar{\alpha}}, \Pi_{\mathrm{ev}} \rangle| = |\langle \Delta_{\bar{\alpha}}^{0,1}, \Pi_{\mathrm{ev}}^{0,1} \rangle| =$$

$$= |\langle \Delta_{\bar{\alpha}}^{1,0}, \Pi_{\mathrm{odd}}^{1,0} \rangle| = |\langle \Delta_{\bar{\alpha}}, \Pi_{\mathrm{odd}} \rangle| = |\langle \Delta_{\bar{\alpha}}^{0,1}, \Pi_{\mathrm{odd}}^{0,1} \rangle|.$$

The proof will be given in § 5.

Corollary. *The following four conditions are equivalent:*

a) the even (respectively, odd) Petrovskii class corresponding to the morsification f_λ of the function f is equal to zero;

b) the even (respectively, odd) Petrovskii class corresponding to the morsification $f_\lambda^{(2,0)}$ of the stabilization $f^{(2,0)}$ of the function f is equal to zero;

c) the even (respectively, odd) Petrovskii class corresponding to the morsification $f_\lambda^{(0,2)}$ of the stabilization $f^{(0,2)}$ of the function f is equal to zero;

d) the **odd** *(respectively,* **even***) Petrovskii class corresponding to the morsification $f_\lambda^{(1,1)}$ of the stabilization $f^{(1,1)}$ of the function f is equal to zero.*

These assertions follow immediately from the previous Theorems 4, 5.

This corollary justifies the following notion.

Definition 3. The *bistable equivalence* of germs of real functions is the minimal equivalence relation such that

a) D_0-equivalent (that is, obtained from one another by a smooth local diffeomorphism in the argument space) germs of functions are bistably equivalent;

b) any function f is bistably equivalent to both its stabilizations $f^{(2,0)}$ and $f^{(0,2)}$ (and is not, in general, bistably equivalent to $f^{(p,q)}$ where either p or q is odd).

(In particular, the numbers of arguments on which the bistably equivalent functions can depend are of the same parity.)

By the previous corollary, there is a natural correspondence between the local lacunas of bistably equivalent functions.

1.8. Multiplication by -1.

Theorem 6. *The perturbation $f_\lambda \equiv F(\cdot,\lambda)$ of the function f and the perturbation $-f_\lambda \equiv -F(\cdot,\lambda)$ of the function $-f$ simultaneously do or do not belong to even (respectively, odd) local lacunas of these deformations.*

This follows immediately from the definitions.

§ 2. Local lacunas close to singularities from the classification tables

2.1. Numbers of local lacunas close to tabulated singularities.

In the next two Tables 12, 13, we give the numbers of formal lacunas in the bases of versal deformations of many different classes of real singularities. These tables cover, in particular, all simple singularities and all singularities of corank 2 whose Milnor numbers do not exceed 11. In each of these tables the first column contains the standard notation of the corresponding singularity (see the Classification Tables 1, 4 – 7 of Chapter I, § 5). Columns 2 – 5 contain the numbers of local lacunas depending on the parity of the dimension n of the argument space and on the parity of the positive inertia index i_+ of the quadratic part of this function (i.e., of the quadratic form Q in Tables 1, 4 – 7). More exactly, these columns present the numbers of formal lacunas of "regular parity" – of even lacunas if n is even, and of odd lacunas if n is odd. By the Corollary of Theorem 1.5, to obtain the numbers of lacunas of "irregular parity" we need to change column 2 to column 3 and vice versa, and also permute the columns 4 and 5.

Theorem 1. *1. For each of the classes of singularities $A_1, \ldots, E_8$, $P_8^1, \ldots, Z_{11}$ given in the left columns of Tables 12 and 13, the number of formal lacunas of regular parity in the base of arbitrary versal deformation of any singularity of this class is equal to the number indicated in the corresponding column (2, 3, 4 or 5 depending on the parity of n and i_+) of this table or satisfies the inequality indicated there.*

2. The same assertion is true for any (not necessarily versal) deformation in the following cases: a) in all cases when the value 0 (but not the expression "≥ 0") is indicated; b) for the singularities $\pm X_9, \pm Y_{k,\mathbf{R}}$ in the case when n is even and i_+ is even (column 2); c) for the same singularities $\pm X_9, \pm Y_{k,\mathbf{R}}$ in the case of column 4 if we replace ≥ 2 by ≥ 1; d) for the singularity $\pm Y^{++}_{2k,2l}$ with any n and i_+.

For the justification of this theorem see subsection 2.2 and § 6 below.

Table 12. Numbers of local lacunas of regular parity for simple singularities

Singularity class	n even i_+ even	n even i_+ odd	n odd i_+ even	n odd i_+ odd
A_1	2	0	1	1
A_{2k}, $k \geq 1$	0	0	1	0
$\pm A_{2k+1}$, $k \geq 1$	0	1	1	1
D_4^-	0	3	1	1
D_{2k}^+, $k \geq 2$	0	0	1	1
D_{2k}^-, $k \geq 3$	0	2	1	1
$\pm D_{2k+1}$, $k \geq 2$	0	0	1	1
$\pm E_6$	0	0	1	1
E_7	0	0	1	1
E_8	0	0	1	1

Corollary. *1. For any class of singularities $A_1, \ldots, E_8, P_8^1, \ldots, Z_{11}$ given in the left columns of Tables 12, 13, the number of local lacunas close to any stable singular point of the same type of a wave front in $\mathbf{R}^{n+2}$ is equal to the number indicated in the corresponding column of this table (2, 3, 4 or 5 depending on the parity of n and of the positive inertia index i_+ of the quadratic part of the corresponding generating function) or satisfies the inequality indicated there.*

2. The same assertion is true for any (not necessarily stable) singularity of the wave front in the four cases indicated in assertion 2 of Theorem 1.

Conjecture. *In Table 13 all "≥ 0" may be replaced by 0.*

2.2. Realization of lacunas.

In this subsection, for every formal lacuna corresponding to a nonzero value in columns 2 – 5 of Tables 12, 13 we present a perturbation of the corresponding singularity that belongs to this lacuna. The proof of the fact that for all these perturbations the corresponding Petrovskii classes vanish follows from formulae (5) – (9) and from the calculation of the intersection matrices of the corresponding complex singularities. For all singularities of corank 2 this calculation is done by the method of Gusein–Zade – A'Campo (see Chapter I, § 4). For the only singularity of corank 3, P_8^1, we use the Gabrielov's method (see [Gabrielov 73]) of calculation of the Dynkin diagrams of the singularities that split into a sum of functions of separated sets of variables. Indeed, the singularity class P_8^1 contains the function

Table 13. Numbers of local lacunas for certain unimodular singularities

Singularity class	n even, i_+ even	n even, i_+ odd	n odd, i_+ even	n odd, i_+ odd
P_8^1	0	0	≥ 1	0
$\pm X_9$	≥ 1	0	≥ 2	0
X_9^1	0	0	0	0
X_9^2	0	≥ 2	0	0
J_{10}^3	≥ 0	≥ 1	0	0
J_{10}^1	≥ 0	≥ 0	0	0
$\pm J_{2k}^3$, $k \geq 6$	0 for k even	≥ 1	0 for $k \geq 8$	0
$\pm J_{2k}^1$, $k \geq 6$	–	0 for k even	0 for $k \geq 7$	0
$\pm J_{11}$	≥ 0	≥ 0	≥ 0	0
$\pm J_{2k+1}$, $k \geq 6$	–	–	0 for $k \geq 7$	0
$\pm Y_{5,\mathbf{R}}$	≥ 1	0	≥ 2	≥ 0
$\pm Y_{k,\mathbf{R}}$, $k \geq 7$	≥ 1	0	≥ 2	0
$\pm Y_{2k,2l}^{++}$, $k \geq 2 < l$	≥ 1	0 for $k \cdot l$ even	≥ 1	0
$\pm Y_{2k,2l}^{+-}$, $k \geq 2 \leq l$	0 for l even	0 for k even, ≥ 0 for $k = 3, l = 2$	0 for $k + l \geq 6$, ≥ 0 for $k + l = 5$	0
$\pm Y_{2k,2l}^{--}$, $k \geq 2 < l$	0 for $k \cdot l$ even	≥ 1; ≥ 2 for $k = 2$	0 for $k + l > 6$; ≥ 0 if $k + l = 5$	0
$\pm Y_{2k,2l+1}^{+}$, $k \geq 2 \leq l$	≥ 0 for $k = l = 2$	0 for k even	0 for $k + l > 4$; ≥ 0 for $k = l = 2$	0
$\pm Y_{2k,2l+1}^{-}$, $k \geq 2 \leq l$	0 for k even	≥ 0 for $k = l = 2$	0 for $k + l > 5$, ≥ 0 for $k = l = 2$	0
$\pm Y_{2k+1,2l+1}$, $k \geq 2 \leq l$	≥ 0 for $k = l = 2$	≥ 0 for $k = l = 2$	0 for $k + l > 4$, ≥ 0 for $k = l = 2$	0
Z_{11}	0	≥ 0	≥ 0	≥ 0

$x^3 + y^3 + z^3$; since this class is connected, this method also gives the intersection matrices for all other functions of it.

In the case of simple singularities the fact that there are no formal lacunas except for those indicated below, will be proved in § 6. Almost all zeros (but not the expressions "≥ 0") in Table 13 follow immediately from Proposition 1.3: the only exceptions are the zeros in columns 2, 3 for the singularity P_8^1, which were proved by the computer program from our Appendix.

Remark. By the exact sequence of the pair $(V_\lambda, \partial V_\lambda)$, the nontriviality of the class $\partial\Pi \in \tilde{H}_{n-2}(\partial V_\lambda)$ (used in Proposition 1.3) is equivalent to the following condition: the vector of intersetion indices $\langle \Pi, \Delta_i \rangle$ of the Petrovskii class with all vanishing cycles is not an *integer* linear combination of columns of the intersection matrix $\langle \Delta_j, \Delta_i \rangle$. Exactly this condition is checked in the corresponding part of the program described in § 8.

A convention. In this subsection, instead of the necessary perturbations of functions $f(x_1, \ldots, x_c) + Q(x_{c+1}, \ldots, x_n)$, $c = \text{corank}(f)$, we present the corresponding perturbations f_λ of the functions f such that the perturbation $f_\lambda + Q$ lies in the desired lacuna.

ε is always a sufficiently small positive number.

The case A$_1$ was considered in detail in Chapter IV, § 3 and § 1.4 of this chapter.

A$_{2k}$. If n is odd and i_+ is even, then the desired perturbation f_λ is any polynomial with $2k+1$ real roots.

$\pm$A$_{2k+1}$. If i_+ is odd and n is arbitrary, then the conditions of Theorem 1.2 are satisfied: the singularity is stably equivalent to one with an extremum point, and the perturbation $f + \varepsilon$ (in the case $+A_{2k+1}$) or $f - \varepsilon$ (in the case $-A_{2k+1}$) belongs to a formal lacuna. If i_+ is even and n is odd, then the perturbation f_λ of the function $f = x^{2k+2}$ belongs to a formal lacuna only if it has exactly $2k+2$ real roots.

Example. Close to a standard singularity A_3 a front is diffeomorphic to the product of a linear space and a swallow-tail (see Figure 3b). For odd i_+ and any n this front is sharp only from the side of the component 3, while for even i_+ and odd n the component 2 is the only formal lacuna. For even n and i_+ there are no lacunas close to this singularity.

For all singularities of corank 2 from our tables the formal lacunas are either counted by Theorem 1.2 (i.e., the corresponding singularities are stably equivalent to extrema), or are constructed as follows. First we construct an appropriate perturbation f_λ of the original function $f = f(x_1, x_2)$, which is a sabirizarion of it; see

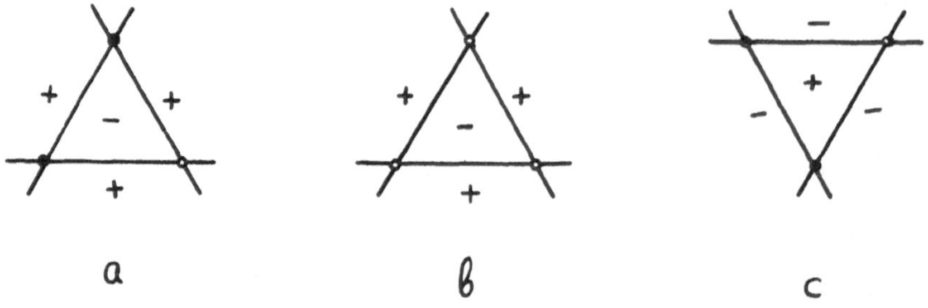

Fig. 39. Sharp morsifications for singularities D_4^-, ...

Chapter I, § 4. These sabirizations are shown in Figures 39–47: the signs + and - denote the signs of the function f_λ in different components of the complement of the zero level. Of course, these perturbations are discriminant ones: all the saddlepoints have zero critical value. Then we improve this sabirization slightly in such a way that these critical values move in the prescribed directions in $\mathbf{R}^1$ (such a move is possible by Proposition I.6.2): in Figures 39–47 the saddlepoints whose critical values should be increased are marked by white circles, and the saddlepoints that should be lowered are marked by black ones.

One more reduction: in all cases when the symbol $\pm$ stands before the notation of the singularity (i.e. this singularity has two real forms which differ from one another by multiplication by -1), then by Theorem 1.6 we can consider only one of them (and we always consider the "+" version).

The case $\mathbf{D}_4^-$. If n is even and i_+ is odd, see Figure 39a. Here any one of the three saddlepoints can be white, and the other two must be black. In § 6 we prove that these three perturbations lie in different components of the complement of the discriminant.

If n and i_+ are odd, see Figure 39b.

If n is odd and i_+ is even, see Figure 39c.

$\mathbf{D}_{2k}^-, k \geq 3$. n even, i_+ odd $\Rightarrow$ see Figure 40a. Reflecting this picture through the horizontal axis we get a picture of another perturbation of the same singularity, which lies in a formal lacuna. In § 6 we prove that these two perturbations belong to different formal lacunas.

n odd, i_+ odd $\Rightarrow$ see Figure 40b.

n odd, i_+ even $\Rightarrow$ Figure 40b should be reflected through a vertical line, the signs + changed to $-$, white circles to black ones and vice versa.

$\mathbf{D}_{2k}^+$. n odd, i_+ odd $\Rightarrow$ see Figure 41.

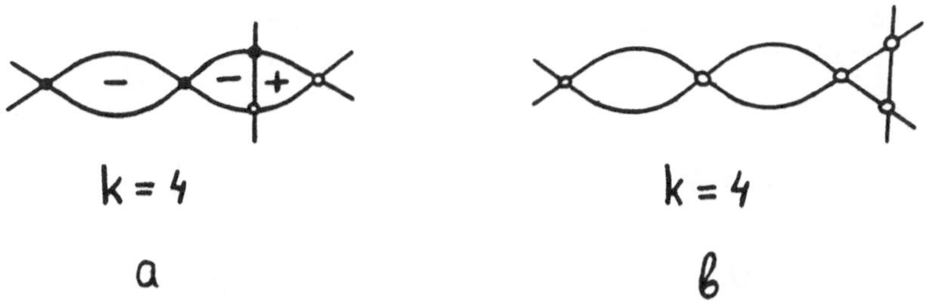

Fig. 40. ... for singularities D^-_{2k}, $k > 2$, ...

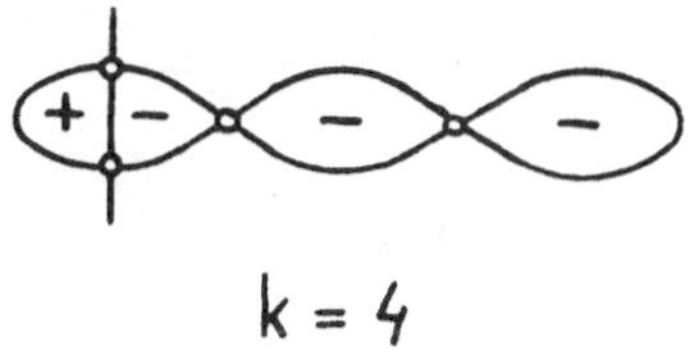

Fig. 41. ... for $D^+_{2k}, k \geq 2$ ...

The case of odd n and even i_+ can be reduced to this one just as for the singularities D^-_{2k}.

$+\mathbf{D}_{2k+1}$. n odd, i_+ odd $\Rightarrow$ see Figure 42a.

n odd, i_+ even $\Rightarrow$ see Figure 42b.

$+\mathbf{E}_6$. n odd, i_+ odd $\Rightarrow$ see Figure 43a.

n odd, i_+ even $\Rightarrow$ see Figure 43b.

$\mathbf{E}_7$. n odd, i_+ odd $\Rightarrow$ see Figure 43c;

the case of even i_+ reduces to this one by multiplication by -1 and reflection in the axis $x_1 = 0$.

$\mathbf{E}_8$. n odd, i_+ odd $\Rightarrow$ see Figure 43d; the case of even i_+ reduces to this one by multiplication by -1 and central symmetry of the argument space.

Singularities $+\mathbf{X}_9$ and $+\mathbf{Y}_{k,\mathbf{R}}$. The function f has a minimum at 0, therefore for even i_+ and arbitrary n the perturbation $f + \varepsilon$ belongs to a lacuna. In the case of odd n a point of one other lacuna is given by Figure 44; the number of petals in this picture is equal to 4 for X_9 and to k for $Y_{k,\mathbf{R}}$.

Proposition 1. *The lacuna shown in Figure 44 is new (i.e. the component of the complement of Σ containing the perturbation shown in this picture does not*

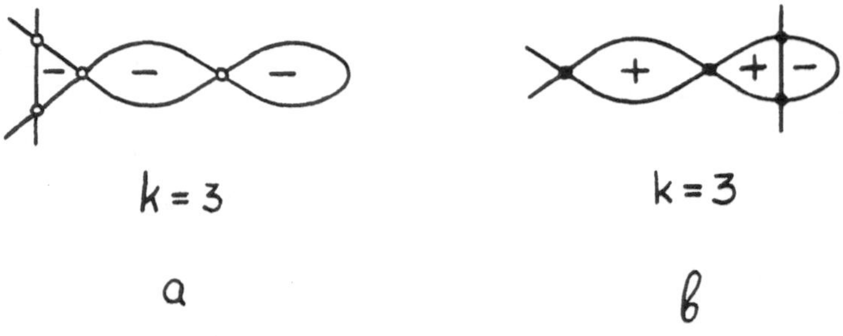

Fig. 42. ... for $+D_{2k+1}$, ...

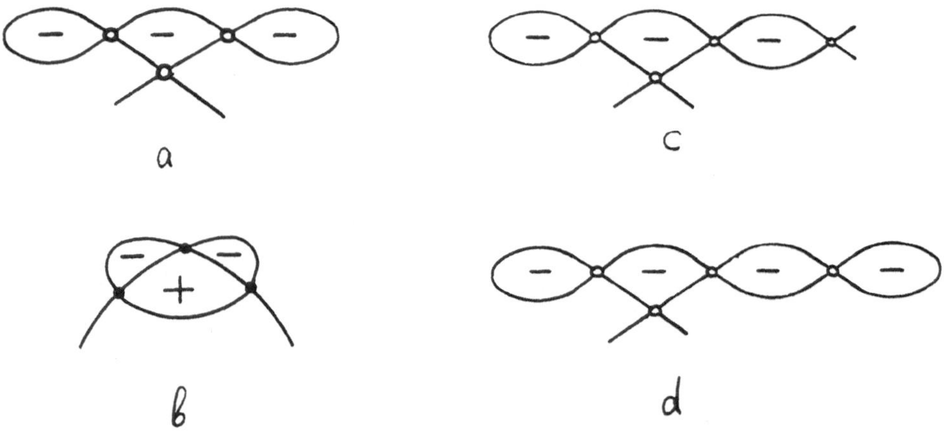

Fig. 43. ... and for singularities $+E_6$, E_7, E_8

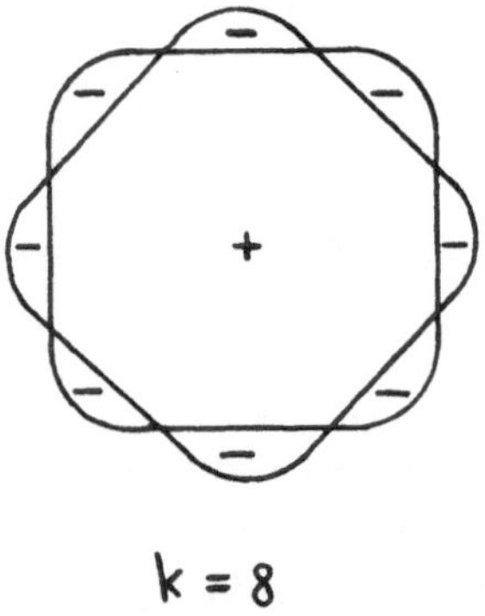

Fig. 44. An additional lacuna for $+Y_{k,\mathbf{R}}$ or $+X_9$

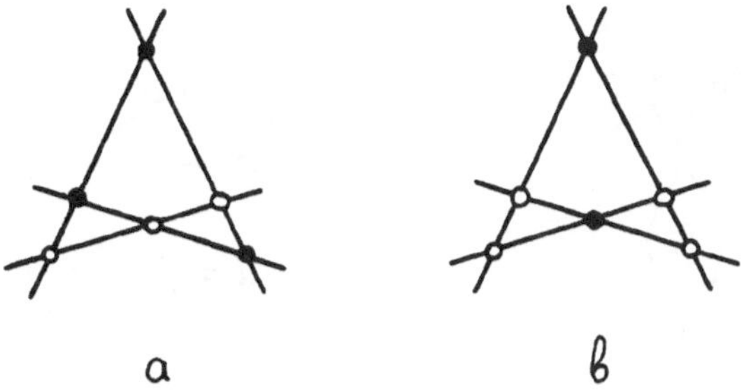

Fig. 45. Lacunas for singularities X_9^2, ...

contain the function $f+\varepsilon$*).*

Proof. It is sufficient to prove that the corresponding two components of the complement of the discriminant variety of some arbitrary stabilizations of our singularity and deformation are different. For the stabilization with even n and even i_+ this difference follows from the fact that the stabilization of the perturbation from Figure 44 does not belong to a lacuna, while the stabilization of the perturbation $f+\varepsilon$ does belong.

$\mathbf{X}_9^2$. n even, i_+ odd $\Rightarrow$ see Figures 45a, 45b.

Proposition 2. *The formal lacunas containing the perturbations from Figures 45a,b are different.*

The proof is based on the following invariant of a component of the complement of the discriminant.

Definition 1. The *negative index* of a component of the complement of Σ is the number of all critical points of any morsification from this component with even Morse indices and negative critical value minus the number of critical points with odd indices and negative values.

Lemma 1. *The negative index is a well-defined function of a component of the complement of the discriminant (i.e., it is the same for all morsifications from it).*

The proof is obvious.

It is easy to see that the indices of the two morsifications from Figures 45a,b are different.

$+\mathbf{Y}_{2k,2l}^{--}$. n even, i_+ odd $\Rightarrow$ see Figure 46a.

In the case $k=2$ see additionally Figure 46b; the fact that these lacunas are different follows again from the comparision of their indices.

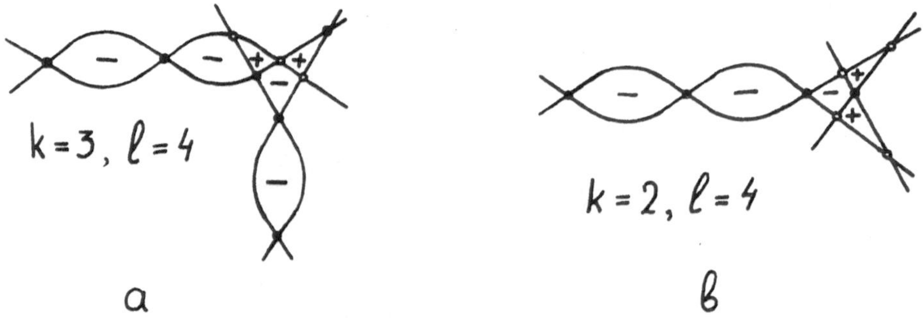

Fig. 46. ... $+Y^{-,-}_{2k,2l}$, ...

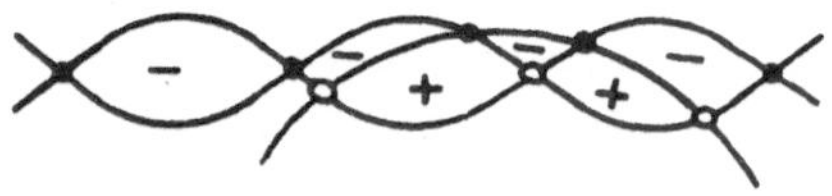

Fig. 47. ... and J^3_{2k}

$\mathbf{J}^3_{2k}$. n even, i_+ odd $\Rightarrow$ see Figure 47.

$+\mathbf{Y}^{++}_{2k,2l}$. See § 1.5.

Finally, let us consider the singularity P^1_8, which is the only singularity of corank 3 in Table 13. This class contains, in particular, functions of the form

$$x^3 + y^3 + z^3 + \alpha xyz, \quad a > -3. \tag{14}$$

Consider any function of this form with $\alpha \in (-3, 6)$. Then the desired perturbations are given by the one-parameter family

$$f_{(\tau)} = f - \tau(x + y + z) - c \cdot \tau^{3/2},$$

where $\tau > 0$, and c is an arbitrary constant from the interval

$$\Big(\frac{2}{9}(\alpha + 3)\sqrt{\frac{3}{\alpha^2 - 3\alpha + 9}},\ 2\sqrt{\frac{1}{a + 3}}\Big)$$

or from the interval opposite to this. The function $f - \tau(x + y + z)$ has two critical points with $i_+ = 0$ and $i_+ = 3$ and critical values equal respectively to $\pm 2\sqrt{\frac{1}{a+3}}\tau^{3/2}$, and two triples of critical points with $i_+ = 1$ and $i_+ = 2$ and critical values equal to $\pm\frac{2}{9}(\alpha + 3)\sqrt{\frac{3}{\alpha^2 - 3\alpha + 9}}\tau^{3/2}$; the constant c above is chosen so that the critical value at one point of the first pair of points is of one sign, and the remaining seven points are of the other. I do not know whether the perturbations $f_{(\tau)}$ corresponding to values of c from these two intervals lie in one formal lacuna or in two different lacunas. The fact that this (or these) lacuna does not vanish for $\alpha > 6$ follows from Theorem I.5.3.

§ 3. Calculation of the even local Petrovskii class

Let $f : (\mathbf{C}^n, 0) \to (\mathbf{C}, 0)$ be a holomorphic function that is real (i.e., takes real values on $\mathbf{R}^n$) and has an isolated singularity at 0, so that its Milnor number $\mu(f)$ is finite. Let B be a small disc with centre at 0, and f_λ a nondiscriminant real strict morsification of f that is very close to f (even in comparison with the diameter of B).

Consider the set Re $V_\lambda(0)$ of real values of the manifold $V_\lambda(0) \equiv f_\lambda^{-1}(0) \cap B$ and orient it in such a way that at any point of it the frame $\{\operatorname{grad} f_\lambda$; a tangent frame to Re $V_\lambda(0)$ positively oriented with respect to this orientation $\}$ defines the distinguished orientation of $\mathbf{R}^n$. The manifold Re $V_\lambda(0)$ thus oriented defines an element Π_{ev} in the group $\tilde{H}_{n-1}(V_\lambda(0), \partial V_\lambda(0))$.

In this section we calculate the intersection indices of this cycle with the vanishing cycles defined by the conventions of § 1.6.

3.1. The cycles $\Pi_{\text{ev}}(a)$, $\Delta_\alpha(a)$, $\Delta_\alpha^+(a)$.

Simultaneously with the local Petrovskii cycle $\Pi_{\text{ev}} \in \tilde{H}_{n-1}(V_\lambda(0), \partial V_\lambda(0))$ for any noncritical real value a of the morsification f_λ we consider the relative cycle in $V_\lambda(a) \equiv f_\lambda^{-1}(a) \cap B$ defined by the set of real points and oriented as above. The homology class of this cycle is denoted by $\Pi_{\text{ev}}(a)$, in particular, $\Pi_{\text{ev}} \equiv \Pi_{\text{ev}}(0)$.

The (absolute) homology of the manifold $V_\lambda(a)$ is generated by the vanishing cycles $\Delta_\alpha(a)$ defined in exactly the same way as the cycles Δ_α in § 1.6 (and coincide with the cycles $\Delta_{\alpha-a}$ of the morsification $f_\lambda - a$ of the same function f). Simultaneously with them we shall consider the vanishing cycles $\Delta_\alpha^+(a)$ defined as follows.

If the critical value α is nonreal, then $\Delta_\alpha^+(a) = \Delta_\alpha(a)$.

If ε is very small, then the oriented cycle $\Delta_\alpha^+(\alpha+\varepsilon) \subset V_\lambda(\alpha+\varepsilon)$ is defined just as the cycle $\Delta_\alpha(\alpha-\varepsilon) \subset V_\lambda(\alpha-\varepsilon)$ in § 1.6, only replacing the sequence of vectors (4) by the sequence

$$\partial/\partial z_1, \ldots, \partial/\partial z_r, i\partial/\partial z_{r+1}, \ldots, i\partial/\partial z_n. \tag{15}$$

For arbitrary noncritical real a the cycle Δ_α^+ corresponding to a critical value α is obtained from $\Delta_\alpha^+(\alpha+\varepsilon)$ by the Gauss–Manin transportation over a path connecting a and $\alpha+\varepsilon$ and going along the real axis in the half-plane $\{\text{Im} > 0\}$.

Lemma 1. *For any noncritical real value a and any real critical value*

$$\alpha,\ \Delta_\alpha^+(a) = (-1)^{n-r+1}\Delta_\alpha(a),$$

where r is the positive inertia index of the quadratic part of the corresponding Morse critical point.

The proof is trivial.

Theorem 1. *For any real critical value α of the morsification f_λ, and sufficiently small positive ε,*

$$\langle \Delta_\alpha^+(\alpha+\varepsilon), \Pi_{ev}(\alpha+\varepsilon)\rangle = (-1)^{n(n-1)/2}\chi(S^{r-1}), \tag{16}$$

$$\langle \Delta_\alpha(\alpha-\varepsilon), \Pi_{ev}(\alpha-\varepsilon)\rangle = (-1)^{n(n-1)/2}\chi(S^{n-r-1}), \tag{17}$$

where χ is the notation for the Euler characteristic, r is the same as in Lemma 1, and $\chi(S^{-1}) \equiv 0$.

Proof of equality (16). In the case $r = 0$ this equality is obvious, thus we assume that $r > 0$.

The cycles $\Delta_\alpha^+(\alpha+\varepsilon)$ and $\Pi_{ev}(\alpha+\varepsilon)$ meet along the $(r-1)$-dimensional sphere S^{r-1}, distinguished in the canonical local coordinates $z_1, \dots, z_n$ in $\mathbf{R}^n$ by the equations $z_{r+1} = \cdots = z_n = 0$, $z_1^2 + \cdots + z_r^2 = \varepsilon$. The vector-field $i \cdot \text{grad}(z_1|_{V_\lambda(\alpha+\varepsilon)}) \subset T_*(V_\lambda(\alpha+\varepsilon))$ is transversal to the sphere $\Delta_\alpha^+(\alpha+\varepsilon)$ at all points of this subsphere S^{r-1} where this field is nonzero (i.e., everywhere but two points $(\pm\sqrt{\varepsilon}, 0, \dots, 0)$). A small shift of the cycle $\Delta_\alpha^+(\alpha+\varepsilon)$ along the corresponding flow makes it transversal to the cycle of real points at these two points, and moves it into the imaginary domain close to all other points of S^{r-1}. The formula (16) follows from a trivial counting of the local orientations of these two cycles close to these two points.

The equality (17) is proved in exactly the same way.

This theorem gives us the intersection indices of the even Petrovskii class with vanishing cycles corresponding to the greatest negative and the smallest positive critical values of f_λ.

3.2. Counting the jump over a critical value.

Notation. Denote by J_α the identification of the homology groups of the spaces $V_\lambda(\alpha+\varepsilon)$ and $V_\lambda(\alpha-\varepsilon)$ (both absolute and relative) given by the arc

$$\alpha + \varepsilon \cdot e^{is}, \quad s \in [0, \pi]. \tag{18}$$

Theorem 2. $\Pi_{ev}(\alpha - \varepsilon) = J_\alpha \Pi_{ev}(\alpha + \varepsilon) + \Delta_\alpha(\alpha - \varepsilon)$.

A preliminary remark. It is obvious that $\Pi_{ev}(\alpha - \varepsilon)$ and $J_\alpha \Pi_{ev}(\alpha - \varepsilon)$ differ by an absolute cycle which is a multiple of $\Delta_\alpha(\alpha - \varepsilon)$: $\Pi_{ev}(\alpha - \varepsilon) - J_\alpha \Pi_{ev}(\alpha - \varepsilon) = \theta \cdot \Delta_\alpha(\alpha - \varepsilon)$. This follows from the fact that the fibration defined by the function f_λ on the union of manifolds of noncritical levels very close to α of this function in B can be trivialized everywhere outside a small neighbourhood of the corresponding critical point. This trivialization can be chosen in such a way that it respects complex conjugation (cf. the proof of Theorem III.3.2). In particular, for all real a close to α the sets of real points of the manifolds $V_\lambda(a)$ are identified with each other outside this small neighbourhood. Therefore the difference $\Pi_{ev}(\alpha - \varepsilon) - J_\alpha \Pi_{ev}(\alpha - \varepsilon)$ is contained in this neighbourhood, and thus is homological to a multiple of $\Delta_\alpha(\alpha - \varepsilon)$.

Moreover, it is clear that the coefficient θ depends only on n and r: $\theta = \theta(n, r)$, and we only need to prove that $\theta(n, r)$ is always equal to 1.

The proof of Theorem 2 is based on the study of a model family of functions (the standard "birth-death surgery"):

$$\phi_t = z_1^3 - 3tz_1 + z_2^2 + \cdots + z_r^2 - z_{r+1}^2 - \cdots - z_n^2, \quad t \in [-1, 1]. \tag{19}$$

The function ϕ_1 has two real critical points with coordinates $z_1 = \pm 1, z_2 = \cdots = z_n = 0$, their critical values are equal to ∓ 2, and the positive inertia indices are equal to r and $r - 1$, respectively. Let us fix a large positive number Λ. Let Q be a disc in $\mathbf{C}^n$ whose radius is much greater than Λ. The family (19) will be considered as a family of functions defined only on this disc. The group $\tilde{H}_{n-1}(\phi_1^{-1}(-\Lambda))$ is generated by two vanishing cycles $\Delta_{+2}(-\Lambda)$, $\Delta_{-2}(-\Lambda)$ (defined as in subsection 3.1), and the group $\tilde{H}_{n-1}(\phi_{-1}^{-1}(-\Lambda))$ by the vanishing cycles $\Delta_{+2i}(-\Lambda)$, $\Delta_{-2i}(-\Lambda)$.

Lemma 2. $\langle \Delta_{+2}(-\Lambda), \Delta_{-2}(-\Lambda) \rangle = (-1)^{(n+2)(n+1)/2}$.

Lemma 1 reduces this lemma to the calculation of the intersection indices of cycles $\Delta_{+2}(0), \Delta_{-2}^{+}(0)$, which is elementary. Indeed, for these cycles we can choose the cycles obtained from cycles $\Delta_{+2}(2 - \varepsilon), \Delta_{-2}^{+}(-2 + \varepsilon)$ by the action of the vector-fields $-\text{grad}\ \phi_1$, $\text{grad}\ \phi_1$, respectively, and these cycles intersect only at the point 0, while the orientation (and the tangent plane) of the first of them is given by the sequence of vectors

$$i\partial/\partial z_2, \ldots, i\partial/\partial z_r, \partial/\partial z_{r+1}, \ldots, \partial/\partial z_n$$

and the orientation of the second is given by the sequence

$$\partial/\partial z_2, \ldots, \partial/\partial z_r, i\partial/\partial z_{r+1}, \ldots, i\partial/\partial z_n.$$

Now the answer follows from the comparison of the concatenation of these two sequences with the complex orientation of the fibre ϕ_1^{-1} given by the sequence

$$\partial/\partial z_2, i\partial/\partial z_2, \ldots, \partial/\partial z_n, i\partial/\partial z_n.$$

Let us consider the relative cycles $\nabla^{(1)}(\pm\Lambda), \nabla^{(-1)}(\pm\Lambda)$ given by the sets of real points in the manifolds $\phi_1^{-1}(\pm\Lambda)$, $\phi_{-1}^{-1}(\pm\Lambda)$, respectively. The family of functions (19) realizes the isotopies

$$\kappa : \phi_1^{-1}(+\Lambda) \leftrightarrow \phi_{-1}^{-1}(+\Lambda), \quad \kappa : \phi_1^{-1}(-\Lambda) \leftrightarrow \phi_{-1}^{-1}(-\Lambda).$$

Lemma 3. *1)*

$$\kappa(\nabla^{(1)}(+\Lambda)) \sim \nabla^{(-1)}(+\Lambda), \quad \kappa(\nabla^{(1)}(-\Lambda)) \sim \nabla^{(-1)}(-\Lambda),$$

where $\sim$ is the notation for the homology;

2) for some choice of the orientations of the cycles $\Delta_{+2i}(-\Lambda)$ and *$\Delta_{-2i}(-\Lambda)$,*

$$\Delta_{-2i}(-\Lambda) \sim \kappa(\Delta_2(-\Lambda)), \; \Delta_{2i}(-\Lambda) \sim \kappa(\Delta_2(-\Lambda) + \Delta_{-2}(-\Lambda)).$$

Assertion 1 is obvious. To prove assertion 2 let us connect ϕ_1 and ϕ_{-1} in the space of (complex) Morse functions by a path consisting of the functions

$$\tilde{\phi}_s = z_1^3 - 3e^{is}z_1 + z_2^2 + \cdots + z_r^2 - z_{r+1}^2 - \cdots - z_n^2, \quad s \in [0, \pi]. \tag{20}$$

Along this path the critical values of the functions $\tilde{\phi}_s$ rotate around the origin by the angle $3\pi/2$. The corresponding isotopy of the pair {the plane $\mathbf{C}^1$; critical values of the function $\tilde{\phi}_s$}, which is constant outside some large disc in $\mathbf{C}^1$, deforms the standard distinguished paths connecting $-\Lambda$ with $+2$ and -2 into the paths 1 and 2 in Figure 48, respectively. Now assertion 2 of Lemma 3 follows immediately from the Picard–Lefschetz theorem and the previous Lemma 2.

The proof of Theorem 2 will be carried out by induction over the indices r. For $r = 0$ the theorem is obvious: if the level goes over the value at a maximum point, then the vanishing of the corresponding vanishing cycle is especially explicit. Suppose that the theorem is proved for real Morse critical points whose positive inertia indices equal $r - 1$.

Let us transport the relative cycle $\nabla^{(1)}(+\Lambda)$ over the dashed path in Figure 49 from the manifold $\phi_1^{-1}(+\Lambda)$ into $\phi_1^{-1}(-\Lambda)$ in such a way that close to the boundary of the disc Q this transportation respects the chosen trivialization of the fibration

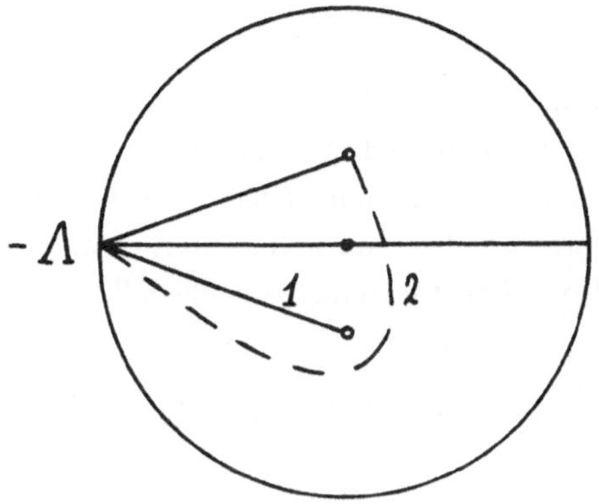

Fig. 48. Distinguished system of paths for morsification ϕ_{-1}

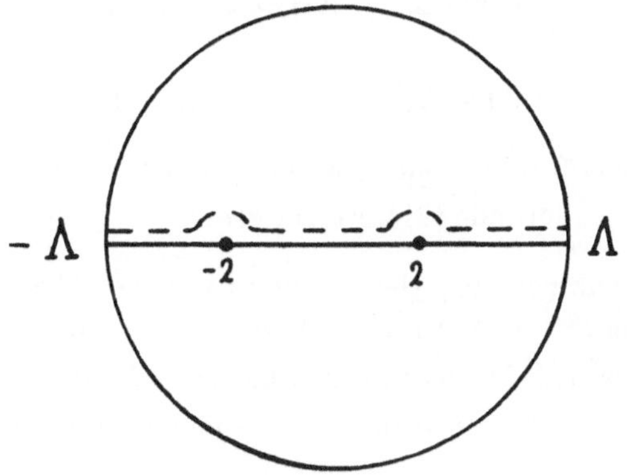

Fig. 49. An identification of negative and positive level varieties for a model morsification

given by the function ϕ_1. By the induction hypothesis, the class of the resulting relative cycle $\nabla! \subset \phi_1^{-1}(-\Lambda)$ in the group $\tilde{H}_{n-1}(\phi_1^{-1}(-\Lambda), \partial Q)$ is equal to

$$\nabla^{(1)}(-\Lambda) + j(\Delta_2(-\Lambda) + \theta(n,r) \cdot \Delta_{-2}(-\Lambda)) \tag{21}$$

(where j is the obvious map of the absolute homology into the relative). On the other hand, our path can be deformed into the arc $\{\Lambda \cdot e^{is},\ s \in [0,\pi]\}$ of the large circle. Let us include our trivialization of the fibre bundle $\phi_1|_{\partial Q}$ in a family of similar trivializations, corresponding to the functions ϕ_t and respecting the complex conjugation. Let us also define the relative cycle $\nabla'! \subset \phi_{-1}^{-1}(-\Lambda)$ obtained from $\nabla^{-1}(+\Lambda)$ by transport over the large arc.

Lemma 4. $\kappa(\nabla^{(1)}(-\Lambda) - \nabla!) \sim (\nabla^{(1)}(-\Lambda) - \nabla'!)$ *in the homology of the space* $\phi_{-1}^{-1}(-\Lambda)$.

This statement is obvious.

But the function φ_{-1} has no real critical values, therefore the cycles $\nabla^{(-1)}(-\Lambda)$ and $\nabla^{(-1)}(+\Lambda)$ are obtained from one another by the isotopy over the real axis. Hence $\nabla'!$ is nothing but the image of the cycle $\nabla^{(-1)}(-\Lambda)$ under the Picard–Lefschetz monodromy defined by the *simple loop* C (see Figure 9) around the distinguished segment $[-\Lambda, 2i]$; see Figure 48. By the Picard–Lefschetz formula and the previous lemma we get

$$\kappa(\Delta_2(-\Lambda) + \theta(n,r) \cdot \Delta_{-2}(-\Lambda)) \sim \nabla^{(-1)}(-\Lambda) - \nabla'! =$$

$$= -\mathrm{Var}_C \nabla^{(-1)}(-\Lambda) = \pm\langle \nabla^{(-1)}(-\Lambda), \Delta_{2i}(-\Lambda)\rangle \cdot \Delta_{2i}(-\Lambda).$$

By assertion 2 of Lemma 3, this cycle is proportional to $\kappa(\Delta_2(-\Lambda) + \Delta_{-2}(-\Lambda))$. Comparing this cycle with formula (21), we get $\theta(n,r) = 1$, q.e.d.

Theorems 1, 2 imply immediately the formulae (5), (7) of Theorem 1.3. Consider for instance the formula (5). If α is the smallest positive critical value of f_λ, then this formula coincides with (17), because in this case $\langle \Delta_\alpha(0), \Pi_{ev}(0)\rangle = \langle \Delta_\alpha(\alpha - \varepsilon), \Pi_{ev}(\alpha - \varepsilon)\rangle$. If there is one critical value α' between α and 0, then

$$\langle \Delta_\alpha(0), \Pi_{ev}(0)\rangle = \langle \Delta_\alpha(\alpha' - \varepsilon), \Pi_{ev}(\alpha' - \varepsilon)\rangle =$$

$$= \langle \Delta_\alpha(\alpha' - \varepsilon), (J_{\alpha'}\Pi_{ev}(\alpha' + \varepsilon) + \Delta_{\alpha'}(\alpha' - \varepsilon))\rangle =$$

$$\langle \Delta_\alpha(\alpha - \varepsilon), \Pi_{ev}(\alpha - \varepsilon)\rangle + \langle \Delta_\alpha, \Delta_{\alpha'}\rangle;$$

by (17) the last expression is equal to the right-hand side of formula (5).

3.3. Proof of the last assertion of Theorem 1.3 for the even Petrovskii class.

We can assume that the cycle $\Delta_\alpha \subset V_\lambda(0)$ is transversal to the set of real points, and at any intersection point the tangent plane $T.\Delta_\alpha \subset T.V_\lambda(0)$ does not contain the lines invariant under complex conjugation. Since the cycle $\Delta_{\bar{\alpha}}$ is complex conjugate to Δ_α, to any intersection point of Δ_α and $\Delta_{\bar{\alpha}}$ there corresponds a complex conjugate intersection point, which coincides with it only if it is simultaneously an intersection point with the real set. This implies the last identity in (9) modulo 2. The first identity in (9) follows from the invariance of the Petrovskii cycle under complex conjugation; see Proposition 1.2. Cf. Lemma 3 in [Arnold 71].

3.4. Proof of formula (9) for a simple singularity.

By Corollary 2 of Chapter I, § 6, the intersection index $\langle \Delta_\alpha, \Delta_{\bar{\alpha}} \rangle$ can be equal only to 0, 1 or -1. Consider the path in the base of the versal deformation that was used in the proof of this Corollary, i.e. the homotopy $f_{\lambda(\tau)}$, $\tau \in [0,1]$, of our function f_λ, along which all the critical values of $f_{\lambda(\tau)}$ except α and $\bar{\alpha}$ remain unmoved, and these two values move towards the value 0 along the corresponding two paths of the distinguished system. These two distinguished paths are complex conjugate, and hence our homotopy lies in the real part of the base of the versal deformation. Obviously during this homotopy none of the three intersection indices (9) changes, therefore we can consider only the last small segment of it.

Suppose that $\langle \Delta_\alpha, \Delta_{\bar{\alpha}} \rangle$ is equal to 1 or -1. We can assume that our two paths of the distinguished system coincide with the two halves of the imaginary axis close to the point 0. Then the last part of our homotopy is locally equivalent to the segment of the family (19) corresponding to the values $t \in [-\delta, 0]$, δ sufficiently small. Then our assertion follows from the local properties of this family, see Lemmas 2 and 3 and Theorem 1.

If $\langle \Delta_\alpha, \Delta_{\bar{\alpha}} \rangle = 0$, then in the last instant of our homotopy we get a function with two complex conjugate Morse singular points with critical value 0, in particular, these two points are nonreal. Then for close instants of the homotopy the corresponding vanishing spheres do not meet the set of real points; of course, this implies the triviality of the corresponding intersection indices $\langle \Delta_\alpha, \Pi_{ev} \rangle$, $\langle \Delta_{\bar{\alpha}}, \Pi_{ev} \rangle$.

§ 4. Calculation of the odd local Petrovskii class

Recall that the odd Petrovskii class is defined as the preimage under the tube isomorphism of the element $b = b_\lambda \in \tilde{H}_n(B - V_\lambda(0), \partial B)$; see § 1.1.

By Proposition 7 of Chapter I, § 3, the diagram (I.7′) is commutative, therefore it is sufficient to calculate the intersection indices of b_λ with the cones $\delta_\alpha \in \check{H}_n(B, V_\lambda(0))$ over the vanishing cycles Δ_α; see Chapter I, § 3.5.

As in the previous section, together with these cycles for every noncritical real value a of the function f_λ we consider the relative cycle $b(a) = b_\lambda(a) \in \check{H}_n(B - V_\lambda(a), \partial B)$ and the cones $\delta_\alpha(a), \delta_\alpha^+(a) \in \check{H}_n(B, V_\lambda(a))$ defined through the vanishing cycles $\Delta_\alpha(a), \Delta_\alpha^+(a)$ (see § 3.1) in the same way as the cones $\delta_\alpha \equiv \delta_\alpha(0)$ were defined through the cycles $\Delta_\alpha \equiv \Delta_\alpha(0)$.

Of course, the diagram (7′) of Chapter I, mentioned in Proposition I.3.7, remains commutative if we change V_λ to $V_\lambda(a)$ everywhere in it.

Theorem 1. *If α is a real critical value of a Morse function f_λ at a real critical point, whose positive inertia index is equal to r, and the segment $[\alpha - \varepsilon, \alpha + \varepsilon]$ contains no other critical values, then*

$$\langle \delta_\alpha^+(\alpha + \varepsilon, b(\alpha + \varepsilon)\rangle = (-1)^{n(n+1)/2}\chi(S^r), \tag{22}$$

$$\langle \delta_\alpha(\alpha - \varepsilon, b(\alpha - \varepsilon)\rangle = (-1)^{1+n(n+1)/2}\chi(S^{n-r}). \tag{23}$$

This immediately implies formulae (6), (8) in the case when α is the smallest positive or the greatest negative critical value of f_λ.

4.1. Proof of Theorem 1.

Instead of the cycles $b(a)$ we shall consider the cycle $\tilde{b}$ homologous to it in $\check{H}_n(B - V_\lambda(a), \partial B)$ that does not depend on a and is defined as follows. Let us divide the cycle $b(a)$ into two cycles $b_+(a)$, $b_-(a)$ participating in the definition of it, so that

$$b_+(a) = \frac{1}{2}(b(a) \cap \mathbf{R}^n) + b(a) \cap \{z | f_\lambda(z) > 0\},$$

$$b_-(a) = \frac{1}{2}(b(a) \cap \mathbf{R}^n) + b(a) \cap \{z | f_\lambda(z) < 0\}.$$

Let γ be a positive constant that is smaller than the absolute values of imaginary parts of all nonreal critical values of f_λ Consider two (discontinuous) vector-fields which coincide with $\operatorname{grad} \operatorname{Im} f_\lambda$ and $-\operatorname{grad} \operatorname{Im} f_\lambda$ in the domain where $|\operatorname{Im} f_\lambda| \leq \gamma$ and are zero outside this domain. Let us act by these two fields on the cycles $b_+(a)$ and $b_-(a)$, respectively. After this action these cycles become two varieties $\tilde{b}_+$ (respectively, $\tilde{b}_-$), which lie along the plane $\mathbf{R}^n$ in the set where $\operatorname{Im} f_\lambda = \gamma$ (respectively, $\operatorname{Im} f_\lambda = -\gamma$); moreover, any real critical point makes an additional contribution to $\tilde{b}_+$ and $\tilde{b}_-$ in the form of the discs lying in the positive (respectively, negative) separatrix manifolds of these vector-fields. Define the cycle $\tilde{b}$ as the sum

$\tilde{b}_+ + \tilde{b}_-$. Obviously for any noncritical a the cycle $\tilde{b}$ is homologous to $b(a)$ in $B - V_\lambda(a) \bmod \partial B$.

In the proof of Theorem 1 we can assume that the critical value α is equal to 0, and the function f_λ is written in local (close to the corresponding critical point) coordinates in the canonical form

$$z_1^2 + \cdots + z_r^2 - z_{r-1}^2 - \cdots - z_n^2.$$

Then for the cone $\delta_\alpha^+(\alpha + \varepsilon) \equiv \delta_0^+(\varepsilon)$ we can take the disc of radius $\varepsilon^{1/2}$ in the real span of the vectors (15), oriented by this sequence of vectors. Close to the critical point the cycle $\tilde{b}_+$ coincides with the positive separatrix manifold of the field grad $\mathrm{Im}(f_\lambda)$, which is the real span of the vectors

$$(1+i)\partial/\partial z_1, \ldots, (1+i)\partial/\partial z_r, (1-i)\partial/\partial z_{r+1}, \ldots, (1-i)\partial/\partial z_n \tag{24}$$

and is oriented by this sequence of vectors. Since all the remaining part of the cycle $\tilde{b}_+$ close to this critical point does not intersect the set where f_λ takes real values, the intersection index $\langle \delta_0^+(\varepsilon), b_+(\varepsilon) \rangle \equiv \langle \delta_0^+(\varepsilon), \tilde{b}_+ \rangle$ is equal to the intersection index of the oriented planes (15), (24) with respect to complex orientation, hence it is equal to $(-1)^{r+n(n+1)/2}$.

In exactly the same way, close to the point 0 the chain $\tilde{b}_-$ coincides with the plane that is the real span of the ordered sequence of vectors

$$(1-i)\partial/\partial z_1, \ldots, (1-i)\partial/\partial z_r, (1+i)\partial/\partial z_{r+1}, \ldots, (1+i)\partial/\partial z_n, \tag{25}$$

and the chain $\delta_\alpha(-\varepsilon)$ coincides with the plane that is the real span of the frame (4); here by the definitions of the orientations of the cycles $\Delta(a), \delta(a)$ given in § 1.6 of this Chapter and § 3.6 of Chapter I, the orientation of $\delta_\alpha(-\varepsilon)$ differs from the orientation defined by this sequence (4). In the same way as above, from this we obtain

$$\langle \delta_0^+(\varepsilon), \tilde{b}_- \rangle = (-1)^{n(n+1)/2}, \quad \langle \delta_0(-\varepsilon), \tilde{b}_+ \rangle = (-1)^{1+n(n+1)/2},$$

$$\langle \delta_0(-\varepsilon), \tilde{b}_- \rangle = (-1)^{1+r+n(n-1)/2},$$

this gives formulae (22), (23), and Theorem 1 is proved.

4.2. Counting the jumps over the critical values.

To prove formulae (6), (8), we must now determine the transformation of the element $b(a)$ of the group $(\tilde{H}_n(B, V_\lambda(a)))^*$ when a jumps over a critical value.

Let α be a real critical value of the function f_λ. For every element Γ of the group $\tilde{H}_n(B, V_\lambda(\alpha+\varepsilon))$ we define the class $\tilde{J}_\alpha\Gamma \in \tilde{H}_n(B, V_\lambda(\alpha-\varepsilon))$ as the class of the relative cycle obtained from any cycle δ representing Γ by adding the chain swept out by the cycle $\partial\delta \subset V_\lambda(\alpha+\varepsilon)$ in transportation over the arc (18).

Of course, the operator $\tilde{J}_\alpha$ is an isomorphism and the diagram

$$\begin{array}{ccc} \tilde{H}_n(B, V_\lambda(\alpha+\varepsilon)) & \to & \tilde{H}_{n-1}(V_\lambda(\alpha+\varepsilon)) \\ \downarrow \tilde{J}_\alpha & & \downarrow J_\alpha \\ \tilde{H}_n(B, V_\lambda(\alpha-\varepsilon)) & \to & \tilde{H}_{n-1}(V_\lambda(\alpha-\varepsilon)), \end{array} \tag{26}$$

in which the horizontal arrows are the boundary homomorphisms, is commutative. In particular, by Lemma 3.1,

$$\tilde{J}_\alpha\delta_\alpha^+(\alpha+\varepsilon) = (-1)^{n-r-1}\delta_\alpha(\alpha-\varepsilon),$$

where r is the positive inertia index of the corresponding critical point.

Theorem 2. *For any class* $\Gamma \in \tilde{H}_n(B, V_\lambda(\alpha+\varepsilon))$,

$$\langle \tilde{J}_\alpha\Gamma, b(\alpha-\varepsilon)\rangle = \langle \Gamma, b(\alpha+\varepsilon)\rangle + (-1)^r\langle \partial\Gamma, \Delta_\alpha^+(\alpha+\varepsilon)\rangle,$$

where $\Delta_\alpha^+(\alpha+\varepsilon)$ *is the vanishing cycle defined in* § *3.1.*

Corollary. $\Pi_{\text{odd}}(\alpha-\varepsilon) = J_\alpha\Pi_{\text{odd}}(\alpha+\varepsilon) - j(\Delta_\alpha(\alpha-\varepsilon))$; *compare with Theorem 3.2.*

This corollary follows immediately from Theorem 2, Lemma 3.1, and Proposition I.3.7.

4.3. Proof of Theorem 2.

We can assume that ε is smaller than the number γ that participated in the construction of the chain $\tilde{b}$. Then the difference $\langle \tilde{J}_\alpha\Gamma, b(\alpha-\varepsilon)\rangle - \langle \Gamma, b(\alpha+\varepsilon)\rangle$ is equal to the intersection index of the relative cycle in

$$B \bmod [V_\lambda(\alpha+\varepsilon) \cup V_\lambda(\alpha-\varepsilon) \cup \{z \mid |\operatorname{Im} f_\lambda(z)| \geq \gamma\}]$$

swept out by the chain $\partial\Gamma$ in transport over the arc (18), and the separatrix manifold (24). As in Proposition I.3.7, it is easy to prove that this index is equal (up to multiplication by $(-1)^n$), to the intersection index in $V_\lambda(\alpha+i\cdot\varepsilon)$ of the cycle $J_\alpha^{1/2}(\partial\Gamma)$ obtained from $\partial\Gamma$ by transportation over a half of the arc (18) and the small sphere $\Delta_\alpha^{\text{Im}}$ that appears in the intersection of the plane (24) with $V_\lambda(\alpha+i\cdot\varepsilon)$ and is oriented by means of a differential form ω such that the orientation $\{d\operatorname{Im} f_\lambda \wedge \omega\}$ of the linear span of vectors (24) coincides with the orientation given by this sequence of vectors. But

$$\langle J_\alpha^{1/2}(\partial\Gamma), \Delta_\alpha^{\mathrm{Im}}\rangle = \langle \partial\Gamma, (J_\alpha^{1/2})^{-1}\Delta_\alpha^{\mathrm{Im}}\rangle,$$

and, since it is easy to calculate that

$$(J_\alpha^{1/2})^{-1}\Delta_\alpha^{\mathrm{Im}} = (-1)^{n-r}\Delta_\alpha^+,$$

Theorem 2 is proved.

The formulae (6), (8) reduce to Theorems 1, 2 in the same way as the formulae (5), (7) were reduced in the last paragraph of § 3.2 to Theorems 3.1, 3.2.

The last two assertions of Theorem 1.3, concerning the intersection indices $\langle \Delta_\alpha, \Pi_{\mathrm{odd}}\rangle$, $\langle \Delta_{\bar\alpha}, \Pi_{\mathrm{odd}}\rangle$ for imaginary α, follow from similar assertions about Π_{ev} (proved in § 3) and from the following theorem.

Theorem 3. *Let α be a nonreal critical value of f_λ, and $\Delta_\alpha(a) \subset V_\lambda(a)$ the corresponding vanishing cycle, defined by a path of the distinguished basis from § 1.6; let $\delta_\alpha(a)$ be the corresponding thimble (see Chapter I, § 3.6). Then*

$$\langle \delta_\alpha(a), b(a)\rangle = (-1)^{n-1}\langle \Delta_\alpha(a), \Pi_{\mathrm{ev}}(a)\rangle \quad \text{if } \mathrm{Im}\,\alpha > 0,$$

$$\langle \delta_\alpha(a), b(a)\rangle = (-1)^{n}\langle \Delta_\alpha(a), \Pi_{\mathrm{ev}}(a)\rangle \quad \text{if } \mathrm{Im}\,\alpha > 0.$$

Since $b(a) = b_+(a) + b_-(a) \sim \tilde b_+ + \tilde b_-$ (see the proof of Theorem 1), and the cones $\delta_\alpha(a)$ corresponding to the upper (respectively, lower) half-plane do not intersect $\tilde b_-$ (respectively, $\tilde b_+$), this theorem follows immediately from Proposition I.3.7 and from the following statement.

Proposition 1. $t(\Pi_{\mathrm{ev}}(a)) = -b_+(a) + b_-(a) \sim -\tilde b_+ + \tilde b_-$.

This proposition follows immediately from the definition of the cycles $b_+(a)$, $b_-(a)$.

Remark. This proposition allows us to re-prove the results of § 3 about the formulae (5), (7) in the same way as we have done in the present section for the formulae (6), (8).

Theorem 3 is completely proved.

§ 5. Stabilization of the local Petrovskii classes. Proof of Theorem 1.5

The equality of the absolute values of all the four intersection indices, placed one above the other in the statement of this theorem, follows from Theorems 1.3 and

4.3. Thus we only need to prove the two equalities from the upper row. We shall demonstrate the proof of only the second of them; the first one is proved in exactly the same way. We suppose that $\operatorname{Im}\alpha > 0$. Let $\{z(\tau),\ \tau \in [0,1]\}$ be the path in $\mathbf{C}^1$ that connects 0 with α and defines the vanishing cycle Δ_α, in particular, $z(0) = 0, z(1) = \alpha$, and $\operatorname{Im} z(\tau) > 0$ for $\tau \neq 0$. Without loss of generality we shall assume that $z(\tau) = i\tau$ for sufficiently small τ.

The cycle $\Delta_\alpha^{0,1} \subset V_\lambda^{0,1}(0)$ can be realized as follows. The cycle $\Delta_\alpha \subset V_\lambda(0)$ is the final position of a family of cycles $\Delta_\alpha(\tau) \subset V_\lambda(z(\tau)) \equiv f_\lambda^{-1}(z(\tau)) \cap B$. Define a cycle $\Delta_\alpha^{0,1} \in \mathbf{C}^{n+1}$ as the set of pairs $\{x \in \mathbf{C}^n, y \in \mathbf{C}^1\}$ such that there exists $\tau \in [0,1]$ such that $x \in \Delta_\alpha(\tau)$ and $y^2 = z(\tau)$.

Obviously this cycle lies in $V_\lambda^{0,1}$; it is easy to see that it vanishes over the critical value α of the function $f^{0,1}$ after transportation over the paths $\{z(\tau)\}$. The intersection of $\Delta^{0,1}$ with the plane $\mathbf{C}^n \equiv \{x, y | y = 0\}$ coincides with the cycle Δ_α, and close to this plane the cycle $\Delta^{0,1}$ is diffeomorphic to the direct product of Δ_α and the interval $[-\epsilon e^{\pi i/4}, \epsilon e^{\pi i/4}]$, where the projection onto the second factor is given by the coordinate function y. Let us orient $\Delta_\alpha^{0,1}$ in such a way that in this neighbourhood the orientation coincides with the natural orientation of the direct product (where the segment $[-\epsilon e^{\pi i/4}, \epsilon e^{\pi i/4}]$ is oriented from $-\epsilon$ to ϵ). The set of real points $\Pi_{\text{ev}}^{0,1} \subset V_\lambda^{0,1}$ close to $\mathbf{C}^n$ is also a direct product $\Pi_{\text{ev}} \times [-\epsilon, \epsilon]$, where the obvious direct product orientation on this manifold is compatible with the orientations of the cycles Π_{ev} and $\Pi_{\text{ev}}^{0,1}$ specified in the definition of Petrovskii classes; see § 1.

The cycles $\Delta_\alpha^{0,1}$ and $\Pi_{\text{ev}}^{0,1}$ intersect at the point $(x, y) \in \mathbf{C}^{n+1}$ if and only if $y = 0$ and x is an intersection point of Δ_α and ∇. It is easy to calculate that the local intersection indices of these pairs of cycles at the corresponding points differ by the constant factor $(-1)^n$, and Theorem 1.5 is proved.

§ 6. Local lacunas close to simple singularities

In this section we prove that all the formal lacunas in the bases of versal deformations of simple singularities are exhausted by those demonstrated in § 2, and that the three formal lacunas for the singularity D_4^- (respectively, two lacunas for D_{2k}^-, $k > 2$) indicated there are different in the case of even n and odd i_+.

Throughout this section we shall assume that the simple singularity is given by the corresponding normal form from Table 1, and its versal deformation is the corresponding quasihomogeneous deformation from Table 3. According to the Remark 1 of Chapter I, § 5, we shall assume that all our functions f_λ are defined everywhere in $\mathbf{C}^n$ (and not in a small neighbourhood of the origin) and for all (not only sufficiently small) λ.

ε is everywhere a sufficiently small positive number.

6.1. Characterizing functions of the components of the complement of the discriminant variety of a simple singularity.

Consider any such polynomial function f with simple singularity at 0. Let μ be its Milnor number, and denote by $\mathbf{R}^\mu$ the base of the corresponding versal deformation F from Table 3.

Lemma 1. *In any component of the complement of the discriminant $\Sigma \equiv \Sigma(F)$ in the parameter space $\mathbf{R}^\mu$ there is a λ such that the function f_λ is Morse and has only real critical points.*

Proof. First of all, in any component of the complement of Σ there is a morsification f_λ having at least one critical point. Indeed, any component has a boundary that belongs to Σ. The generic points of the boundary are the functions having exactly one Morse real singular point with zero critical value: all the more complicated discriminant functions form a set of codimension 2 and cannot separate our component from the others. Close to such a generic discriminant point f'_λ the discriminant is smooth, and its complement consists of exactly two components (one of which is ours). One of these components contains the function $f'_\lambda + \varepsilon$, and the other contains the function $f'_\lambda - \varepsilon$; both these functions have at least one real critical point, and the lemma is proved.

Let f_λ be some nondiscriminant strictly Morse function with at least one real critical point, and suppose that it has at least one pair of complex conjugate critical points with critical values α and $\bar{\alpha}$. Let $a \neq 0$ be any real number that is not a critical value of f_λ. Consider any distinguished system of paths connecting a with all critical values of f_λ and satisfying the conditions (i), (ii) of § 1.6. Suppose that the intersection index $\langle \Delta_\alpha(a), \Delta_{\bar{\alpha}}(a)\rangle$ of the corresponding vanishing cycles in $V_\lambda(a)$ is equal to 1 or -1. Then the path in $\mathbf{R}^\mu$ described in § 3.4 in the proof of formula (9) joins the function f_λ to a function which lies in the same component of the complement of the discriminant and has only one non-Morse critical point of type A_2 at a real point. An appropriate small real perturbation of this function splits this critical point into two real points; the number of real critical points of the Morse function thus obtained is more than that for the initial morsification f_λ.

Suppose now that $\langle \Delta_\alpha(a), \Delta_{\bar{\alpha}}(a)\rangle = 0$ for all nonreal α. Let c be the real critical value closest to a such that for some nonreal critical value α we have $\langle \Delta_\alpha(a), \Delta_c(a)\rangle \neq 0$ (and hence, by Corollary I.6.2, the numbers $|\langle \Delta_\alpha(a), \Delta_c(a)\rangle|$ and $|\langle \Delta_{\bar{\alpha}}(a), \Delta_c(a)\rangle|$ are equal to 1). (Such a c exists because the Dynkin diagram is connected; see Proposition 4 of Chapter I, § 3.) Set $c' = c + \varepsilon$ if $c > a$ and $c' = c - \varepsilon$ if $c < a$. Let us replace a by c', and replace the paths of the distinguished system that defined the elements $\Delta_\alpha(a)$, $\Delta_{\bar{\alpha}}(a)$ by the paths obtained from them by adding the complex conjugate paths connecting a with c' in the upper and lower half-planes and going along the real axis. It follows from the Picard–Lefschetz formula that in this case the intersection index $\langle \Delta_\alpha(c'), \Delta_\delta(c') \rangle$ of the corresponding basis elements is equal to 1 or -1, and we get the situation considered above.

Definition 1. A real function f with a critical point at 0 satisfies the *even Petrovskii plus-condition* (respectively, *even minus-condition*) if the local even Petrovskii class corresponding to the function $f - \varepsilon$ (respectively, $f + \varepsilon$) is homologous to zero, or, which is the same, the class $\Pi_{\text{ev}}(\varepsilon) \in \tilde{H}_{n-1}(V_\lambda(\varepsilon), \partial V_\lambda(\varepsilon))$ (respectively, $\Pi_{\text{ev}}(-\varepsilon) \in \tilde{H}_{n-1}(V_\lambda(-\varepsilon), \partial V_\lambda(-\varepsilon)))$ defined in § 3.1 is trivial. The definition of *odd* Petrovskii plus- and minus-conditions is obtained from this one by replacing all even Petrovskii classes by odd ones.

Lemma 2. *The Petrovskii plus- (or minus-) condition is satisfied or not satisfied simultaneously for all function-germs from any class of bistable equivalence.*

This lemma follows immediately from Corollary 1.2.

Let $f_\lambda \in \mathbf{R}^\mu - \Sigma$ be a morsification of some simple singularity that belongs to the corresponding deformation from Table 3 and is such that all μ critical points of f_λ are real. Using the Lyashko–Looijenga covering, we can deform f_λ inside the same component of the complement of Σ to a function $\tilde{f}$ having either two critical values, 1 and -1, or only one of them in such a way that all the intermediate functions $f_{\lambda(t)}$, $t \in (0, 1)$, of this deformation except for the last one are Morse and have only real critical points, and $f_{\lambda(0)} \equiv f_\lambda$. At the last instant $t = 1$ of this move some critical points can coincide; since any simple singularity can split only into simple ones, all the critical points of the function $\tilde{f} \equiv f_{\lambda(t)}$ will be simple (and real).

Definition 2. The function $\tilde{f}$ thus obtained is called the *characterizing function* of the original morsification f_λ (and of the entire component of the set of morsifications of f that contains f_λ).

Denote by $\Xi_i^+(\tilde{f})$ and $\Xi_j^-(\tilde{f})$ the classes of bistable equivalence of all critical points with zero critical value of the functions $\tilde{f} - 1$ ($\tilde{f} + 1$, respectively).

Consider the Dynkin diagram of the morsification f_λ defined by the distinguished system of paths of the form indicated in § 1.6.

Lemma 3. *The collection of Dynkin diagrams of the singularities $\Xi_i^+(\tilde{f})$ and $\Xi_j^-(\tilde{f})$ coincides with the collection of connected graphs obtained from this Dynkin diagram by removing all edges connecting the nodes corresponding to critical values of different signs.*

This is proved essentially in [Lyashko 76] (see the proof of Theorem 2 there).

Theorem 1. *The component of the complement of Σ that contains $\tilde{f}$ is an (even or odd) formal lacuna if and only if for all singularities $\Xi_i^+(\tilde{f})$ the Petrovskii* *minus-condition (of the same parity) is satisfied, and all singularities* $\Xi_j^-(\tilde{f})$ *satisfy the plus-condition.*

This theorem follows immediately from Theorem 1.3 applied to any small morsification of $\tilde{f}$ having only real critical points.

Thus we have to study simple singularities satisfying the Petrovskii plus- and minus-conditions, and then to determine which singularities can be split into these in a proper way.

Theorem 2. *Among the simple function singularities the even Petrovskii minus- (respectively, plus-) condition is satisfied only by functions that are bistably equivalent to functions having a minimum (respectively, maximum) point at 0, namely, by functions bistably equivalent to the functions* $x^{2k} + y^2$ *(respectively,* $-x^{2k} - y^2$*) if* n *is even and to* x^{2k} *(respectively,* $-x^{2k}$*) if* n *is odd.*

Theorem 3. *Among the simple function singularities the odd Petrovskii minus- (respectively, plus-) condition is satisfied only by functions bistably equivalent to functions of the form* $\phi(x,y) - z^2$ *(respectively,* $\phi(x,y) + z^2$*), where* ϕ *has a minimum (respectively, maximum) point at 0: namely, by functions bistably equivalent to the functions* $x^{2k} + y^2 - z^2$ *(respectively,* $-x^{2k} - y^2 + z^2$*) if* n *is odd, and to* $x^{2k} - y^2$ *(respectively,* $-x^{2k} + y^2$*) if* n *is even.*

The proofs of these two theorems follow immediately from Theorems I.4.3 and 1.3 and the Corollary of Theorem 1.5.

6.2. Even formal lacunas for simple singularities.

Everywhere in this subsection the number of variables n is assumed to be even. If n is odd, then the Corollary of Theorem 1.5 reduces the problem of enumeration of even lacunas to a similar problem on odd lacunas for a function of $n+2$ variables, which will be solved in subsection 6.3 below.

Proposition 1. *If the singularity of a function* f *at 0 is simple but is not bistably equivalent to a singularity of the form* $\pm(x_1^{2k} + y^2)$ *or* $\pm(x^2y - y^{2k-1} + z^2 - w^2)$*, then the space of its versal deformation does not contain even formal lacunas.*

Proof. For any perturbation f_λ of an isolated real singular point 0, the sum of the indices of the vector-field grad f_λ over all real critical points close to 0 is equal to the index of the field grad f at the point 0. Therefore it follows from Theorem 2 and from the fact that the coranks of all simple singularities are ≤ 2 that if L is a formal lacuna in $\mathbf{R}^\mu - \Sigma$, and $\tilde{f}$ any characterizing function of it, then the number of critical points of $\tilde{f}$ is equal to the index of grad f. But the indices of all simple singularities belong to the segment $[-2, 2]$. Rejecting all singularities whose indices

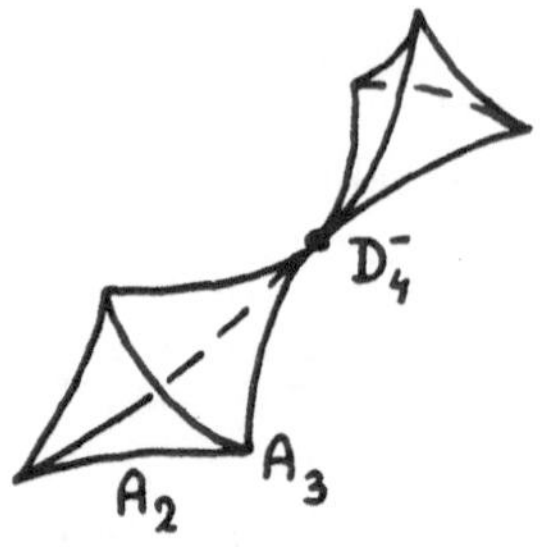

Fig. 50. Pyramid: the caustic for the singularity D_4^-

are equal to $-2, -1, 0$, and also the singularities of index $+1$ that are not bistably equivalent to an extremum, we get the desired result; thus we have justified all zeros in columns 2, 3 of Table 12.

Some formal lacunas for the remaining singularities were demonstrated in § 2. We only need to prove that these singularities have no extra lacunas.

For the Morse singularities A_1 everything is wellknown; see for instance [ABG 73], [Gårding 72], [Leray 62], [Borovikov 59] etc. These results imply our assertion.

6.2.1. Functions A_{2k-1} bistably equivalent to the functions $\pm(x_1^{2k}+y^2)$, $k > 1$.

In this case $\operatorname{ind}\operatorname{grad} f = 1$, and hence the characterizing function $\tilde{f}$ has a unique critical point whose Milnor number is the same as for f. It follows from the simplicity of the singularity f that all such functions of our versal deformation (see Table 3) have the form $f + \lambda_0$ and belong to Σ if and only if $\lambda_0 = 0$. By Theorem 2, the desired function can lie on this line only on one side of f, and we get the assertion of Theorem 2.1 for functions of type A_{2k-1}.

6.2.2. Functions of class D_4^- bistably equivalent to singularities of the form $\pm(x^2y - y^3 + z^2 - w^2)$.

For these singularities we shall prove the following statement in addition to the assertions of Theorem 2.1.

Proposition 2. *For any real strictly Morse function f_λ that is a perturbation of the above-mentioned function of class D_4^- and is of the form shown in Table 3, the set of all constants c such that the function $f_\lambda - c$ belongs to an even formal lacuna is a nonempty interval bounded from both sides.*

Fig. 51. Sliced pyramid and Maxwell set

Proof. In the case D_4^-, the versal deformation from Table 3 depends on four parameters $\lambda_0, \ldots, \lambda_3$. Rejecting the parameter λ_0, we get a three-parameter family of functions; obviously it is sufficient to prove the assertion of Proposition 2 for functions of this family only. The set of those of them that have at least one non-Morse point is shaped as shown in Figure 50, is homeomorphic to the standard cone $x^2 + y^2 = x^2$, and is called a *pyramid*; see for example [AVGL 89]. The transversal slice of this cone is shown in Figure 51. The dashed line on this figure shows the *Maxwell set*, i.e., the set of Morse but not strictly Morse functions having several critical points on the same critical level. The interior points of the curvilinear triangle correspond to functions having three saddlepoints and one minimum (respectively, maximum) point, depending on the sliced half of the pyramid; the critical value at this point is smaller (respectively, greater) than all values at the saddlepoints. See also Figure 39.

Consider the case when this is a minimum point (the case of a maximum point can be considered in exactly the same way). Then the edges of the triangle correspond to functions $\varphi_\lambda(x, y)$ having a critical point of class A_3, and, more exactly, the variant of this class consisting of singularities which, after direct summation with the form $z^2 - w^2$ of additional variables, satisfies the (even) Petrovskii plus-condition. Suppose that we go out from the triangle through a smooth piece of its boundary. At the instant of going out, one saddlepoint and the minimum coalesce and go into the imaginary domain.

Let the parameter $\lambda = (\lambda_0, \lambda_1, \lambda_2)$ corresponding to a strictly Morse function belong to this slice. It follows immediately from Theorem 1.3 that $\varphi_\lambda - c$ lies in an even lacuna if and only if c satisfies the following conditions.

If λ belongs to the interior part of the triangle, then c must lie in the interval between two maximal critical values (between the minimal ones if we consider the slice containing functions with one maximum point).

If λ lies outside the triangle, then φ_λ has exactly two critical values, and c must lie between them.

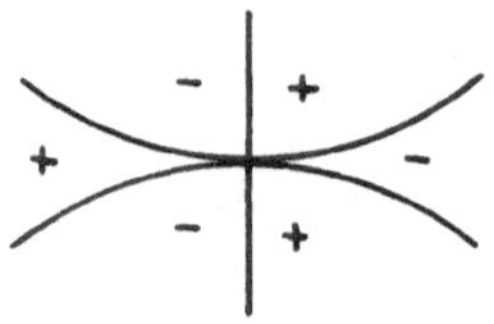

Fig. 52. Zero level set of a singularity D_{2k}^-, $k > 2$

This proves Proposition 2.

When λ goes over a segment of the Maxwell set, which lies in Figure 51 between an edge and the centre of the triangle, such an interval of admissible values of c does not vanish, because on this segment the second and third critical values become equal, and these values lie on the same side of our interval. But when we traverse the other part of the Maxwell set (which is represented in Figure 51 by three infinite rays issuing from the centre of the triangle), then our interval collapses, and hence in the space of the whole versal deformation from Table 3 there exists no curve lying in an even formal lacuna in such a way that its projection along the axis λ_0 traverses this part of the Maxwell set. Since the pair $\{\mathbf{R}^3;$ the closure of this part $\}$ is homeomorphic to the direct product of its slice (shown in Figure 51) and a line, Theorem 2.1 is proved in the case of singularities D_4^- and even n.

6.2.3. Singularities D_{2k}^-, $k > 2$.

Let $\varphi = x_1^2 x_2 - x_2^{2k-1}$, and let $\{\varphi_\lambda\}$ be the versal deformation of φ indicated in Table 3. The set $\{\varphi = 0\}$ is shown in Figure 52. During the proof of Theorem 2.1 for the singularities D_{2k}^- we shall say for simplicity that a subset $L \subset \{\varphi_\lambda\}$ is an (even) formal lacuna if in the space of the stabilized versal deformation $\{f_\lambda \equiv \varphi_\lambda + x_3^2 - x_4^2\}$ of the function f the set of functions $\{f_\lambda | \lambda \in L\}$ is an even formal lacuna.

Let L be a formal lacuna in $\{\varphi_\lambda\}$, and $\tilde{\varphi}$ an arbitrary characterizing function of it; see Definition 2.

Since $\operatorname{ind}\operatorname{grad} f = 2$, $\tilde{\varphi}$ has exactly two critical points; it follows from the connectedness of the Dynkin diagrams that its critical values are distinct (and hence are equal to 1 and -1). According to Theorems 1 and 2, close to one of them $\tilde{\varphi}$ is D-equivalent to the function $x^{2l} - y^2 + 1$, and close to the other to the function $-x^{2s} + y^2 - 1$, where $l + s = k + 1$. By the definition of versal deformations, there is a function φ_1 arbitrarily close to $\tilde{\varphi}$ in the space of the versal deformation that has l saddlepoints with critical value 1, $l - 1$ maxima close to these saddlepoints with critical value $1 + \eta$ (where $0 < \eta << 1$), s saddlepoints with value -1 and $s - 1$

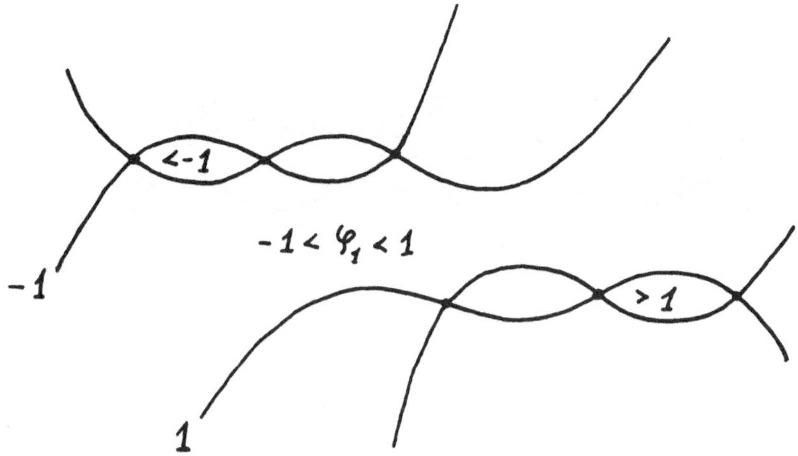

Fig. 53. Topology of level sets of characterizing functions for the singularity D_{2k}^{-} ($k = 5$)

minima with value $-1 - \eta$. The topology of the mutual disposition of the level sets $\{\varphi_1 = \pm 1\}$ is as shown in Figure 53 in the special case $l = s = 3$: this follows from the geometry of the level sets $\tilde{\varphi} = \pm 1$ (see Figure 52) and from the fact that the set $\{x \in \mathbf{R}^2 | \varphi(x) = 0\}$ (and hence also the set $\{x \in \mathbf{R}^2 | \varphi(x) = 0\}$) goes to infinity along exactly 6 branches. Using the Lyashko–Looijenga covering, we can include φ_1 in a one-parameter family of functions φ_τ, $\tau \in [0, 1]$, having only the critical values $\tau + \eta, \tau, -\tau, -\tau - \eta$. As τ tends to 0, to any self-intersection point of the set $\{\varphi_\tau = -\tau\}$ there tends a smooth branch of the line $\{\varphi_\tau = \tau\}$ and vice versa.

Proposition 3. *The number of self-intersection points of the set $\{\varphi_\tau = -\tau\}$ to which there tend the components of the set $\{\varphi_\tau = \tau\}$ that bound a maximum point is equal to 1.*

Proof. Suppose that the number of such self-intersection points is greater than one. Using the Lyashko–Looijenga covering, we can move the function φ_τ inside our versal deformation in such a way that during this movement the values of the functions at almost all critical points will be constant, and only the critical values corresponding to these self-intersection points move from $-\tau$ to $+\tau/2$. Consider the Dynkin diagram of φ calculated by the methods of Chapter I, § 4, and forget all its vertices corresponding to the negative critical values of the resulting function, together with the segments entering these vertices. It is easy to calculate that the resulting graph is connected. Move this function in such a way that all negative critical values stay unchanged, and the positive ones attract to the value τ. Then

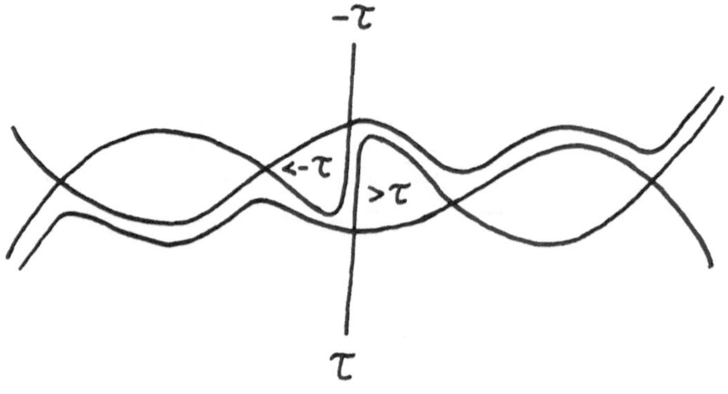

Fig. 54.

all the critical points corresponding to the latter values coalesce together in a unique critical point (cf. [Lyashko 76]) and the local index of the gradient field of this critical point does not exceed -3. This is impossible, because the decompositions of simple singularities can give only simple singularities, and the absolute values of indices of simple singularities cannot exceed 2.

The number of the desired self-intersection points of the set $\{\varphi_\tau = -\tau\}$ cannot be equal to 0, otherwise we see by the methods of Chapter I, § 4, that the Dynkin diagram of the singularity φ is not connected; Proposition 3 is proved.

Permuting the sets $\{\varphi_\tau = -\tau\}$ and $\{\varphi_\tau = \tau\}$ in the statement of this Proposition, we deduce that immediately before the coalescing the union of these two sets is shaped as shown in Figure 54, and hence the set $\{\varphi_0 = 0\}$ is homeomorphic to the one from Figure 40a with appropriate k.

Proposition 4. *In the base of the versal deformation of φ there is at least one formal lacuna.*

Indeed, this lacuna is shown in Figure 40a.

Proposition 5. *The number of formal lacunas in the base of the versal deformation of φ is even.*

Proof. The base of the deformation indicated in Table 3 has an involution, induced by the reflection of the space of arguments with respect to the line $\{x_1 = 0\}$; this involution acts on $\mathbf{R}^\mu$ by changing the sign of the parameter λ_1. The fixed set of this involution is the hyperplane given by $\lambda_1 = 0$. Thus we only need to prove that this plane does not intersect the lacunas. In any component of the intersection of the plane $\{\lambda_1 = 0\}$ with the complement of Σ there is a characterizing function: indeed, the homotopy of functions used in the proof of the existence of characterizing functions is invariant under the reflection in this plane. If this component belongs to a formal lacuna, then this characterizing function must again have a point of type A_{2l-1} and a point of type A_{2s-1}, where $l + s = k + 1$. Elementary calculations show that all functions of this versal deformation with $\lambda_1 = 0$ and a pair of such critical points are of the form

$$x_1^2 x_2 - (x_2 + (2k-2)t^{2k-2}(x_2 - t) + c,$$

where t and c are parameters. If $t > 0$, then such a characterizing function has points close to which it is diffeomorphic to $x^{2k-2} - y^2 - 1$ and $x^4 - y^2 + 1$; if $t < 0$, then it has points of types $-x^{2k-2} + y^2 + 1$ and $-x^4 + y^2 - 1$. Such combinations of singular points are not satisfactory for us (see Theorem 2), and Proposition 5 is proved.

Proposition 6. *The number of formal lacunas in the base of the versal deformation of φ does not exceed 2.*

Proof. Above we have proved that any formal lacuna of our deformation contains a function φ_τ such that the picture of its real level sets is topologically equivalent to the ones shown in Figsures 53, 54. Let us check which of these functions lie in the same lacuna. We say that our function φ_τ is of type (l, s) if it has $l - 1$ maxima and $s - 1$ minima: for instance, in Figure 54 a function of type (3,3) is shown. Suppose that $l > 1$. As above, let us move φ_τ in the class of real functions in such a way that the critical values of almost all critical points stay unchanged, and only the critical value corresponding to the extreme left self-intersection point of the set $\{\varphi_\tau = \tau\}$ in Figure 54, and the critical value $\tau + \eta$ corresponding to its neighbouring maximum, coalesce in a point with critical value $\tau + \eta/2$, come out into the complex domain, and go there along the paths connecting the points $\tau+\eta/2$ and $\tau - \eta/2$ in the half-planes $\{\mathrm{Im} > 0\}$ and $\{\mathrm{Im} < 0\}$. The intersection matrix of vanishing cycles of the limit sabirization shown in Figure 40a (and hence also of the neighbouring morsification φ_τ) can be calculated by the method of Chapter I, § 4. This calculation together with the Picard-Lefschetz formula implies that the critical points corresponding to the two critical values that we move coalesce at the final instant of this homotopy, and we get two new real critical points whose values will be close to $-\tau - \eta/2$. As above, we move these critical values to $-\tau$ and $-\tau - \eta$ respectively and get a function φ' of type $(l - 1, s + 1)$. Iterating this process, we deduce, in particular, that in our formal lacuna there is a characterizing function $\tilde{\varphi}$ having a critical point of type A_{2k-1}, close to which $\tilde{\varphi}$ is D-equivalent to the function

$$-x^{2k} + y^2 + 1. \qquad (27)$$

As in part B_2 of this subsection, consider the deformation of the original singularity $\varphi \in D_{2k}^-$ obtained from the one indicated in Table 3 by forgetting the parameter

λ_0. In the space of this truncated deformation consider the values of the parameter $\lambda' = (\lambda_1, \ldots, \lambda_{2k-1})$ such that the function $\varphi_{\lambda'}$ has a point equivalent to (27) (up to addition of a constant). It is easy to calculate (see for example [Vassiliev 88 & 93]) that this set belongs to two smooth curves passing through the origin, and on either of these curves it occupies a half of it; here it is essential that $k > 2$. For any function $\varphi_{\lambda'}$ that belongs to such a half-curve, the set of values λ_0 such that the function $\varphi_{\lambda'} + \lambda_0 + z^2 - w^2$ satisfies the even local Petrovskiĭ condition forms an interval. We have proved that in the space of the deformation from Table 3 there is a set having exactly two connected components and such that any formal lacuna contains at least one of these components. Proposition 6 is completely proved, and hence also the assertions of Theorem 2.1 concerning the even formal lacunas.

6.3. Odd formal lacunas for simple singularities.

Here we enumerate all odd lacunas for singularities of an odd number of variables. If n is even, then the last statement of Corollary 1.2 reduces the similar problem to the enumeration of even lacunas for a singularity of an even number $n+2$ of variables, which was accomplished in the previous subsection 6.2.

By Corollary 1.2 and Theorem 1.6 it is sufficient to consider only the functions of the form

$$f_\lambda \equiv \varphi_\lambda(z_1, z_2) + z_3^2,$$

where $\{\varphi_\lambda\}$ is a versal deformation of a simple singularity φ.

Definition 3. A polynomial $\tilde{\varphi}(x_1, x_2)$ is called *two-index* if it has only real critical points and at all these points it is D-equivalent either to $x^2 + y^2 + c$, $c < 0$, or to $x^{2k} - y^2 + c$, $c > 0$ (where k and c depend on these points).

Proposition 7. *Suppose that the morsification $f_\lambda = \varphi_\lambda + x_3^2$ of a simple real singularity $f(x_1, x_2, x_3) \equiv \varphi(x_1, x_2) + x_3^2$ belongs to the deformation of $f \equiv f_0$ given in Table 3 and belongs to a formal lacuna, and $\tilde{f} \equiv \tilde{\varphi} + x_3^2$ is the characterizing function corresponding to this morsification. Then either $\tilde{\varphi}$ has only one critical point and is D-equivalent to the function $-x^{2k} - y^2 - 1$ at it (and hence, similarly to part B_1, φ is D-equivalent to $-x^{2k} - y^2$) or $\tilde{\varphi}$ is a two-index polynomial.*

Proof. It follows from Theorems 1 and 3 that at any critical point with critical value 1 $\tilde{\varphi}$ is D-equivalent to $x^{2k} - y^2 + 1$, and at any critical point with value -1 it is D-equivalent to either $-x^{2k} - y^2 - 1$ or $x^2 + y^2 - 1$. We need to prove that if there is at least one point D-equivalent to $-x^{2k} - y^2 - 1$, then $\tilde{\varphi}$ has no other critical points. Suppose this is false. Let us slightly perturb all critical points in such a way that all points of type $x^2 + y^2 - 1$ of the resulting function φ_1 become critical points with critical value $-1 - \varepsilon$, all points of type $-x^{2k} - y^2 - 1$ split into $k-1$ saddlepoints with critical value -1 and k maxima with value $-1+\varepsilon$, and all points of type $x^{2k}-y^2+1$ split into k saddlepoints with value 1 and $k-1$ maxima with value $1+\varepsilon$. Using the Lyashko–Looijenga covering, let us try to include this function φ_1 in a continuous family of functions φ_t, $t \in (0,1]$, such that the critical points of these five groups at the instant t become critical points with values $-t-\varepsilon$, $-t$, $-t+\varepsilon$, t, $t+\varepsilon$ respectively. The only obstruction to such a movement could consist in the fact that some of the critical points of the function φ_t with critical value $-t+\varepsilon$ do not miss the points having critical values t at the instant $t=\varepsilon/2$. This situation is impossible, because the former points are the maxima, and the latter are the saddlepoints. At the final instant $t=0$ two groups of points, the third and the fifth ones, get the critical value ε, and the points of the second and the fourth groups get the value 0. But all the points of groups 3 and 5 are maxima, and the points of groups 2 and 4 are saddlepoints, hence none of them coalesce at the instant $t=0$, and the function φ_0 is a sabirization of φ. Then any group of critical points appearing from a critical point of $\tilde{\varphi}$, D-equivalent to $-x^{2k}-y^2-1$, is not connected by separatrices with other critical points. Hence, by Theorem I.4.2, the Dynkin diagram of the function φ is not connected, a contradiction.

It remains to count all two-index characterizing functions that can appear in the versal deformations of simple singularities. First, we count all two-index sabirizations of simple singularities, and then reduce the general problem to this one.

Theorem 4. *If the real function $\varphi(x_1,x_2)$ has a simple singularity at 0, then*

1) in the base of its versal deformation $\{\varphi_\lambda\}$ there is at most one connected component of the complement of Σ that contains the two-index functions, all of whose critical points are Morse;

2) there is one such component if and only if φ is D-equivalent at 0 to one of the polynomials

$$\pm x^{k+1}+y^2 \in A_k,\ k\geq 1, \quad x^2y-y^{2k-1}\in D_{2k}^-, \quad x^2y+y^{2k}\in +D_{2k+1},$$

$$x^3+y^4\in +E_6, \quad x^3+xy^3\in E_7, \quad x^3+y^5\in E_8;$$

3) such a component consists entirely of two-index functions having only Morse critical points;

4) the critical points of the functions of this component admit an ordering depending continously on the points of this component and such that the map of this component into $\mathbf{R}^\mu$, defined by this ordering and sending any function into the set of its (ordered) critical values, is a diffeomorphism of this component onto an open coordinate octant in $\mathbf{R}^\mu$. In particular, this component contains only one characterizing function.

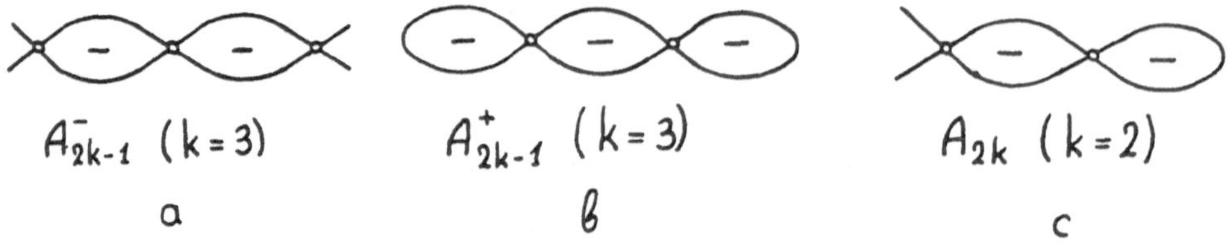

Fig. 55. Two-index morsifications of singularities A_μ

Proof. For a function to go out from the set of Morse two-index functions, two of its critical points of different indices must meet together. In the case of two-index Morse functions, the critical values of such critical points are separated by the number 0, which is prohibited for the critical values of functions from the complement of Σ. This proves assertion 3.

The two-index morsifications of all singularities indicated in assertion 2 are shown in Figures 55, 39b, 40b, 42a, 43a, 43c, 43d (recall that the white circles at the intersection points of all these pictures mean that the corresponding critical value 0 should be slightly increased). Let us prove that all two-index morsifications of simple singularities lie in the same components of the complements of discriminants.

Lemma 4. *If $\tilde{\varphi}$ is a two-index morsification of a simple singularity $\varphi(x_1, x_2)$, and $\tilde{f} \equiv \tilde{\varphi} + x_3^2$, then the involution of complex conjugation acts trivially on the homology of the corresponding nonsingular level manifold $\tilde{V} \equiv \tilde{f}^{-1}(0) \cap B_\varepsilon$.*

Indeed, all the corresponding vanishing cycles $\Delta_c^+(0)$, $c < 0$, and $\Delta_c(0)$, $c > 0$, generating the group $\tilde{H}_2(\tilde{V})$ and constructed in § 3.1, are mapped by this involution into themselves, and this map preserves their orientations.

Now assertions 1 and 2 of Theorem 4 follow from Theorem I.7.2 and Proposition I.7.3.

Finally, assertion 4 follows from the contractibility of any component of the complement of the discriminant of a simple singularity (see Theorem I.7.1), assertions 1 and 3, and Proposition I.6.2 about the Lyashko–Looijenga covering. Theorem 4 is completely proved.

Let us return to the case of a general (maybe, not Morse) two-index characterizing function $\tilde{\varphi}$.

Lemma 5. *In arbitrarily small neighbourhood of the function $\tilde{\varphi}$ there is a function $\tilde{\varphi}'$ which is obtained from $\tilde{\varphi}$ by decomposition of any point of it that is D-equivalent to $x^{2k} - y^2 + 1$ for some k into k saddlepoints and $k - 1$ maxima, such*

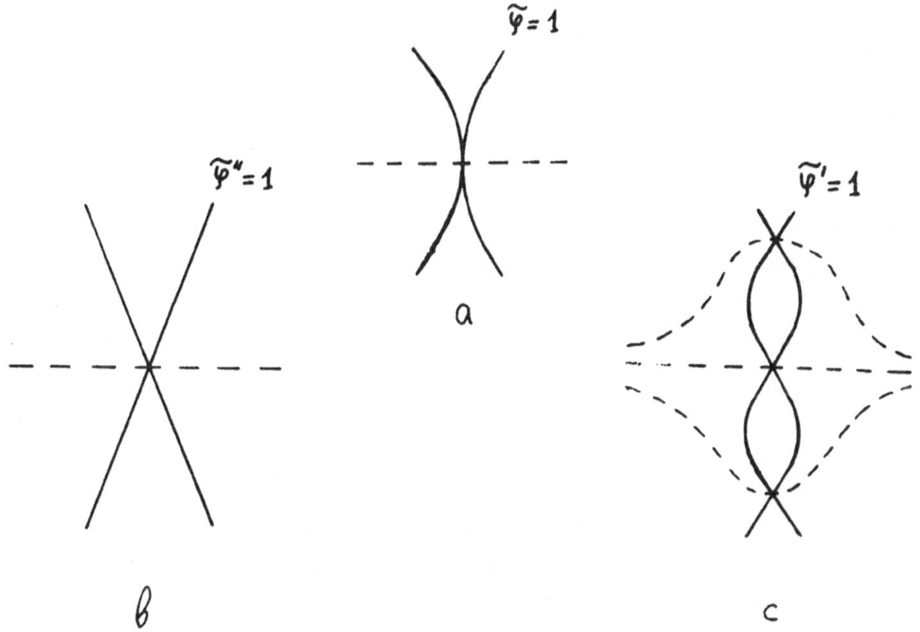

Fig. 56. Perturbations of non-Morse points of two-index functions

that all saddlepoints obtained by decompositions of all these critical points lie on the level set $\tilde{\varphi}' = 1$, *and all the maxima on a slightly higher level.*

(Such a decomposition of a critical point of $\tilde{\varphi}$ is illustrated by passing from Figure 56a to Figure 56b.)

Indeed, by the versality of our deformation $\{\varphi_\lambda\}$ all non-Morse critical points of $\tilde{\varphi}$ may be morsovized (and even sabirized) independently; using additionally the Lyashko–Looijenga covering, we ensure the desired critical values.

Simultaneously we consider one more perturbation $\tilde{\varphi}''$ of the function $\tilde{\varphi}$, such that any critical point, D-equivalent to $x^{2k} - y^2 + 1$ for some k, splits into $2k - 2$ nonreal critical points with nonreal values and one real saddlepoint on the level $\tilde{\varphi}'' = 1$; see Figure 56c. This perturbation exists by the same arguments as $\tilde{\varphi}'$. The function $\tilde{\varphi}' - 1$ is a sabirization of φ. Consider the real separatrix diagram of this function, i.e., the graph whose vertices correspond to critical points of $\tilde{\varphi}' - 1$, and any two vertices, one of which corresponds to a saddlepoint and the other to an extremum, may be connected by segments that are in one-to-one correspondence with the separatrices of the gradient vector-field connecting these critical points. This diagram is connected (otherwise the Dynkin graph of the singularity φ computed in terms of this diagram will be disconnected). Comparing the real separatrix diagrams of the functions $\tilde{\varphi}'$ and $\tilde{\varphi}''$ (see the dashed lines in Figure 56), we see that the separatrices connecting the real critical points of the function $\tilde{\varphi}''$ also form a connected graph. Using the Lyashko–Loojienga covering, we include the function $\tilde{\varphi}''$ in a family of real functions $\tilde{\varphi}''_s \subset \{\varphi_\lambda\}$, $s \in [0,1]$, such that

a) $\tilde{\varphi}'' \equiv \tilde{\varphi}''_1$;

b) for $s \neq 0$ all critical values of $\tilde{\varphi}''_s$ are nonzero;

c) the nonreal critical values of the functions $\tilde{\varphi}''_s$ do not depend on s, the values at all saddlepoints are equal to s, and all the other real critical values also tend to 0 as $s \to 0$.

By the previous remark about the connectedness of the separatrix diagram of $\tilde{\varphi}''$, all the real critical points coincide at the instant $s = 0$ and form a simple singularity; let Ξ be the class of this singularity. For small values of s the function $\tilde{\varphi}''_s$ close to the point of coincidence is a two-index morsification of the singularity of type Ξ. It follows from Theorem 4 that in this case the picture of the sets of negative and positive values of the function $\tilde{\varphi}''_s - s$ is homeomorphic to one from Figures 55, 39b, 40b, 42a, 43a, 43c, or 43d, and since for $s \in (0,1]$ the real Morse structure of functions $\tilde{\varphi}''_s$ does not change the same is also true for $s = 1$, that is, for the function $\tilde{\varphi}''$.

All we need to do now is to study which of the saddlepoints in these pictures can be locally replaced by singularities D-equivalent to $x^{2k} - y^2 + c$ or, which is the same by the existence of the perturbation $\tilde{\varphi}'$, for which of these saddlepoints we can make the surgery connecting Figures 56b and 56c in such a way that the picture of the level sets thus obtained could correspond to a simple singularity. The last condition can easily be checked: indeed, we can calculate the Dynkin diagram of the singularity φ by the resulting picture, and the Dynkin diagrams of the simple singularities are very special.

We cannot consider the cases $\Xi = E_7$ or E_8, since there are no simple singularities that adjoin these with a jump of Milnor numbers equal to two or more.

Lemma 6. *It is impossible to delete a part of the vertices of an arbitrary Dynkin diagram of an arbitrary simple singularity (together with all edges entering these vertices) in such a way that the resulting diagram will be one of the diagrams of Figure 57.*

Indeed, a) if we delete a part of the vertices together with all their incident edges, then the rest will be a collection of Dynkin diagrams of simple singularities, see [Lyashko 76]; b) the quadratic form defined by the intersection indices in the middle (= two-dimensional) homology of the nonsingular level manifold of a simple singularity of three variables is negative definite (see Proposition I.5.2); c) all the

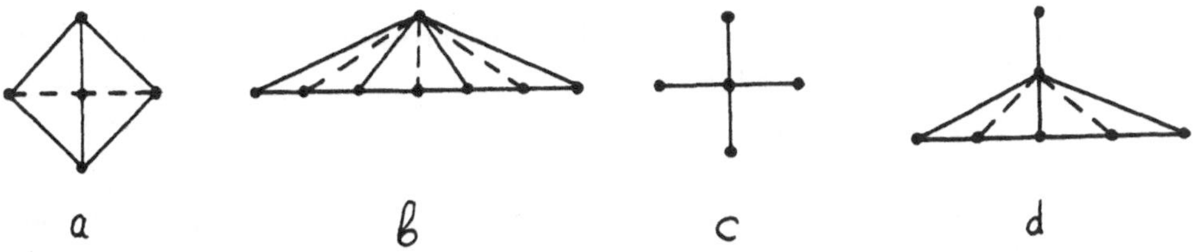

Fig. 57. Forbidden subdiagrams of Dynkin diagrams of simple singularities

quadratic forms defined by the diagrams of Figure 57 are degenerate.

If $\tilde{\varphi}''$ has no other real critical points besides the only saddlepoint which we perturb, then $\tilde{\varphi}$ has a unique critical point at which it can be D-equivalent to an arbitrary function of the form $x^{2k} - y^2 + 1$, $k > 1$; in this case φ is D-equivalent to $x^{2k} - y^2$. This case is considered in Theorem 1.2. In what follows we shall assume that $\tilde{\varphi}''$ has other real critical points.

Let us call a *birth of* A_{2k-1} a local surgery $\tilde{\varphi}'' \leftrightarrow \tilde{\varphi}'$ equivalent to the one relating Figures 56b and 56c and corresponding to a critical point of the characterizing function $\tilde{\varphi}$ at which this function is D-equivalent to $x^{2k} - y^2 + 1$. It follows from the assertion of Lemma 6 about diagram 57a that we cannot execute the surgeries of a birth of A_{2k-1}, $k \geq 2$, over the saddlepoints *both* of whose lower separatrices terminate at some minimum points. It follows from the assertion about diagram 57b that if there is at least one such separatrix, then only the birth of A_3 of A_5 is permitted, i.e. the corresponding saddlepoint can be changed either by two saddles and one maximum or by three saddles and two maxima. It follows from the assertion about diagram 57c that even the surgeries of these two types cannot be executed over the saddlepoints of singularities $\Xi = +D_{2k-1}, D_4^-, +E_6$, and over either of the two extreme right (in Figure 40b) saddlepoints of a singularity D_{2k}^-, $k \geq 3$.

Suppose that such a surgery is admissible for the extreme left saddlepoint for D_{2k}^-, $k \geq 3$. Using the Lyashko–Looijenga covering, it is easy to prove that in this case the singularity φ has a decomposition having two points of type D_4. According to [Lyashko 76], a singularity having such a decomposition can only be of type E_8. But if $k = 3$ and a singularity A_3 is born, then the intersection form of the resulting singularity is again degenerate, while if $k \geq 4$ or A_5 is born, then the Milnor number of the singularity thus obtained is more than 8.

In exactly the same way it can be proved that we cannot execute the surgery of the birth of A_{2k-1}, $k \geq 2$, over both marginal saddlepoints of the singularity A_{2l-1}^-, $l \geq 3$. In the case $l = 2$ the same follows from the assertion of Lemma 6 about diagram 57c.

Thus, all we can accomplish are the births of A_3 or A_5 at one boundary saddlepoint for $\Xi = A_{2l}, A^-_{2l-1}$. It follows from the assertion about the 57d that the birth of A_5 is admissible only if $\Xi = A_2$. This case gives us the picture 43b. The surgery "birth of A_3" is admissible for all $\Xi = A_{2l}$ or A^-_{2l-1} and gives the functions $\tilde{\varphi}'$ whose level sets $\{\tilde{\varphi}' = 1\}$ are homeomorphic to the ones in Figures 41 and 42b (with appropriate k).

Thus we have obtained two infinite series and one picture, which pretend that they depict (up to a homeomorphism) a partition of the plane into the sets of positive and negative values of the sabirizations $\tilde{\varphi}' - 1$ of certain simple singularities.

Theorem 5. *A simple singularity of a function of two real variables has a sabirization outlined in Figure 41 (respectively, 42b, 43b) if and only if this singularity is of class D^+_{2k} (respectively, $-D_{2k+1}$, $-E_6$), i.e. is D-equivalent to the function $x^2y + y^{2k-1}$ (respectively, $x^2y - y^{2k}$, $x^3 - y^4$).*

Proof. For the "if" part see § 2; let us prove "only if". Suppose that two singularities of functions of two variables have sabirizations with homeomorphic partitions of the plane into the sets of positive and negative values. This homeomorphism defines a correspondence between the critical points of these sabirizations, and hence, for a real number a greater than all critical values of these sabirizations, an isomorphism between the homology groups of the corresponding Milnor fibres $V(a)$: this isomorphism is given by the correspondence of vanishing cycles. The complex conjugation and the Weyl groups of these singularities act in the same way on these homology groups. Therefore it follows from the "if" part and from Proposition I.7.3 that the singularities whose sabirizations are shown in Figure 41 (respectively, 42b, 43b) cannot belong to the classes D^-_{2k} (respectively, $+D_{2k+1}$, $+E_6$). Also they cannot belong to any of the other classes because of the nonequivalence of Dynkin diagrams, and Theorem 5 is proved.

It remains to prove that in the parameter space of the versal deformation of a singularity of every one of these types D^+_{2k}, $-D_{2k+1}$, $-E_6$ there is exactly one formal lacuna.

For the singularities D^+_{2k} and $-D_{2k+1}$ the proof will be carried out by induction.

Base of induction: singularity D^+_4. As in the proof of Proposition 2, consider its truncated three-parameter versal deformation (without the constant term). The set of functions from this deformation that have non-Morse points is diffeomorphic to the surface shown in Figure 58 (the *purse*, see for example [AVGL 89]). The upper (in this picture) component of the complement of this surface consists of functions that have four real critical points (two saddlepoints, one maximum and one minimum), the lower one consists of functions that do not have real critical

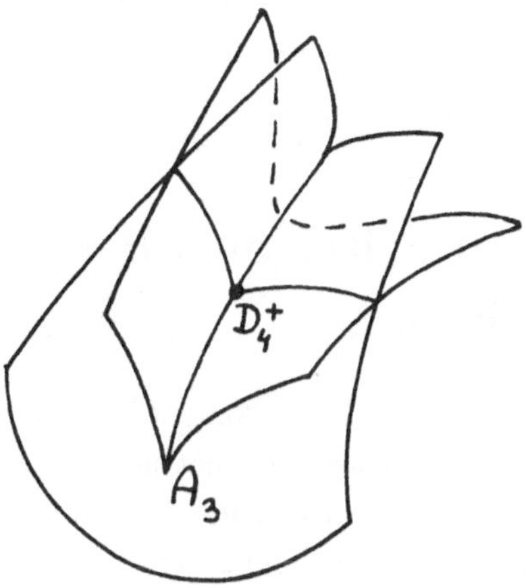

Fig. 58. Purse: the caustic for the singularity D_4^+

points at all, and the two symmetric intermediate components consist of functions with one saddlepoint and one extremum, minimum or maximum depending on the component.

The upper component contains, in particular, the function $x^3 - x + y^3 - y$. Its Dynkin diagram can be calculated by the methods of Chapter I, § 4. It follows from this calculation and from Theorem 1.3 that the projection map from the space of the complete (four-parameter) versal deformation into our truncated one, defined by forgetting the constant term, maps the union of formal lacunas into the union of a) the upper component, b) the one intermediate component that corresponds to functio.., having a saddlepoint and a minimum, and c) the piece of the surface between these two components. The restriction of this projection to the union of formal lacunas is a locally trivial bundle whose fibre is an interval, and our assertion for the case D_4^+ is proved.

Induction step. It was proved above that in every formal lacuna of a versal deformation of a function of the class D_{2k}^+, $k > 2$, there is a function $\tilde{\varphi}'$ whose level curve $\tilde{\varphi}' = 1$ is homeomorphic to one shown in Figure 41. Fix the critical value of this function corresponding to the extreme right minimum in this picture and move all other critical values to zero. Lifting this move to the Lyashko–Looijenga covering, we get a path in the space of the versal deformation that lies completely in our formal lacuna and approaches a point of the class $-D_{2k-1}$. The set of all points of this class in the space of the versal deformation from Table 3 is a ray given by the conditions $\lambda_0 = \cdots = \lambda_{2k-2} = \lambda_{2k} = 0, \lambda_{2k-1} < 0$. Since any transversal manifold to this ray is a versal deformation for the corresponding function of the class $-D_{2k-1}$, the desired assertion for D_{2k}^+ follows from the inductive hypothesis for $-D_{2k-1}$.

The reduction of the case $-D_{2k-1}$ to D_{2k-2}^+ and of $-E_6$ to $-A_5$ is carried out in exactly the same way.

Theorem 2.1 is completely proved.

§ 7. Geometrical criterion for sharpness close to simple singularities

In this section we prove Theorem 1.1. Its "only if" part is obvious (even close to the nonsimple singularities), so we prove only that nontriviality of the Petrovskii class implies that one of the two geometrical conditions of this theorem is not satisfied.

Theorem 6.1 and the Lyashko–Looijenga covering reduce this assertion to its special case which concerns the components of the complement of $\Sigma(F)$ containing the functions $f_0 + \varepsilon$ and $f_0 - \varepsilon$, where f is the original singularity. Namely, it is sufficient to prove the following theorem.

Theorem 1. *Suppose that a simple singularity f does not satisfy the plus- (respectively, minus-) Petrovskii condition. Then the component of the complement of $\Sigma(F)$ that contains the function $f - \varepsilon$ (respectively, $f + \varepsilon$) does not satisfy at least one of the two geometrical conditions from Theorem 1.1.*

Proof. We can assume that f is some function in Table 1, and F the corresponding versal deformation from Table 3, in particular, the number of parameters of F is equal to the Milnor number $\mu(f)$ of f. Then by Remark I.5.1 we can suppose that $\varepsilon = 1$. We shall consider only the assertion about the Petrovskii plus-condition and the case when n is even: the other three cases reduce to this one by means of the multiplication of f and F by -1 (see Theorem 1.6) and the Stabilization Theorems 1.4, 1.5.

So suppose that n is even and f does not satisfy the Petrovskii plus-condition. Since the corank of a simple singularity does not exceed 2, it is sufficient to consider two cases:

a) $n = 2$, so that $f = f(x, y)$,

b) $n = 4$ and $f = \varphi(x, y) + z^2 - w^2$.

All other cases reduce to these by the Stabilization Theorems 1.4, 1.5.

7.1. The case a).

In this case the singularity of f is neither a maximum point nor a Morse minimum: otherwise f satisfies the Petrovskii plus-condition.

Let f' be a small sabirization of f that has only the critical values $\delta, 0$ and $-\delta$, where $0 < \delta << 1$ (say, the sabirization shown in the corresponding Figure 55, 39b, 40b, 42a, 43a, 43c, or 43d; the requirement about the critical values is ensured by the Lyashko–Looijenga covering). Then the function $f' - 1$ belongs to the same component of $\mathbf{R}^{\mu} - \Sigma(F)$ as $f - 1$. By Theorem I.4.3, there is at least one saddlepoint of the function f' such that at least one of the two upper separatrices of grad f' issuing from this saddlepoint does not terminate at any maximum point. Suppose that there are two such separatrices. Then, using the Lyashko–Looijenga covering, we construct a one-parameter family of functions $f_{\lambda(t)}$, $t \in [0, 1)$, $\lambda(t) \in \mathbf{R}^{\mu}$, such that $f_{\lambda(0)} = f' - 1$ and all but one of the critical values of the functions $f_{\lambda(t)}$ are the same for all t (and hence are equal to $-1+\delta, -1$ and $-1-\delta$), while the critical value at the saddlepoint obtained by a continuous deformation from the distinguished saddlepoint is equal to $-1+t$. (Indeed, the only obstacle to such a homotopy of the function $f' - 1$ could appear from the fact that at the instant $t = \delta$ our saddlepoint meets some other critical points. Such points can only be the maxima connected with our saddle by a trajectory of the gradient flow: such points do not exist by our assumption.)

This family of functions lies in the complement of the discriminant and approaches a nonsingular discriminant point. Since the critical point with value 0 of the corresponding function is a saddlepoint, the Davydova–Borovikov condition is not satisfied at this discriminant point.

Supppose now that only one upper separatrix, issuing from our saddlepoint of $f' - 1$, goes infinitely upwards, while the other terminates at some maximum point (whose critical value is equal to $-1 + \delta$). Consider a continuous family of functions $f_{\lambda(t)}$, $t \in [0, 1)$, $f_{\lambda(0)} = f' - 1$ such that all but two of the critical values of $f_{\lambda(t)}$ stay unchanged, and the critical values at the saddlepoint and the maximum, obtained by the continuous homotopy from the distinguished ones, are equal to $-1 + t$ and $(1 - \delta)(-1 + t)$ respectively. The existence of such a family follows again from the Lyashko–Looijenga covering. For $t \in [0, 1)$ the function $f_{\lambda(t)}$ lies in the complement of the discriminant, and at the instant $t = 1$ the two distinguished critical points coalesce in a point of type A_2 with zero critical value. Thus the family $\{f_{\lambda(t)}\}$ approaches the cuspidal edge of the discriminant. It is easy to verify that this approach is from the "bigger" component 1 in Figure 3 (indeed, when we come to the edge from the "smaller" component, two critical values of different signs coalesce; cf. Definition 2.1 and Lemma 2.1). Thus the second geometrical condition of Theorem 1.1 is not satisfied for our component.

7.2. The case b) : $f = \varphi(x, y) + z^2 - u^2$.

Let φ' be a sabirization of the function φ, having only the critical values $\delta, 0, -\delta$, where $0 < \delta << 1$; let $f' = \varphi' + z^2 - u^2$ and suppose that for this function f' the local Petrovskii condition is not satisfied. We now prove that the component of the complement of $\Sigma(F)$ that contains $f' - 1$ does not satisfy the conditions of Theorem 1.1.

If φ' has at least one maximum point (and hence f' has a point of index 3), then the obstruction to the Davydova–Borovikov condition is constructed in exactly the same way as in subsection A in the case of two upper separatrices: the family $f_{\lambda(t)}$ consists of functions whose $\mu - 1$ critical values are the same, while the value at the distinguished maximum is equal to $(1 - \delta)(-1 + t)$. Therefore it remains to prove only the following statement.

Lemma 1. *Every simple singularity* $(\mathbf{R}^2, 0) \to (\mathbf{R}, 0)$, *except for the singularities of class* $+A_{2k-1}$, *which are D-equivalent to* $x^{2k} + y^2$, *admits a sabirization having a maximum point.*

The proof is immediate; see for example Figures 55, 39 – 43.

Theorem 1.1 is completely proved.

§ 8. A program for counting the topologically different morsifications of a real singularity

The program in its present form (see Appendix) is intended especially for the search of formal lacunas in the bases of versal deformations of real singularities; however it can also be used to answer many other problems about the existence of morsifications with fixed topological properties.

The possible results of the execution of the program are:

a) a proof of the fact that there are no lacunas close to singularities of the investigated class;

b) the indication that there can be lacunas close to the singularities of this class and the list of topological characteristics of some possible morsifications from these lacunas. The question whether a morsification with these characteristics exists is a nonformalized problem, which is left to the investigator; however, in all the cases already calculated the answer to this question has turned out to be affirmative;

c) the computer, with its restricted resources of memory and time, did not find any collection of topological characteristics that some lacunar morsification of the given singularity could have. If these resources are sufficiently large, this is a strong argument towards the absence of lacunas. The strict proof in this case is also left to the investigator (and is a very nontrivial problem). I only formulate a conjecture in this situation: ail the signs "≥ 0" in Table 13 are due to this circumstance.

8.1. The main ideas of the program.

Every morsification f_λ of a given singularity f is characterized by the following set of discrete data:

a) the intersection matrix of vanishing cycles Δ_j, $j = 1, \ldots, \mu(f)$, defined by some system of paths described in § 1.6;

b) the intersection indices of both local Petrovskii classes with these vanishing cycles;

c) the number of negative, positive and imaginary critical values;

d) the *Euler indices* ind(j) of real critical points (recall that the Euler index of an isolated singular point of a real function φ is the index of the mapping grad $\varphi/|\text{grad}\varphi| : S_\epsilon^{n-1} \to S^{n-1}$, where S_ϵ^{n-1} is a small sphere with center at the singular point; in the case of a Morse singularity it is easy to see that ind $= (-1)^{i_-}$).

From any point λ of the complement of the discriminant corresponding to a morsification f_λ we can go to any other by a path along which only the following elementary metamorphoses over the functions are executed:

M1) the largest negative critical value passes through 0 and becomes the smallest positive one;

M2) the operation inverse to M1;

M3) two neighbouring real critical values of the same sign change places, and the corresponding critical points do not meet during this change;

M4) two complex conjugate critical values cross the real axis, and the corresponding critical points do not meet and do not go out from the complex domain;

M5) two neighbouring real critical values coincide and then exit into the complex domain, and the corresponding critical points at one instant coalesce in a point of type A_2 and then also exit from $\mathbf{R}^n$;

M6) the operation inverse to M5;

M7) additionally, we can change the system of paths joining 0 and the nonreal critical values, so that the conditions of § 1.6 on the system of paths remain satisfied. All such changes are generated by the elementary changes α_i and β_{i+1}, see [AVG 84], [Gusein–Zade 77], applied to the paths in the upper half-plane, and the simultaneous mirror images of these changes applied to the lower paths; see Figure 59. Here i runs from 1 to the number of pairs of complex conjugate critical values minus 1.

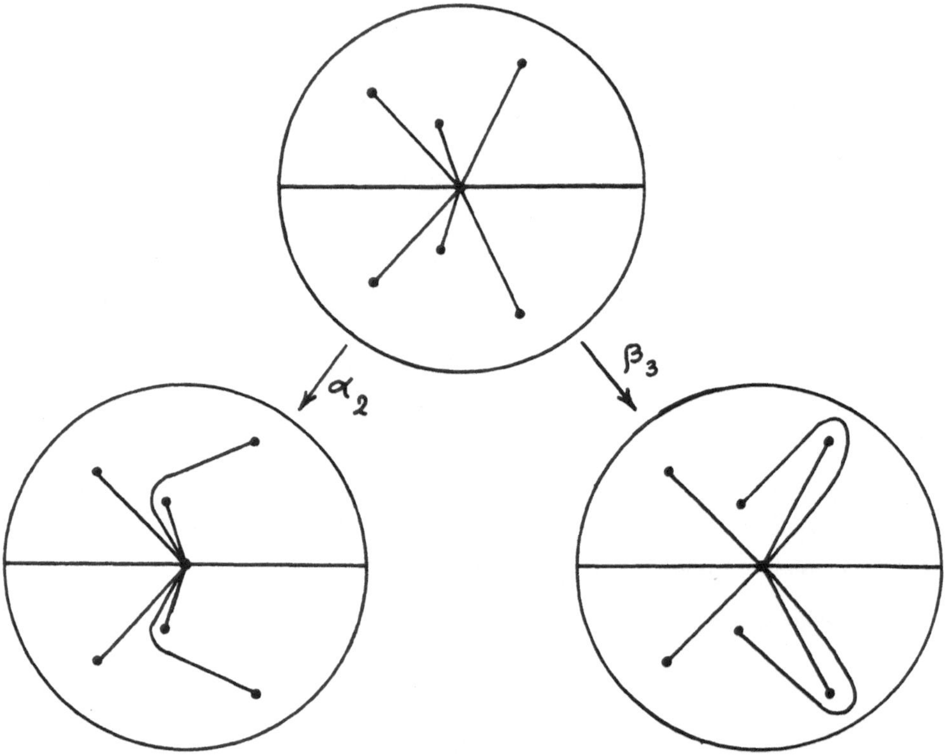

Fig. 59. Transformations α_i and β_{i+1}

Under all these surgeries the topological data a) – d) change in a known way (see § 9.6 below and the comments to the program in the Appendix). Therefore any generic path in the base of the deformation can be simulated by some sequence of elementary transformations, executed over the characteristic data of the original singularity listed above.

The search algorithm for morsifications of f consists in the following. First we define the topological characteristics for some real morsification of f. (This is not a formalized problem: if our morsification is a sabirization of a singularity of corank 2 then it can be solved by the methods of Chapter I, § 4 and Theorem 1.3; on the other hand, the absence of the singularity P_8^2 in Table 13 is due to the fact that I have not solved the similar problem for this singularity.) Then we successively apply all the possible admissible transformations to the set of these characteristics, and we look at the vectors of intersection indices of the vanishing cycles with the Petrovskii classes. If one of these vectors vanishes, the program says that we have found a lacuna of corresponding parity and prints out the current discrete parameters. The recovery of a real morsification with these parameters is a nonformal (but usually not very complicated) problem.

If the Petrovskii class is not equal to zero for any set of indices obtained by this method, this ensures the absence of lacunas. For instance, using this method the absence of lacunas for the singularity P_8^1 in the case of even n has been proved, i.e., the last two zeros in Table 13 are justified that do not follow from Proposition 1.3.

Among the singularities described in Table 13, only for $P_8^1, \pm X_9$ and X_9^2 was the algorithm executed up to the end (in the case X_9^1 the calculation was aborted at the very beginning by a subprogram checking the condition $\partial\Pi = 0$ from Proposition 1.3 for our singularity and for all singularities stably equivalent to it). For any of the remaining singularities with Milnor numbers ≤ 11, about 40000 topologically distinct morsifications were viewed by this program; in the case when this view did not find any lacuna, we put ≥ 0 in Table 13.

An inperfection of the method consists in the fact that we cannot be sure that any chain of admissible combinatorial transformations of the topological characteristics corresponds to some real path in the base of the versal deformation. A one-to-one correspondence between combinatorial and real transformations is ensured in the case of simple singularities by Theorem I.6.1 on the existence of the Lyashko–Looijenga covering, but in this case the complete answer was obtained by theoretical methods: see § 6. However, in selectively checking the intermediate results of the computation we found that any set of discrete indices, obtained by our program, actually corresponded to some real morsification. (Maybe it is the specifics of singularities of corank 2 that constitute a majority of the calculated examples.)

Moreover, the problem of the existence of lacunas is also equivalent to the above computational problem in the case of parabolic singularities P_8, X_9 and J_{10}: indeed, by the Jaworski theorem I.6.2, any combinatorial chain can be realized by a path lying, in general, in a neighbourhood of the whole $\mu = \text{const}$ stratum of the corresponding versal deformation from Table 8; on the other hand, by a theorem of Looijenga (see [Looijenga 77–78]) the discriminant variety is topologically trivial along this stratum.

8.2. Preliminary remarks on the realization of the algorithm.

1) All the explicit formulae for the transformations M1 - M7 follow from the Picard–Lefschetz formula, Theorem 3.2 and Corollary 4.1; they will be listed in § 9.6 below. Among all these transformations only M1 and M2 change the local Petrovskii classes, while under M3–M7 (which do not change a component of the complement of the discriminant) the spaces $\tilde{H}_{n-1}(V_\lambda)$ for the initial and final values of λ are naturally identified by the Gauss–Manin connection. This identification preserves the Petrovskii classes, and the corresponding transformation of the set of discrete characteristics is simply reduced to a change of the basis of vanishing cycles in these spaces. The jump of the Petrovskii class under the operations $M1, M2$ consists in adding to it the vanishing cycle, taken with appropriate sign, corresponding to the critical value which jumps over 0; see Theorem 3.2 and Corollary 4.1.

2) The vanishing cycles Δ_j of a morsification f_λ (and the critical values over which they vanish) are ordered as follows: first the real critical values in order of their increase, then the values with positive imaginary parts in order of increase of the angles at 0 between the ray $(0, -\infty)$ and the tangent direction of the distinguished path along which our cycle vanishes, and finally the values with negative imaginary parts ordered in accordance with the orders of their complex conjugate critical values.

3) The discrete characteristics a) – d) depend on the choice of the orientation of the vanishing cycles. In § 1.6 we have agreed on the orientations of the cycles vanishing at the real critical points, and on the concordance of the cycles vanishing at the complex conjugate points. We exterminate the remaining freedom by choosing such orientations in the following way: this set of orientations must provide the lexicographic maximization of the intersection matrices (over the set of orientations satisfying the above conditions). See statements 624–690 in the Appendix. Since the Dynkin diagram of a singularity is connected, these rules uniquely fix the orientations of all vanishing cycles if there is at least one real critical point. If all of them are imaginary, then there are exactly two optimal (in this sense) choices of orientations, which are opposite to one another. Since both Petrovskii cycles cannot vanish

simultaneously (see Corollary 1.1), one of these choices can be selected in a unique way by the additional condition of the lexicographic maximization of Petrovskii cycles (presented by their intersection indices with our vanishing cycles); see the statements 691–704. After any elementary transformation M1–M7, the algorithm makes a reorientation of the resulting basis of vanishing cycles that ensures such maximization.

4) The operations M3 and M5 are combined into one: the attempt to make two neighbouring real critical values α_i, α_{i+1} coincide. If the intersection index $\langle \Delta_i, \Delta_{i+1} \rangle$ of the corresponding vanishing cycles is equal to zero, then the transformation M3 arises; if

$$\langle \Delta_i, \Delta_{i+1} \rangle = (-1)^{n(n+1)/2} \tag{28}$$

and, moreover, the Euler indices of the corresponding real Morse critical points are different, then M5 arises, and no coincidence occurs in the remaining cases: a "strong repulsing interaction" arises between the critical values. See statements 214–343 in the program. The formula (28) is taken from Lemma 3.2.

(Of course, in reality other obstructions to the operation M5 and M3 also exist, so that the critical values do not coincide even if both conditions (28) and $\epsilon(i) \neq \epsilon(i+1)$ are satisfied: for example, almost all of the modern 3-dimensional topology studies such additional obstructions.)

Similarly, M4 and M6 are combined into the attempt to make two complex conjugate critical values coincide. From the intersection matrix we can determine whether the transformation M4, M6 or repulsion occurs: moreover, in the case of M6 we can predict the Euler indices of the newborn real critical points. For example, if coincidence of the critical values α_i and $\alpha_j \equiv \bar{\alpha}_i$ occurs at a point not separated from 0 by some real critical values, and the paths connecting 0 with these critical values and defining the corresponding vanishing cycles are line segments (that is, do not embrace other critical values), then M4 occurs if and only if $\langle \Delta_i, \Delta_j \rangle = 0$, and M6 occurs if and only if $\langle \Delta_i, \Delta_j \rangle = 1$ or -1. In the last case, by Lemmas 3.2 and 3.3, the Euler index of the newborn critical point with smaller critical value is equal to $(-1)^{1+(n+1)n/2} \langle \Delta_i, \Delta_j \rangle$, where Δ_i is the cycle vanishing over the nonreal critical value with positive imaginary part. See statements 345–511 in the Appendix.

5) The examination of "virtual morsifications", i.e., of collections of topological characteristics that are obtained by the chains of our transformations, is organised in the following way. All these collections are written in a queue in the order of their detection. For all collections from some initial segment of this list, it is known that all the similar collections that can be obtained from them by *one* elementary transformation are already written in the queue.

On any examination step we:

take the first collection not belonging to this initial segment,

apply to it all elementary transformations,

compare all the resulting collections with the collections that already are in the queue. If the collection thus obtained is new, then we write it at the end of the queue, otherwise we ignore it.

6) Moreover, to reduce the number of operations, for every virtual morsification we remember the types of the elementary transformations by which it was obtained or by which an attempt was made to obtain it again. Later we do not apply to this collection all the transformations opposite to these remembered operations. Therefore we can compare the new collections obtained at one step of the examination not with all the queue, but only with the collections that are not contained in the initial segment of it. See the statements containing the parameters $INVC$ and $INVB$.

7) Using the Stabilization Theorems 1.4, 1.5, we examine simultaneously four singularities that are stably but not bistably equivalent to each other. See statements 738–790.

8) The same local lacuna can contain a very large number of topologically distinct morsifications; it is impossible to print out all of them. Therefore we print only one morsification for each value of the invariant from Definition 2.1 (the sum of the Euler indices over critical points with negative critical values). However, in this way we can miss some nonequivalent lacunas.

A more regular algorithm of separating lacunar (with zero Petrovskii class) virtual morsifications from different lacunas could consist in the following: to number all such morsifications in order of their appearance in the queue unless they are obtained from one another by some elementary transformation M3 - M7; in the latter case they get the same numbers, and, moreover, if at some step one discovers that some lacunar morsification can be obtained by any of these operations from some other with a different number, then these morsifications (and all morsifications having the same numbers) get the smallest of their numbers. At the end of the job, the program prints out one lacunar collection of data with each number. However, this procedure is not realized in the program because of the poor attainable computer resources.

8.3. Open problems and possible improvements of the algorithm.

1) What are the obstructions to the realization of any chain of admissible transformations M1–M7 by some path in the space of a versal deformation, and hence to the implementation of any virtual morsification by a real one? According to Proposition 2 of Chapter I, § 6, any small perturbation of the set of critical values of a

morsification f_λ is covered by some perturbation of this morsification. However, the attempt to lift a longer path can press the morsification out from a neighbourhood of the origin in the parameter space of the deformation. One of the reasons for this is the "strong repulsion" from part 4 of subsection 8.2 above. Is it the unique obstruction to the lifting of a generic path from the space of sets of critical values invariant under complex conjugation?

2) The problem of computing the initial data for singularities of corank ≥ 3 is the main obstruction to the extensive exploitation of the program; this problem is stated as follows. Let f be a real singularity. The problem is to present a generic morsification of it and a set of parameters a) – d) for this morsification, computed in the basis of vanishing cycles defined by above-described system of paths and orientations; see § 1.6. (In the case when all the critical points of this morsification are real, the corresponding Petrovskii cycles are computed automatically from the other data; the subroutine LPC that executes this calculation is based on Theorem 1.3. Many strong methods for calculating the Dynkin diagrams were developed in [Gabrielov 73], [Gabrielov 79], but the morsifications and basis systems of vanishing cycles used there are, in general, not the ones needed in our problem.

3) Do there exist real isolated singularities having no morsifications without imaginary critical points?

4) Can one predict the Morse indices of the critical points born during the elementary transformation M6, knowing only the set of indices a) – d) from subsection 8.1 and the Morse indices of the real critical points of the morsification before this transformation? (Now we can only predict their *Euler indices*, i.e., the parities of the Morse indices: of course, if $n = 2$, as in the majority of calculated examples, then these two problems are equivalent.) This is obviously impossible if the morsification before the transformation has no real critical points: indeed, all other algebraic data are the same for bistably equivalent singularities and their perturbations. What is the extra topological information known to the coinciding imaginary critical points that enables them to determine their Morse indices after the birth?

5) The number of virtual morsifications obtained by admissible transformations for complicated singularities exceeds the computer resources, and, moreover, for sufficiently complicated singularities it simply becomes infinite (because of the transformations M7). One way to overcome this difficulty consists in proving some *a priori* estimates of the following type: from any (or from some specific) morsification of a given singularity one can pass to any component of the complement of $\Sigma(F)$ in at most T elementary transformations. The obvious estimates of this kind that follow from Bezout's theorem are unsatisfactory for practical use.

6) Of course, our algorithm (or simple modifications of it) can solve numerous

problems not related to lacunas, for example: does a given singularity have a morsification of some prescribed type (such problems often appear in the theory concerning Hilbert's 16-th problem; see [Gudkov 74], [Kharlamov 86]); can two given morsifications lie in the same component of the complement of the discriminant (for this it is sufficient to drop the operations M1, M2 in the algorithm), and so on.

§ 9. More detailed description of the algorithm

This section is intended for the reader who wants to read the program and, maybe, to adapt it to his own problems.

9.1. The parameters and constants of the program.

The constant characteristics along the program are:

the number N which is equal to $(-1)^n$, the parity of the number of variables;

the number $N2 \equiv (-1)^{n(n-1)/2}$;

M, the Milnor number of the singularity.

The topological characteristics of the current (virtual) morsifications are:

1) the number MIB of pairs of complex conjugate imaginary critical values; the number of real values is equal to $M - 2 * MIB$ and is denoted by MRB;

2) the number $NPOZB$ of negative real critical values;

3) the $M \times M$ intersection matrix $B(I, J)$ of vanishing cycles defined by the current system of paths, satisfying the conditions of § 1.6 and the optimization conditions of part 3 of § 8.2;

4) the Euler indices $INDB(I)$ of critical points: they take values 1 or -1 at real points and are assumed to be equal to zero at the complex ones;

5) even and odd Petrovskii classes, given by the vectors of their intersection indices ($PB2(I)$ and $PB1(I)$, respectively) with the basis vanishing cycles.

The similar characteristics of the new virtual morsification, which appears at the current step of the algorithm, are denoted in exactly the same way, only with the letter C instead of B: MIC, MRC, $NPOZC$, $C(I, J)$, $PC2(I)$ and $PC1(I)$.

The current characteristics of the examination process, KA and KN, are respectively the number of all virtual morsifications in the queue and the number of virtual morsifications over which all standard transformations have already been executed.

We accomplish not all the transformations M1 - M7, but a subset of them, such that all other transformations are compositions of these.

Namely, the transformations M3 and M5 are executed over two real critical points with the two smallest *positive* critical values: all other transformations of these two types can be reduced to these by several operations M1 and M2.

Also in the transformations M4 and M6 the coincidence of two imaginary critical values occurs at the positive point that is not separated from 0 by any other critical value (other transformations of these types are again conjugate to these by the operations M1 and M2). Moreover, we suppose that the paths connecting 0 with these critical values and defining the corresponding vanishing cycles are line segments (i.e., do not embrace other critical values; by the above convention on the ordering of vanishing cycles and critical values, the coinciding critical values are the last ones among these with positive and negative imaginary parts respectively, i.e., their numbers are $MRB + MIB$ and M). This can be ensured by an appropriate change of type M7.

The group of transformations M7 is generated by the elementary operations α_i and $\beta_{i+1} \equiv \alpha_i^{-1}$, $i = 1, \ldots, MIB - 1$, see Figure 59, therefore we consider only such transformations.

9.2. Preparation for the work.

The initial data of the program (in its form presented in the Appendix) are the constants M, N, $N2$, and the list of characteristics of some real morsification having only real critical values, namely, the intersection matrix $C(I, J)$ of its vanishing cycles, defined as in § 1.6 and ordered as in part 2 of § 8.2, and the vector $INDC(I)$ of Euler indices of the corresponding critical points. The intersection matrix should be presented only by *nonzero* entries $C(I, J)$ with $I < J$: all the other entries will be added automatically; also only the indices $INDC(I)$ equal to -1 should be input, and the program assumes all the other indices to be equal to $+1$.

These initial data should be written in the subroutine DATA and hitched at the very end of the program; see statements 945–978 in the Appendix. In the *FORMAT* statement of this subroutine, instead of $Y(1R)$ the notation of the investigated singularity must be written.

In the initial situation, when all the critical points of the current virtual morsification are real, the Petrovskii classes can be explicitly calculated through the above data via formulae (5)–(8) (and this calculation is carried by the subroutine LPC; see statements 839–862 in the Appendix). The value of the parameter $NPOZC$ (the number of negative critical values) is automatically set equal to $[M/2]$. Thus we get all data a) – d) from § 8.1 for a certain morsification of our singularity.

In the first seven lines of the main program we put the following parameters: the actual Milnor number M instead of all numbers 11 indicated in the program of the

Appendix, and the number $NMASS$ of possible virtual morsifications in the queue instead of all 15000. The number $NMASS$ can be estimated by the fact that the memory needed for all the queue satisfies

$$(MEMORY\ SIZE) \simeq (M+4) * 8 * NMASS \text{ bytes.} \tag{29}$$

The constants encoded in these seven lines have the following meaning: for any $K = 1, \ldots, NMASS$, the data $A(J, I, K)$, $P1(I, K)$, $P2(I, K)$, $IND(I, K)$, where $I = 1, 2$; $J = 1, \ldots, M$, are respectively the (converted) intersection matrix, the Petrovskii classes, and the Euler indices of the K-th virtual morsification. $PARAM(K)$ is the converted information about numbers of negative and imaginary critical values. $NAIS(K)$ is the converted information about the types of standard transformations by which the K-th virtual morsification was obtained, or by which an attempt was made to obtain it again; see part 6 of § 8.2.

All other constants not mentioned in this and the previous subsection are used in the intermediate calculations only.

9.3. Two more features.

At any instant, the interval $[0, KN-1]$ of the queue is useless. Therefore, when the queue becomes too large (i.e., exhausts all the memory) we forget all the elements of this interval, and lower the order numbers of all other virtual morsifications by $KN-1$; the program prints the message

$$SHIFT \quad < \text{the value of number } KN-1 > .$$

Any singularity has, besides its own, three more classes of bistable equivalence, which are stably but not bistably equivalent to each other; see Definition 1.3. Namely, if the original function is $f = f(x_1, \ldots, x_n)$, then these classes are represented by the functions $f - x_{n+1}^2 + x_{n+2}^2$, $f - x_{n+1}^2$, $f + x_{n+1}^2$. All four classes of bistable equivalence are investigated simultaneously: indeed, the Petrovskii classes of all of them are reduced to each other by Theorems 1.4, 1.5. If for some of these four stably equivalent virtual morsifications the corresponding Petrovskii class of *regular parity* vanishes, and we are sure that this lacuna is a new one (see part 8 of § 8.2), the program prints the message

$$LACUNA\,!\,,$$

the value of the negative index of this morsification (see Lemma 2.1), the number of the discovered local lacunas with different values of the negative index, the indication

$STEP = 1$ (respectively, $STEP = 2$, $STEP = 3$ or $STEP = 4$) if the current virtual morsification is that of the original function f (respectively, of the function $f - x_{n+1}^2 + x_{n+2}^2$, $f - x_{n+1}^2$, or $f + x_{n+1}^2$), and finally the parameters of the corresponding virtual morsification (for the list of these parameters see the description of subroutine PECH in subsection 9.5 below). At the very beginning the program requests the subroutine BOUNDARY, which checks the conditions of Proposition 1.3 for any of four classes of bistable equivalence. If the boundary of the Petrovskii class for some of them turns out to be nontrival, this subroutine informs you that this singularity cannot have any lacunas at all by printing the message "THE BOUNDARY IS NONTRIVIAL" and the number KSTEP of the class of bistable equivalence.

9.4. Termination of the program.

The program terminates in one of the following cases:

1) $KN > KA$: this means that all virtual morsifications that can be obtained from the initial one have been considered;

2) the resource of memory is exhausted, and the first KN elements of the queue take less than 1/10 of this memory: the iteration of the operation $SHIFT$ becomes non rentable. Of course, the parameter 1/10 here can be changed; see the statement 65 in the Appendix. Also it is possible to use the interchange with the slow access memory to increase the possible size of the queue.

3) the subroutine BOUNDARY says that by Proposition 1.3 none of the four classes of bistable equivalence, stably equivalent to the given one, can have formal lacunas.

4) See item **ECON** of the following subsection.

9.5. Description of the subroutines.

ECON is the program converting the data of virtual morsifications for writing them in the queue. The absolute values of the data $C(I, J), PC1(I), PC2(I)$ must be less than 16. If this condition is not satisfied, then the program writes the message

TOO LARGE ENTRIES

and terminates. In all the cases actually calculated this situation did not appear. The parameter 16 can be increased, but then the conversion index will be worse.

DISSIP is inverse to **ECON**.

PECH is the program printing the parameters of the current virtual morsification and other current parameters. First it writes five numbers in a line: $NOP =$ the number of the last elementary operation (which is equal to i for α_i, $i + [M/2]$ for

β_{i+1}, $M+2$ for M2, $M+3$ for M1, $M+1$ for $M3$ & $M5$, and $[M/2]$ for $M4$ & $M6$); KA; KN; MRC; NPOZC. Then it writes the intersection matrix, odd and even Petrovskii classes, and Euler indices of the critical points. This subroutine works at the very beginning (when it prints out the data of the initial morsification) and after the discovery of any *new* formal lacuna; see the description of subroutine *NLAC* below. In the present form of the program, it prints also the data of the $3000 \cdot k$-th morsification in the queue for any k; see statement 606. I have done it for the selective checking of the fact that all virtual morsifications actually correspond to some real morsifications of our singularities. The parameter 3000 here can be replaced by any other, or this statement under the label 9000 can be replaced by *CONTINUE*.

LPC calculates the intersection indices of both Petrovskii classes with all vanishing cycles corresponding to real critical points. The initial data are: intersection matrix, Euler indices, N, $N2$, M, $NPOZC$. See Theorem 1.3.

BOUNDARY checks the conditions of Proposition 1.3 for the given singularity and for three singularities stably equivalent to it. If it establishes the absence of lacunas for some of these four singularities, it prints at the end of the execution of the program the message(s) "THERE ARE NO LACUNAS < *KSTEP* >", where KSTEP can take values 1, 2, 3, 4; see subsection 9.3 above.

NLAC gives a lower estimate of the number of discovered lacunas. It is based on the invariant used in Lemma 2.1: the sum of the Euler indices over the critical points with negative values. It allows us to print the message *LACUNA* ! only if a virtual morsification with zero Petrovskii class and given value of this invariant appears for the first time, or if such a morsification additionally having only real critical points appears for the first time.

DATA: list of initial data.

9.6. Explicit formulae for the elementary transformations M1 – M7.

Operations α_i and $\beta_{i+1} \equiv \alpha_i^{-1}$ are some standard changes of the i-th and $(i+1)$-th upper and lower basis cycles (in the general numbering these cycles have numbers $MRB+i, MRB+i+1$ and $MRB+MIB+i, MRB+MIB+i+1$, respectively) determined by the simultaneous and symmetric change of the paths connecting 0 to the corresponding two pairs of critical values; see Figure 59. The explicit expression of new basis cycles $\tilde{\Delta}_j$ in terms of the old cycles Δ_j follows immediately from the Picard–Lefschetz formula and is as follows:

$\tilde{\Delta}_j = \Delta_j$ for all $j \neq MRB+i, MRB+i+1, MRB+MIB+i, MRB+MIB+i+1$

and both operations α_i, β_{i+1}:

operator α_i changes the remaining basis cycles by the formulae

$\tilde{\Delta}_{MRB+i+1} = \Delta_{MRB+i}$;

$\tilde{\Delta}_{MRB+i} = \Delta_{MRB+i+1} + N * N2 * B(MRB+i+1, MRB+i) * \Delta_{MRB+i}$;

$\tilde{\Delta}_{MRB+MIB+i+1} = \Delta_{MRB+MIB+i}$;

$\tilde{\Delta}_{MRB+MIB+i} = \Delta_{MRB+MIB+i+1} -$

$-N2 * B(MRB+MIB+i+1, MRB+MIB+i) * \Delta_{MRB+MIB+i}$;

see statements 130–170 in the program;

and operator β_{i+1} changes them in the following way:

$\tilde{\Delta}_{MRB+i} = \Delta_{MRB+i+1}$;

$\tilde{\Delta}_{MRB+i+1} = \Delta_{MRB+i} + N * N2 * B(MRB+i+1, MRB+i) * \Delta_{MRB+i+1}$;

$\tilde{\Delta}_{MRB+MIB+i} = \Delta_{MRB+MIB+i+1}$;

$\tilde{\Delta}_{MRB+MIB+i+1} = \Delta_{MRB+MIB+i} -$

$-N2 * B(MRB+MIB+i+1, MRB+MIB+i) * \Delta_{MRB+MIB+i+1}$;

see statements 172–212.

The operations M3 and M5 (more precisely, their special cases specified in subsection 9.2) are combined into one: the attempt to make two *smallest positive* critical values coincide, i.e., the values over which the basis cycles $\Delta_{NPOZB+1}$ and $\Delta_{NPOZB+2}$ vanish. The result of the coincidence depends on the intersection index $B(NPOZC+1, NPOZC+2)$ of these two cycles.

If $B(NPOZC+1, NPOZC+2) = 0$, then we have transformation M3, the change of orders of critical values. The corresponding change of basis cycles is trivial:

$\tilde{\Delta}_{NPOZC+1} = \Delta_{NPOZC+2}$;

$\tilde{\Delta}_{NPOZC+2} = \Delta_{NPOZC+1}$;

$\tilde{\Delta}_i = \Delta_i$ for $i \neq NPOZC+1, NPOZC+2$;

see statements 229–251 in the program.

If $B(NPOZC+1, NPOZC+2) = N * N2$, then we have the transformation M5, the "death" of two real critical points. The newborn imaginary critical points become the last numbers among the points with critical values in their half-planes, i.e., the numbers $MRC + MIC \equiv MRB + MIB - 1$ for the value in the half-plane

Im $z > 0$ and the number M for the value in the half-plane Im $z < 0$. The explicit formulae for the change of basis cycles follow from the Picard–Lefschetz formula and Lemmas 3.2, 3.3, and are as follows:

$\tilde{\Delta}_i = \Delta_i$ for $i \leq NPOZB$;

$\tilde{\Delta}_i = \Delta_{i+2} + N * N2 * (B(NPOZB+1, i+2)+$
$+B(NPOZB+2, i+2)) * (\Delta(NPOZB+1) + \Delta(NPOZB+2))$
for $i \in [NPOZB+1, MRB-2]$;

$\tilde{\Delta}_i = \Delta_{i+2}$ for $i \in [MRB-1, MRB+MIB-2]$;

$\tilde{\Delta}_{MRB+MIB-1} = \Delta_{NPOZB+1} + \Delta_{NPOZB+2}$;

$\tilde{\Delta}_i = \Delta_{i+1}$ for $i \in [MRB+MIB, M-1]$;

$\tilde{\Delta}_M = -INDB(NPOZB+1) * \Delta_{NPOZB+2}$;

see statements 253–343.

(Recall that $INDB(i)$ is the Euler index of the i-th critical point; the coefficient $-INDB(NPOZB+1)$ in the last line ensures the compatibility under complex conjugation of the two cycles that vanish at the newborn critical points.)

Finally, if $B(NPOZC+1, NPOZC+2) \neq 0$ or $N * N2$, then the coincidence is forbidden.

The operations M4 and M6 in their special form (see subsection 9.1) are also combined into one operation: the attempt to make two imaginary critical values coincide, whose numbers in the general ordering are $MRB+MIB$ and M, at a *real positive* point not separated from 0 by any real critical value. The precise kind of transformation depends again on the intersection index $B(MRB+MIB, M)$ of the corresponding vanishing cycles.

If $B(MRB+MIB, M) = 0$, then the transformation M4 arises: the critical points stay in the imaginary domain but change their critical values. The corresponding change of basis cycles follows immediately from the Picard–Lefschetz formula and is as follows:

$\tilde{\Delta}_i = \Delta_i$ for $i \in [1, NPOZB] \cup [MRB+1, MRB+MIB-1] \cup [MRB+MIB+1, M-1]$;

$\tilde{\Delta}_i = \Delta_i + N * N2 * B(i, MRB+MIB) * \Delta_{MRB+MIB} - N2 * B(i, M) * \Delta_M$
for $i \in [NPOZC+1, MRB]$;

$\tilde{\Delta}_{MRB+MIB} = \Delta_M$;

$\tilde{\Delta}_M = \Delta_{MRB+MIB}$;

see statements 358–396.

If $B(MRB + MIB, M) = 1$ or -1, then two complex conjugate points coincide and two real points appear. The Euler indices of them depend on the sign of $B(MRB + MIB, M)$: the index of the one with the smaller critical value is equal to $-B(MRB + MIB, M) * N2$, and the index of the other is opposite to this one. In the following formulae, which express the elements of the new basis of vanishing cycles in terms of the old ones, two more numbers participate: INP and $INORM$.

INP is nothing but the above index $-B(MRB + MIB, M) * N2$, while the coefficient $INORM = \pm 1$ is chosen in such a way that the orientation of the two cycles $\tilde{\Delta}_i$, $i = NPOZB+1, NPOZB+2$, vanishing at the two newborn real critical points and defined by the second and third formulae of the list below, coincides with the one specified in § 1.6. Namely, if $INP * N = -1$, then this coefficient is equal to $N2*(PB2(MRB+MIB) - INP * PB2(M))/2$; if $INP * N = 1$, then it is equal to $-N2*(PB1(MRB+MIB) + INP * PB1(M))/2$; recall that $PB1(I)$ (respectively, $PB2(I)$) is the intersection index of the odd (respectively, even) Petrovskii cycle with the I-th (old) basis vanishing cycle. So our formulae for the new basis cycles are:

$\tilde{\Delta}_i = \Delta_i$ for $i \in [1, NPOZB]$;

$\tilde{\Delta}_{NPOZB+1} = INORM * (\Delta_{MRB+MIB} + INP * \Delta_M)$;

$\tilde{\Delta}_{NPOZB+2} = -INORM * INP * \Delta_M$;

$\tilde{\Delta}_i = \Delta_{i-2} + N * N2 * B(i-2, MRB + MIB) * \Delta_{MRB+MIB}$ for $i \in [NPOZB + 3, MRB + 2]$;

$\tilde{\Delta}_i = \Delta_{i-2}$ for $i \in [MRB + 3, MRB + MIB + 1]$;

$\tilde{\Delta}_i = \Delta_{i-1}$ for $i \in [MRB + MIB + 2, M]$;

see statements 398–510 of the program.

Again, if $|B(MRB+MIB, M)| \geq 2$, then the coincidence of these critical values is impossible.

The transformations M1 and M2 are the only ones that change the Petrovskii classes themselves, and not only their expression in bases of vanishing cycles. In particular, only these operations can lead us for the first time to a formal lacuna.

The change of basis vanishing cycles for the operation M1 is as follows (see statements 560–584 of the program):

$\tilde{\Delta}_i = \Delta_i$ for $i \in [1, MRB + MIB]$;

$\tilde{\Delta}_i = \Delta_i - N2 * B(i, NPOZB) * \Delta_{NPOZB}$ for $i \in [MRB + MIB + 1, M]$.

The change of Petrovskii classes (i.e., the intersection indices of new Petrovskii cycles $PC1, PC2$ with the new basis vanishing cycles) is given by the following formulae:

$$PC1(i) = PB1(i) - B(i, NPOZB), \quad PC2(i) = PB2(i) + B(i, NPOZB)$$
$$\text{for } i \in [1, MRB + MIB];$$
$$PC1(i) = PB1(i) - B(i, NPOZB) -$$
$$- N2 * B(i, NPOZB) * (PB1(NPOZB) - B(1,1));$$
$$PC2(i) = PB2(i) + B(i, NPOZB) -$$
$$- N2 * B(i, NPOZB) * (PB2(NPOZB) + B(1,1))$$
$$\text{for } i \in [MRB + MIB + 1, M];$$

see statements 585–605.

(Recall that $B(1,1) \equiv B(i,i)$ for any i is equal to 0 if $N = 1$, is equal to 2 if $N = -1$ and $N2 = 1$, and is equal to -2 if $N = -1$, $N2 = -1$; see Proposition I.2.2.)

For the operation M2 (which is inverse to M1) the similar expressions are:

$$\tilde{\Delta}_i = \Delta_i \quad \text{for } i \in [1, MRB + MIB];$$
$$\tilde{\Delta}_i = \Delta_i + N * N2 * B(i, NPOZB+1) * \Delta_{NPOZB+1} \qquad \text{for } i \in [MRB + MIB + 1, M].$$

The intersection indices of the new Petrovskii classes $PC1, PC2$ with the new basis vanishing cycles are expressed in terms of the old ones by the following formulae:

$$PC1(i) = PB1(i) + B(i, NPOZB+1), \quad PC2(i) = PB2(i) - B(i, NPOZB+1)$$
$$\text{for } i \in [1, MRB + MIB];$$
$$PC1(i) = PB1(i) + B(i, NPOZB+1) +$$
$$+ N * N2 * B(i, NPOZB+1) * (PB1(NPOZB+1) + B(1,1));$$
$$PC2(i) = PB2(i) - B(i, NPOZB+1) +$$
$$+ N * N2 * B(i, NPOZB+1) * (PB2(NPOZB+1) - B(1,1))$$
$$\text{for } i \in [MRB + MIB + 1, M];$$

see statements 512–558.

Reminder. All these changes are not final: after any of them, the corresponding data $C(I,J)$, $PC1(I)$ and $PC2(I)$ are subject to some lexicographic maximization which specifies the canonical orientations of cycles vanishing at the imaginary critical points; see part 3 of § 8.2 and statements 624–705.

9.7. An example of the final print-out.

At the end of the Appendix, the print-out of the calculation for the singularity $+E_6$ is presented. It starts with the name of the singularity (which should be written in subroutine DATA instead of $Y(1R)$). The next 14 lines 2–15 contain the standard data (see the description of subroutine PECH) for the initial morsification. The value 0 of the parameter NOP appears only for the initial morsification.

Next two lines contain additional data for the first discovered lacuna: its negative index, the number of this lacuna for this class of bistable equivalence, the number of imaginary critical values, and the index $STEP$ of the class of bistable equivalence. The number 4 in the position $STEP$ denotes that this is a lacuna for the singularity $f(x_1, \ldots, x_n) + x_{n+1}^2$, where f is the original singularity. (In our case $n = 2$.)

In the next 14 lines 18–31 there follow the standard data for the morsification f_λ of the original singularity f such that the morsification $f_\lambda + x_{n+1}^2$ belongs to the discovered lacuna.

The next $2 + 14$ lines contain similar information about the other lacuna (in the deformation of the function $f(x_1, \ldots, x_n) - x_{n+1}^2$). The two lines "THERE ARE NO LACUNAS 1" (respectively, 2) inform us that by Proposition V.1.3 the original singularity f and its stabilization $f - x_{n+1}^2 + x_{n+2}^2$ cannot have local lacunas. The concluding numbers 456 and 457 are the values of the parameters KA and KN at the final instant. They indicate that the queue contains 456 topologically different morsifications, and the "initial segment" of the queue, which consists of virtual morsifications such that all morsifications whose one-step perturbations are also written in the queue, is also equal to 456. Since these numbers coincide, the program terminates; see § 9.4. The number 456 is thus the number of *all* topologically different morsifications of the given singularity. (In general, of *virtual* morsifications, but in our case, when the investigated singularity is simple, all virtual morsifications can surely be realized. Numerous calculations, carried out by our program and partially presented in Table 13, allow us to conjecture that this is also the case for all singularities of corank 2.)

APPENDIX

A FORTRAN program searching for the lacunas and enumerating the morsifications of real function singularities

```
      INTEGER A(11,2,15000),B(11,11),C(11,11), D(11,11), P1(2,15000),   001
     *P2(2,15000),PARAM(15000),PB1(11),PB2(11),PAR,PC1(11),PC2(11),      002
     *PDEL1,PDEL2                                                        003
      DIMENSION NAIS(15000),INDB(11),IND(2,15000),L(14),LA(4),           004
     *INDC(11),NDEL(11),IN(11),LF(11),IC(11),INEUL(22),INEU4(22,4)       005
      M=11                                                               006
      NMASS=15000                                                        007
C                                                                        008
C       IN THE PREVIOUS 7 LINES, ALL NUMBERS 11 SHOULD BE               009
C       REPLACED BY M, THE MILNOR NUMBER OF THE INVESTIGATED            010
C       SINGULARITY, THE NUMBERS 22 BY 2*M, L(14) BY L(M+3),            011
C       AND 15000 BY NMASS, THE POSSIBLE SIZE OF THE QUEUE,             012
C       CALCULATED FROM THE EQUALITY (29) IN CHAPTER V                  013
C                                                                        014
      M3=M+3                                                             015
      MH=M/2                                                             016
      MD=M*2                                                             017
      DO 149 J=1,4                                                       018
      DO 148 I=1,MD                                                      019
  148 INEU4(I,J)=MD                                                      020
  149 LA(J)=0                                                            021
      DO 178 I=1,M                                                       022
      DO 179 J=1,M                                                       023
```

249

```
179 C(I,J)=0                                                     024
    INEUL(I)=MD                                                  025
    INEUL(I+M)=MD                                                026
    IN(I)=0                                                      027
178 INDC(I)=1                                                    028
    CALL ECON(NULL2,NULL1,IN,M)                                  029
    CALL DATA(C,INDC,M,N,N2)                                     030
    MIC=0                                                        031
    MRC=M                                                        032
    NPOZC=MH                                                     033
    INVC=0                                                       034
    KN=0                                                         035
    KA=0                                                         036
    KAO=1                                                        037
    NOP=0                                                        038
    DO 194 I=1,M3                                                039
194 L(I)=1                                                       040
    DO 182 I=1,M                                                 041
    DO 181 J=1,I                                                 042
181 C(I,J)=-N*C(J,I)                                             043
182 C(I,I)=N2*(1-N)                                              044
    CALL LPC(M,C,INDC,N,N2,NPOZC,M,PC1,PC2)                      045
    KSTEP=1                                                      046
    CALL BOUNDARY(M,C,B,PC1,PC2,PB1,N,LA,KSTEP)                  047
    DO 183 I=1,M                                                 048
    DO 184 J=I,M                                                 049
    D(I,J)=-N*C(I,J)                                             050
184 D(J,I)=N*D(I,J)                                              051
    IF(I.GT.NPOZC) GO TO 185                                     052
    IN(I)=-N*PC1(I)                                              053
    NDEL(I)=-N*PC2(I)                                            054
    GO TO 183                                                    055
185 IN(I)=PC1(I)*N+2*N2*INDC(I)                                  056
    NDEL(I)=PC2(I)*N+2*N2*INDC(I)                                057
183 D(I,I)=N*N2*(1+N)                                            058
    N=-N                                                         059
    CALL BOUNDARY(M,D,B,IN,NDEL,PB1,N,LA,KSTEP)                  060
    N=-N                                                         061
    LAM=LA(1)+LA(2)+LA(3)+LA(4)                                  062
    IF(LAM+4) 9000,541,9000                                      063
340 IF(KA.LT.NMASS-MD) GO TO 1114                                064
    IF(KN.LT.NMASS/10.OR.KN.EQ.KA) GO TO 541                     065
```

```
1113 KN1=KN+1
     DO 76 K=KN1,KA
     PARAM(K-KN)=PARAM(K)
     NAIS(K-KN)=NAIS(K)
     DO 76 J=1,2
     P1(J,K-KN)=P1(J,K)
     P2(J,K-KN)=P2(J,K)
     IND(J,K-KN)=IND(J,K)
     DO 76 I=1,M
  76 A(I,J,K-KN)=A(I,J,K)
9631 FORMAT(1H,'SHIFT',I8)
     WRITE (6,9631) KN
6638 FORMAT(10I4)
     WRITE (6,6638) (LA(I),I=1,4)
     KA=KA-KN
     KA0=KA+1
     KN=0
1114 KN=KN+1
     IF(KN.GT.KA) GO TO 541
     CALL DISSIP(P1(2,KN),P1(1,KN),PB1,M)
     CALL DISSIP(P2(2,KN),P2(1,KN),PB2,M)
     CALL DISSIP(IND(2,KN),IND(1,KN),INDB,M)
     DO 79 I=1,M
     CALL DISSIP(A(I,2,KN),A(I,1,KN),IC,M)
     DO 79 J=1,M
  79 B(I,J)=IC(J)
     NPOZB=PARAM(KN)/16
     MIB=PARAM(KN)-16*NPOZB
     MRB=M-2*MIB
     DO 445 NOP=1,M
 445 L(NOP)=1
     L(M+1)=0
     L(M+2)=0
     L(M+3)=0
     IF(MIB.LE.0) GO TO 447
     DO 446 I=1,MIB
     L(I)=0
 446 L(I+MH)=0
     L(MIB)=1
     L(MH)=0
 447 L(MH+MIB)=1
     INVB=NAIS(KN)
```

```
          DO 448 I=1,M3                                       108
          PAR=MOD(INVB,2)                                     109
          L(I)=MAX0(L(I),PAR)                                 110
      448 INVB=INVB/2                                         111
     9201 FORMAT(/I3,I4,I4,I3,I3)                             112
      640 DO 140 NOP=1,M3                                     113
          IF(L(NOP).EQ.0) GO TO 240                           114
      140 CONTINUE                                            115
          GO TO 340                                           116
      240 L(NOP)=1                                            117
          DO 300 I=1,M                                        118
          DO 400 J=1,M                                        119
      400 C(I,J)=B(I,J)                                       120
          PC1(I)=PB1(I)                                       121
          PC2(I)=PB2(I)                                       122
      300 INDC(I)=INDB(I)                                     123
          MIC=MIB                                             124
          MRC=MRB                                             125
          NPOZC=NPOZB                                         126
          IF(NOP-MH) 2000,5000,1040                           127
     1040 IF(NOP-M-1) 3000,4000,1340                          128
     1340 IF(NOP-M-3) 7000,8000,8000                          129
    C                                                         130
    C              ALPHA(NOP)                                 131
    C                                                         132
    C     THE OPERATION ALPHA(J) CHANGES THE J-TH AND (J+1)-TH UPPER   133
    C     (RESP., LOWER) BASIS CYCLES E(J) AND E(J+1) (WHOSE TOTAL     134
    C     NUMBERS ARE  MRB+J AND MRB+J+1 (RESP., MRB+MIB+J AND         135
    C     MRB+MIB+J+1) BY THE FOLLOWING FORMULAE:             136
    C         IN THE UPPER HALF-PLANE:  E(J+1,NEW) = E(J,OLD),     137
    C         E(J,NEW) = E(J+1,OLD) + N*N2*B(J+1,J)*E(J,OLD)   138
    C                                                         139
    C         IN THE LOWER HALF-PLANE:  E(J+1,NEW) = E(J,OLD),     140
    C         E(J,NEW) = E(J+1,OLD) - N2*B(J+1,J)*E(J,OLD),    141
    C         WHERE B(*,*) IS THE INTERSECTION INDEX OF THE    142
    C         CORRESPONDING OLD CYCLES. COMPLEX CONJUGATION    143
    C         TRANSMITS THESE FORMULAE INTO ONE ANOTHER        144
    C                                                         145
     2000 J=NOP+MRB                                           146
          INTI=B(J+1,J)*N*N2                                  147
          DO 2003 I=1,M                                       148
          DO 2003 K=1,M                                       149
```

```
 2003 D(I,K)=B(I,K)                                                   150
 2004 DO 2040 I=1,M                                                   151
      C(I,J+1)=D(I,J)                                                 152
      C(I,J)=D(I,J+1)+INTI*D(I,J)                                     153
      C(J+1,I)=D(J,I)                                                 154
 2040 C(J,I)=D(J+1,I)+INTI*D(J,I)                                     155
      C(J,J+1)=D(J+1,J)+INTI*D(J,J)                                   156
      C(J+1,J)=-C(J,J+1)*N                                            157
      C(J,J)=B(J,J)                                                   158
      C(J+1,J+1)=B(J,J)                                               159
      PC1(J+1)=PB1(J)                                                 160
      PC2(J+1)=PB2(J)                                                 161
      PC1(J)=PB1(J+1)+INTI*PB1(J)                                     162
      PC2(J)=PB2(J+1)+INTI*PB2(J)                                     163
      J=J+MIB                                                         164
      IF(J.GT.M) GO TO 2030                                           165
      DO 2033 I=1,M                                                   166
      DO 2033 K=1,M                                                   167
 2033 D(I,K)=C(I,K)                                                   168
      GO TO 2004                                                      169
 2030 INVC=NOP+MH                                                     170
      GO TO 9000                                                      171
C                                                                     172
C             BETA (NOP - M/2 + 1)                                    173
C                                                                     174
C     THE OPERATION BETA(J+1) CHANGES THE J-TH AND (J+1)-TH UPPER     175
C     (RESP., LOWER) BASIS CYCLES E(J) AND E(J+1) (WHOSE TOTAL        176
C     NUMBERS ARE  MRB+J AND MRB+J+1 (RESP., MRB+MIB+J AND            177
C     MRB+MIB+J+1) BY THE FOLLOWING FORMULAE:                         178
C         IN THE UPPER HALF-PLANE:  E(J,NEW) = E(J+1,OLD),            179
C         E(J+1,NEW) = E(J,OLD) + N*N2*B(J+1,J)*E(J+1,OLD)            180
C                                                                     181
C         IN THE LOWER HALF-PLANE:  E(J,NEW) = E(J+1,OLD),            182
C         E(J+1,NEW) = E(J,OLD) - N2*B(J+1,J)*E(J+1,OLD),            183
C      WHERE B(*,*) IS THE INTERSECTION INDEX OF THE                  184
C      CORRESPONDING OLD CYCLES. COMPLEX CONJUGATION                  185
C      TRANSMITS THESE FORMULAE INTO ONE ANOTHER                      186
C                                                                     187
 3000 J =NOP-MH+MRB                                                   188
      INTI=B(J+1 ,J)*N*N2                                             189
      DO 3003 I=1,M                                                   190
      DO 3003 K=1,M                                                   191
```

```
3003 D(I,K)=B(I,K)                                            192
3004 DO 3040 I=1,M                                            193
     C(I,J)=D(I,J+1)                                          194
     C(I,J+1)=D(I,J)+INTI*D(I,J+1)                            195
     C(J,I)=D(J+1,I)                                          196
3040 C(J+1,I)=D(J,I)+INTI*D(J+1,I)                            197
     C(J,J+1)=D(J+1,J)+INTI*D(J+1,J+1)                        198
     C(J+1,J)=-C(J,J+1)*N                                     199
     C(J,J)=B(J,J)                                            200
     C(J+1,J+1)=B(J,J)                                        201
     PC1(J)=PB1(J+1)                                          202
     PC2(J)=PB2(J+1)                                          203
     PC1(J+1)=PB1(J)+INTI*PB1(J+1)                            204
     PC2(J+1)=PB2(J)+INTI*PB2(J+1)                            205
     J=J+MIB                                                  206
     IF (J.GT.M) GO TO 3030                                   207
     DO 3033 I=1,M                                            208
     DO 3033 K=1,M                                            209
3033 D(I,K)=C(I,K)                                            210
     GO TO 3004                                               211
3030 INVC=NOP-MH                                              212
     GO TO 9000                                               213
C                                                             214
C         COINCIDENCE OF TWO SMALLEST POSITIVE                215
C                                                             216
C                  CRITICAL VALUES                            217
C                                                             218
4000 NJ=NPOZB+1                                               219
     IF(MRB.LE.NJ) GO TO 640                                  220
C                                                             221
C       IF THEIR INTERSECTION INDEX IS NOT EQUAL TO 1, 0 OR -1, 222
C                                                             223
C          OR IS EQUAL TO  -N*N2,  NO COINCIDENCE HAPPENS:    224
C                                                             225
     I=B(NJ,NJ+1)                                             226
     IF(I.NE.0.AND.I.NE.N*N2) GO TO 640                       227
     IF(I.NE.0) GO TO 4080                                    228
C                                                             229
C      IF THE INTERSECTION INDEX IS EQUAL TO  0, THEN THE     230
C              CHANGE OF ALL PARAMETERS IS TRIVIAL:           231
C                                                             232
4020 DO 4050 I=1,M                                            233
```

```
      C(I,NJ)=B(I,NJ+1)                                     234
      C(I,NJ+1)=B(I,NJ)                                     235
      C(NJ,I)=B(NJ+1,I)                                     236
 4050 C(NJ+1,I)=B(NJ,I)                                     237
      PC1(NJ)=PB1(NJ+1)                                     238
      PC2(NJ)=PB2(NJ+1)                                     239
      PC1(NJ+1)=PB1(NJ)                                     240
      PC2(NJ+1)=PB2(NJ)                                     241
      C(NJ,NJ)=B(NJ,NJ)                                     242
      C(NJ+1,NJ+1)=B(NJ,NJ)                                 243
      C(NJ,NJ+1)=0                                          244
      C(NJ+1,NJ)=0                                          245
      INDC(NJ)=INDB(NJ+1)                                   246
      INDC(NJ+1)=INDB(NJ)                                   247
C                                                           248
C        THIS OPERATION IS SELF-INVERSE!                    249
C                                                           250
      INVC=NOP                                              251
      GO TO 9000                                            252
C                                                           253
C     IF THE INTERSECTION INDEX IS EQUAL TO N*N2 AND THE EULER INDICES 254
C     ARE DIFFERENT, THEN THE CRITICAL VALUES COME OUT IN THE COMPLEX   255
C     DOMAIN; WE CHOOSE THE CORRESPONDING PATHS SO THAT THESE VALUES    256
C     BECOME THE LAST (MIB+1)-ST IN THE CORRESPONDING COLLECTIONS       257
C     OF NONREAL CRITICAL VALUES. BY THE PICARD-LEFSCHETZ FORMULA,      258
C     THE NEW BASIS VANISHING CYCLES E'(I) ARE EXPRESSED IN TERMS OF    259
C     THE OLD ONES E(I) BY THE FOLLOWING FORMULAE: E'(I) =              260
C        E(I)                FOR  I  FROM THE SEGMENT [1,NPOZB];        261
C        E(I+2)+N*N2*(B(NJ,I+2)+B(NJ+1,I+2))*(E(NJ)+E(NJ+1))           262
C                            FOR  I FROM THE SEGMENT [NJ=NPOZC+1,MRB-2];263
C        E(I+2)              FOR  I FROM  [MRB-1,MRB+MIB-2];            264
C        E(NJ)+E(NJ+1)       FOR  I = MRB+MIB-1;                        265
C        E(I+1)              FOR  I FROM [MRB+MIB,M-1];                 266
C        -INDB(NJ)*E(NJ+1)   FOR I=M.                                   267
C                                                           268
C        THE COEFFICIENT -INDB(NJ) IN THE LAST EXPRESSION IS DUE        269
C        TO THE CHOICE OF ORIENTATIONS OF THE CYCLES OBTAINED AFTER     270
C        THE COINCIDENCE WHICH ENSURES THAT COMPLEX CONJUGATION         271
C        TRANSMITS THEM INTO EACH OTHER                                 272
C                                                           273
 4080 IF(INDB(NJ).EQ.INDB(NJ+1)) GO TO 640                  274
      MRB1=MRB+1                                            275
```

```
      MRIB=MRB+MIB
      M1=M-1
      NJ2=NJ+2
      DO 4135 I=1,M
 4135 NDEL(I)=B(NJ,I)+B(NJ+1,I)
      IF(NPOZB.LE.0) GO TO 4136
      DO 4090 I=1,NPOZB
      IF(NJ2.GT.MRB) GO TO 4119
      DO 4110 J=NJ2,MRB
 4110 C(I,J-2)=B(I,J)-N2*NDEL(J)*NDEL(I)
 4119 IF(MIB.LE.0) GO TO 4121
      DO 4120 J=MRB1,MRIB
      C(I,J-2)=B(I,J)
 4120 C(I,J+MIB-1)=B(I,J+MIB)
 4121 C(I,MRIB-1)=-N*NDEL(I)
 4090 C(I,M)=-INDB(NJ)*B(I,NJ+1)
 4136 PDEL1=PB1(NJ)+PB1(NJ+1)
      PDEL2=PB2(NJ)+PB2(NJ+1)
      IF(NJ2.GT.MRB) GO TO 4141
      DO 4140 I=NJ2,MRB
      DO 4150 J=I,MRB
 4150 C(I-2,J-2)=B(I,J)
      IF(MIB.LE.0) GO TO 4161
      DO 4160 J=MRB1,MRIB
 4160 C(I-2,J-2)=B(I,J)+N*N2*NDEL(J)*NDEL(I)
      DO 4170 J=MRIB,M1
 4170 C(I-2,J)=B(I,J+1)+N*N2*NDEL(I)*NDEL(J+1)
 4161 C(I-2,MRIB-1)=-NDEL(I)
      C(I-2,M)=-INDB(NJ)*(B(I,NJ+1)+N*N2*NDEL(I)*NDEL(NJ+1))
      PC1(I-2)=PB1(I)+N*N2*NDEL(I)*PDEL1
 4140 PC2(I-2)=PB2(I)+N*N2*NDEL(I)*PDEL2
 4141 IF(MIB.LE.0) GO TO 4207
      DO 4200 I=MRB1,MRIB
      DO 4210 J=I,MRIB
 4210 C(I-2,J-2)=B(I,J)
      C(I-2,MRIB-1)=-N*NDEL(I)
      DO 4220 J=MRIB,M1
 4220 C(I-2,J)=B(I,J+1)
      C(I-2,M)=-INDB(NJ)*B(I,NJ+1)
      PC1(I-2)=PB1(I)
 4200 PC2(I-2)=PB2(I)
      DO 4240 J=MRIB,M1
```

```
4240 C(MRIB-1,J)=NDEL(J+1)                                  318
4207 C(MRIB-1,MRIB-1)=B(1,1)                                319
     C(MRIB-1,M)=-INDB(NJ)*NDEL(NJ+1)                       320
     PC1(MRIB-1)=PDEL1                                      321
     PC2(MRIB-1)=PDEL2                                      322
     IF(MIB.LE.0) GO TO 4307                                323
     DO 4300 I=MRIB,M1                                      324
     DO 4310 J=I,M1                                         325
4310 C(I,J)=B(I+1,J+1)                                      326
     C(I,M)=-INDB(NJ)*B(I+1,NJ+1)                           327
     PC1(I)=PB1(I+1)                                        328
4300 PC2(I)=PB2(I+1)                                        329
4307 C(M,M)=B(M,M)                                          330
     PC1(M)=-INDB(NJ)*PB1(NJ+1)                             331
     PC2(M)=-INDB(NJ)*PB2(NJ+1)                             332
     DO 4400 I=1,M                                          333
     DO 4400 J=1,I                                          334
4400 C(I,J)=-N*C(J,I)                                       335
     MRC=MRB-2                                              336
     MIC=MIB+1                                              337
     INVC=MH                                                338
     IF(NJ.GT.MRB) GO TO 9000                               339
     DO 4500 I=NJ,MRB                                       340
4500 INDC(I)=INDB(I+2)                                      341
     INDC(MRC+1)=0                                          342
     INDC(MRC+2)=0                                          343
     GO TO 9000                                             344
C                                                           345
C    COINCIDENCE OF TWO COMPLEX CONJUGATE CRITICAL VALUES,  346
C    WHICH ARE THE LATEST IN THEIR SERIES (I.E., HAVE       347
C    TOTAL NUMBERS  MRB+MIB  AND  M), IN THE POSITIVE NONCRITICAL  348
C    INTERVAL IMMEDIATELY  OVER  0                          349
C                                                           350
5000 NJ=NPOZB+1                                             351
     MRIB=MRB+MIB                                           352
C                                                           353
C    IF THE INTERSECTION INDEX IS NOT EQUAL TO 1, 0 OR -1,  354
C              NO COINCIDENCE HAPPENS:                      355
C                                                           356
     IF(B(MRIB,M)**2-1) 5010,5020,640                       357
C                                                           358
C    IF THE INTERSECTION INDEX IS EQUAL TO  0, THEN THE CHANGE OF  359
```

```
C          THE BASIS IS AS FOLLOWS: THE CYCLES E(MRB+MIB) AND E(M)        360
C          INTERCHANGGE, ALL CYCLES E(I,NEW) FOR I FROM THE SEGMENT       361
C          [NJ,MRB] ARE EQUAL TO E(I,OLD) +                               362
C             + N*N2*B(I,MRB+MIB)*E(MRB+MIB,OLD) - N2*B(I,M)*E(M,OLD),    363
C      AND ALL OTHER CYCLES STAY UNCHANGED. THEREFORE THE INTERSECTION    364
C          MATRIX AND VALUES OF PETROVSKII CLASSES CHANGE AS FOLLOWS:     365
C                                                                         366
 5010 IF(NJ.GT.MRB) GO TO 5065                                           367
      DO 5035 I=NJ,MRB                                                    368
      DO 5040 J=1,M                                                       369
 5040 C(I,J)=B(I,J)+N*N2*B(I,MRIB)*B(MRIB,J)-N2*B(I,M)*B(M,J)            370
      PC1(I)=PB1(I)+N*N2*B(I,MRIB)*PB1(MRIB)-N2*B(I,M)*PB1(M)            371
 5035 PC2(I)=PB2(I)+N*N2*B(I,MRIB)*PB2(MRIB)-N2*B(I,M)*PB2(M)            372
      IF(NPOZB.LE.0) GO TO 5038                                           373
      DO 5050 I=1,NPOZB                                                   374
      DO 5050 J=NJ,MRB                                                    375
 5050 C(I,J)=-N*C(J,I)                                                    376
 5038 DO 5060 I=NJ,MRB                                                    377
      DO 5060 J=I,MRB                                                     378
 5060 C(I,J)=B(I,J)                                                       379
 5065 DO 5070 I=1,M                                                       380
      D(I,M)=C(I,M)                                                       381
      C(I,M)=C(I,MRIB)                                                    382
 5070 C(I,MRIB)=D(I,M)                                                    383
      PC1(MRIB)=PB1(M)                                                    384
      PC2(MRIB)=PB2(M)                                                    385
      PC1(M)=PB1(MRIB)                                                    386
      PC2(M)=PB2(MRIB)                                                    387
      DO 5080 J=MRIB,M                                                    388
 5080 C(MRIB,J)=-N*C(J,MRIB)                                              389
      C(MRIB,MRIB)=B(1,1)                                                 390
      C(M,M)=B(1,1)                                                       391
      C(MRIB,M)=B(M,MRIB)                                                 392
      DO 5090 I=1,M                                                       393
      DO 5090 J=1,I                                                       394
 5090 C(I,J)=-N*C(J,I)                                                    395
      INVC=NOP                                                            396
      GO TO 9000                                                          397
C                                                                         398
C      IF THE INTERSECTION INDEX IS EQUAL TO 1 OR -1, THEN THE COMPLEX    399
C      CRITICAL POINTS WITH NUMBERS  MRB+MIB  AND M COINCIDE AT A         400
C      POINT OF TYPE A_2 AND BECOME A PAIR OF REAL ONES. THE             401
```

```
C        EULER INDEX OF THE NEWBORN CRITICAL POINT WITH THE SMALLER    402
C        CRITICAL VALUE IS EQUAL TO   INP = -N2*B(MRB+MIB,M), THAT OF THE 403
C        OTHER IS EQUAL TO  -INP.                                        404
C               THE ELEMENTS OF THE NEW VANISHING BASIS ARE             405
C        EXPRESSED IN TERMS OF THE OLD ONES BY THE FOLLOWING FORMULAE:   406
 5020 CONTINUE                                                          407
C                                                                       408
C        E(I,NEW)=E(I,OLD)      FOR I FROM [1,NPOZB];                    409
C        E(NPOZB+1,NEW) = INORM*(E(MRB+MIB,OLD)+INP*E(M,OLD)),          410
C                          WHERE THE COEFFICIENT INORM=1 OR -1 IS       411
C                          CHOSEN SO THAT THE ORIENTATION OF THE        412
C                          TWO NEWBORN REAL CYCLES COINCIDES WITH       413
C                          THE ONE SPECIFIED IN SUBSECTION V.1.F; THE   414
C                          PRECISE VALUE OF THIS COEFFICIENT FOLLOWS    415
C                          FROM THE FORMULAE (V.5), (V.7);             416
C        E(NPOZB+2,NEW) = -INORM*INP*E(M,OLD);                          417
C                                                                       418
C        E(I,NEW) = E(I-2,OLD) + N*N2*B(I-2,MRB+MIB)*E(MRB+MIB,OLD)     419
C                          FOR I FROM [NPOZB+3, MRB+2];                  420
C        E(I,NEW) = E(I-2,OLD)    FOR I FROM [MRB+3,MRB+MIB+1];         421
C        E(I,NEW) = E(I-1,OLD)    FOR I FROM [MRB+MIB+2,M].             422
C                                                                       423
C        THEREFORE THE NEW INTERSECTION MATRIX AND VALUES               424
C          OF PETROVSKII CYCLES ARE CALCULATED AS FOLLOWS:              425
C                                                                       426
      NJ2=NJ+2                                                          427
      MRB2=MRB+2                                                        428
      MRB3=MRB+3                                                        429
      MRIB1=MRIB+1                                                      430
      MRIB2=MRIB+2                                                      431
      INP=-B(MRIB,M)*N2                                                 432
      IF(INP*N.EQ.1) GO TO 6016                                         433
      PC2(NJ)=PB2(MRIB)+INP*PB2(M)                                      434
      INORM=N2*PC2(NJ)/2                                                435
      GO TO 6017                                                        436
 6016 PC1(NJ)=PB1(MRIB)+INP*PB1(M)                                      437
      INORM=-N2*PC1(NJ)/2                                               438
 6017 IF(INORM*INORM.NE.1) GO TO 640                                    439
      IF(NPOZB.LE.0) GO TO 6051                                         440
      DO 6050 I=1,NPOZB                                                 441
      C(I,NJ)=INORM*(B(I,MRIB)+INP*B(I,M))                              442
      C(I,NJ+1)=-INP*INORM*B(I,M)                                       443
```

```
      IF(NJ.GT.MRB) GO TO 6039                                      444
      DO 6030 J=NJ2,MRB2                                            445
 6030 C(I,J)=B(I,J-2)+N*N2*B(J-2,MRIB)*B(I,MRIB)                    446
 6039 IF(2.GT.MIB) GO TO 6050                                       447
      DO 6040 J=MRB3,MRIB1                                          448
 6040 C(I,J)=B(I,J-2)                                               449
      DO 6047 J=MRIB2,M                                             450
 6047 C(I,J)=B(I,J-1)                                               451
 6050 CONTINUE                                                      452
 6051 C(NJ,NJ)=B(1,1)                                               453
      C(NJ,NJ+1)=N*N2                                               454
      IF(NJ.GT.MRB) GO TO 6068                                      455
      DO 6060 J=NJ2,MRB2                                            456
 6060 C(NJ,J)=INP*INORM*B(M,J-2)                                    457
 6068 IF(2.GT.MIB) GO TO 6071                                       458
      DO 6070 J=MRB3,MRIB1                                          459
 6070 C(NJ,J)=INORM*(B(MRIB,J-2)+INP*B(M,J-2))                      460
      DO 6072 J=MRIB2,M                                             461
 6072 C(NJ,J)=INORM*(B(MRIB,J-1)+INP*B(M,J-1))                      462
 6071 PC1(NJ)=INORM*(PB1(MRIB)+INP*PB1(M))                          463
      PC2(NJ)=INORM*(PB2(MRIB)+INP*PB2(M))                          464
      C(NJ+1,NJ+1)=B(1,1)                                           465
      IF(NJ.GT.MRB) GO TO 6098                                      466
      DO 6090 J=NJ2,MRB2                                            467
 6090 C(NJ+1,J)=-INORM*(INP*B(M,J-2)+B(J-2,MRIB))                   468
 6098 IF(2.GT.MIB) GO TO 6108                                       469
      DO 6100 J=MRB3,MRIB1                                          470
 6100 C(NJ+1,J)=-INP*INORM*B(M,J-2)                                 471
      DO 6110 J=MRIB2,M                                             472
 6110 C(NJ+1,J)=-INP*INORM*B(M,J-1)                                 473
 6108 PC1(NJ+1)=-INP*INORM*PB1(M)                                   474
      PC2(NJ+1)=-INP*INORM*PB2(M)                                   475
      IF(NJ.GT.MRB) GO TO 6021                                      476
      DO 6120 I=NJ2,MRB2                                            477
      DO 6130 J=I,MRB2                                              478
 6130 C(I,J)=B(I-2,J-2)                                             479
      IF(2.GT.MIB) GO TO 6151                                       480
      DO 6140 J=MRB3,MRIB1                                          481
 6140 C(I,J)=B(I-2,J-2)+N*N2*B(I-2,MRIB)*B(MRIB,J-2)                482
      DO 6150 J=MRIB2,M                                             483
 6150 C(I,J)=B(I-2,J-1)+N*N2*B(I-2,MRIB)*B(MRIB,J-1)                484
 6151 PC1(I)=PB1(I-2)   +N*N2*B(I-2,MRIB)*PB1(MRIB)                 485
```

```
6120 PC2(I)=PB2(I-2)   +N*N2*B(I-2,MRIB)*PB2(MRIB)              486
6021 IF(2.GT.MIB) GO TO 6191                                   487
     DO 6160 I=MRB3,MRIB1                                      488
     DO 6170 J=I,MRIB1                                         489
6170 C(I,J)=B(I-2,J-2)                                         490
     DO 6180 J=MRIB2,M                                         491
6180 C(I,J)=B(I-2,J-1)                                         492
     PC1(I)=PB1(I-2)                                           493
6160 PC2(I)=PB2(I-2)                                           494
     DO 6190 I=MRIB2,M                                         495
     DO 6200 J=1,M                                             496
6200 C(I,J)=B(I-1,J-1)                                         497
     PC1(I)=PB1(I-1)                                           498
6190 PC2(I)=PB2(I-1)                                           499
6191 DO 6220 I=1,M                                             500
     DO 6220 J=1,I                                             501
6220 C(I,J)=-N*C(J,I)                                          502
     INVC=M+1                                                  503
     MRC=MRB+2                                                 504
     MIC=MIB-1                                                 505
     INDC(NJ)=INP                                              506
     INDC(NJ+1)=-INP                                           507
     IF(NJ2.GT.MRC) GO TO 9000                                 508
     DO 6210 I=NJ2,MRC                                         509
6210 INDC(I)=INDB(I-2)                                         510
     GO TO 9000                                                511
7000 IF(NPOZB.GE.MRB) GO TO 640                                512
C                                                              513
C        JUMP OF THE SMALLEST POSITIVE CRITICAL VALUE OVER     514
C        ZERO (AFTER WHICH IT BECOMES THE GREATEST NEGATIVE)   515
C                                                              516
C        IN THE NEW BASIS, ALL ELEMENTS E(I) WITH  I   FROM    517
C        THE SEGMENT [1,MRB+MIB]  ARE NATURALLY IDENTIFIED     518
C        WITH THE OLD ONES, WHILE FOR I FROM [MRB+MIB+1,M]     519
C        THEY ARE IDENTIFIED WITH  THE CYCLES   E(I,OLD) +     520
C          +N*N2*B(I,NPOZB+1)*E(NPOZB+1,OLD).                  521
C        THEREFORE THE INTERSECTION MATRIX CHANGES AS FOLLOWS: 522
C                                                              523
     NJ=NPOZB+1                                                524
     MRIB=MRB+MIB                                              525
     MRIB1=MRIB+1                                              526
     IF(MIB.EQ.0) GO TO 7023                                   527
```

```
      DO 7020 I=1,M                                          528
      DO 7021 J=MRIB1,M                                      529
7021 C(I,J)=B(I,J)+N*N2*B(I,NJ)*B(J,NJ)                      530
7020 CONTINUE                                                531
      DO 7030 I=MRIB1,M                                      532
      DO 7030 J=I,M                                          533
7030 C(I,J)=B(I,J)                                           534
7023 DO 7040 I=1,M                                           535
      DO 7040 J=1,I                                          536
7040 C(I,J)=-N*C(J,I)                                        537
C                                                            538
C     ACCORDING TO THEOREM V.3.2, THE VALUE OF THE NEW EVEN  539
C     PETROVSKII CYCLE P2 ON THE VANISHING CYCLES E(I), I FROM THE  540
C     SEGMENT [1,MRIB], IS EQUAL TO THE EXPRESSION UNDER THE LABEL  541
C     7050. IT FOLLOWS FROM THEOREM V.4.2 AND COROLLARY V.4.1,  542
C     THAT THE ODD PETROVSKII CYCLE P1 CHANGES AS INDICATED  543
C     IN THE STATEMENT BEFORE THE LABEL 7050                 544
C                                                            545
      DO 7050 I=1,MRIB                                       546
      PC1(I)=PB1(I)+B(I,NJ)                                  547
7050 PC2(I)=PB2(I)-B(I,NJ)                                   548
C                                                            549
C     FOR  I  FROM THE SEGMENT [MRIB+1,M], SIMILARLY AND     550
C     BY THE PICARD-LEFSCHETZ FORMULA,                       551
C                                                            552
      IF(MIB.EQ.0) GO TO 7061                                553
      DO 7060 I=MRIB1,M                                      554
      PC1(I)=PB1(I)+B(I,NJ)+N*N2*B(I,NJ)*(PB1(NJ)+B(1,1))    555
7060 PC2(I)=PB2(I)-B(I,NJ)+N*N2*B(I,NJ)*(PB2(NJ)-B(1,1))     556
7061 NPOZC=NPOZB+1                                           557
      INVC=NOP+1                                             558
      GO TO 9000                                             559
8000 IF(NPOZB.EQ.0) GO TO 640                                560
C                                                            561
C     JUMP OF THE GREATEST NEGATIVE CRITICAL VALUE OVER      562
C     ZERO (AFTER WHICH IT BECOMES THE SMALLEST POSITIVE)    563
C                                                            564
C      IN THE NEW BASIS, ALL ELEMENTS E(I) WITH  I  FROM     565
C     THE SEGMENT [1,MRB+MIB]  ARE NATURALLY IDENTIFIED      566
C     WITH THE OLD ONES, WHILE FOR I FROM [MRB+MIB+1,M]      567
C     THEY ARE IDENTIFIED WITH  THE CYCLES    E(I,OLD) -     568
C      - N2*B(I,NPOZB)*E(NPOZB,OLD).                         569
```

```
C       THEREFORE THE INTERSECTION MATRIX CHANGES AS FOLLOWS:            570
C                                                                        571
        MRIB=MRB+MIB                                                     572
        MRIB1=MRIB+1                                                     573
        IF(MIB.EQ.0) GO TO 8041                                         574
        DO 8021 I=1,M                                                    575
        DO 8020 J=MRIB1,M                                                576
 8020 C(I,J)=B(I,J)-N2*B(I,NPOZB)*B(J,NPOZB)                             577
 8021 CONTINUE                                                           578
        DO 8030 I=MRIB1,M                                                579
        DO 8030 J=I,M                                                    580
 8030 C(I,J)=B(I,J)                                                      581
 8041 DO 8040 I=1,M                                                      582
        DO 8040 J=1,I                                                    583
 8040 C(I,J)=-N*C(J,I)                                                   584
C                                                                        585
C       ACCORDING TO THEOREM V.3.2, THE VALUE OF THE NEW EVEN            586
C       PETROVSKII CYCLE P2 ON THE VANISHING CYCLES E(I), I FROM THE     587
C       SEGMENT [1,MRIB], IS EQUAL TO THE EXPRESSION UNDER THE LABEL     588
C       7050. IT FOLLOWS FROM THEOREM V.4.2 AND COROLLARY V.4.1,         589
C       THAT THE ODD PETROVSKII CYCLE P1 CHANGES AS INDICATED            590
C       IN THE STATEMENT BEFORE THE LABEL 7050                           591
C                                                                        592
        DO 8050 I=1,MRIB                                                 593
        PC1(I)=PB1(I)-B(I,NPOZB)                                         594
 8050 PC2(I)=PB2(I)+B(I,NPOZB)                                           595
C                                                                        596
C       FOR  I  FROM THE SEGMENT [MRIB+1,M], SIMILARLY AND               597
C       BY THE PICARD-LEFSCHETZ FORMULA,                                 598
C                                                                        599
        IF(MIC.LE.0) GO TO 8062                                         600
        DO 8060 I=MRIB1,M                                                601
        PC1(I)=PB1(I)-B(I,NPOZB)-N2*B(I,NPOZB)*(PB1(NPOZB)-B(1,1))       602
 8060 PC2(I)=PB2(I)+B(I,NPOZB)-N2*B(I,NPOZB)*(PB2(NPOZB)+B(1,1))         603
 8062 NPOZC=NPOZB-1                                                      604
        INVC=NOP-1                                                       605
 9000 IF(KA.NE.3000*(KA/3000)) GO TO 1640                               606
        CALL PECH(M,C,PC1,PC2,INDC,NOP,KA,KN,MRC,NPOZC)                 607
C                                                                        608
C       CHECK WHETHER THE ELEMENTS OF THE INTERSECTION MATRICES          609
C           AND PETROVSKII COVECTORS ARE NOT TOO LARGE                   610
C                                                                        611
```

```
7344 FORMAT(1H,'TOO LARGE ENTIRES',I7)                           612
1640 DO 7341 I=1,M                                               613
     DO 7342 J=1,I                                               614
     IF(C(I,J)**2.GT.150) GO TO 7343                             615
7342 CONTINUE                                                    616
     IF(PC1(I)**2.GT.150) GO TO 7343                             617
     IF(INDC(I)**2.GT.4) GO TO 7343                              618
     IF(PC2(I)**2.LE.150) GO TO 7341                             619
7343 WRITE(6,7344) KA                                            620
     CALL PECH(M,C,PC1,PC2,INDC,NOP,KA,KN,MRC,NPOZC)             621
     GO TO 541                                                   622
7341 CONTINUE                                                    623
C                                                                624
C     OPTIMIZATION OF THE INTERSECTION MATRIX OVER ALL COMPATIBLE 625
C                                                                626
C          ORIENTATIONS OF COMPLEX VANISHING CYCLES             627
C                                                                628
C     IS INTENDED FOR THE COLLECTIONS OF INDICES THAT DIFFER ONLY 629
C     IN THE CHOICE OF ORIENTATIONS OF SUCH CYCLES DO NOT TAKE  630
C     DIFFERENT PLACES IN THE QUEUE.                            631
C        IS ORGANIZED AS FOLLOWS. AT ANY INSTANT, THE SET OF ALL 632
C     VANISHING CYCLES IS DIVIDED INTO THE ONES WITH FIXED OR   633
C     QUESTIONABLE ORIENTATIONS. AT THE FIRST INSTANT, ALL      634
C     THE IMAGINARY CYCLES ARE QUESTIONABLE. IN THE EXCEPTIONAL 635
C     CASE, WHEN THERE ARE NO REAL CRITICAL POINTS, THE FIRST   636
C     PAIR OF COMPLEX CONJUGATE POINTS (I.E., THESE WITH NUMBERS 637
C                                                                638
     IF(MIC.LE.0) GO TO 531                                      639
C                                                                640
C     1 AND MIC+1) IS DECLARED TO BE FIXED. AT ANY STEP WE TAKE 641
C     THE FIRST QUESTIONABLE CYCLE E(J) SUCH THAT AT LEAST ONE  642
C     OF ITS INTERSECTION INDICES WITH THE FIXED CYCLES IS NONZERO, 643
C     AND CHOOSE THE ORIENTATION OF THIS CYCLE (AND THUS ALSO OF ITS 644
C     COMPLEX CONJUGATE) WHICH LEXICOGRAPHICALLY OPTIMIZES THE SET 645
C     OF SUCH INTERSECTION INDICES WITH THE FIXED CYCLES. THEN  646
C     WE FIX THIS ORIENTATION OF THIS CYCLE AND ITS COMPLEX CONJUGATE,647
C     THUS THESE TWO CYCLES ARE TRANSFERRED INTO THE FIXED CATEGORY. 648
C          BY THE CONNECTEDNESS OF THE DYNKIN DIAGRAM, THIS PROCESS 649
C     LEADS TO UNIQUE ORIENTATION OF ALL VANISHING CYCLES (MAYBE 650
C     UP TO SIMULTANEOUS CHANGE OF ALL OF THEM), IN PARTICULAR, THE 651
C     NEW OPTIMIZED INTERSECTION MATRIX IS UNIQUELY DETERMINED  652
C                                                                653
```

```
        MRIC=MRC+MIC                                                       654
        IF(MRC.LE.0) MRC=1                                                 655
        MRC1=MRC+1                                                         656
        DO 1610 J=MRC1,M                                                   657
 1610 LF(J)=0                                                              658
        DO 1620 I=1,MRC                                                    659
 1620 LF(I)=1                                                              660
        IF(M.LE.2*MIC) LF(MIC+1)=1                                         661
 1656 DO 1655 J=MRC1,M                                                     662
        DO 1650 I=1,M                                                      663
        IF(LF(I).NE.1.OR.LF(J).NE.0) GO TO 1650                           664
        IF(C(I,J)) 1670,1650,1680                                          665
 1670 DO 1675 K=1,M                                                        666
        C(J,K)=-C(J,K)                                                     667
 1675 C(K,J)=-C(K,J)                                                       668
        PC1(J)=-PC1(J)                                                     669
        PC2(J)=-PC2(J)                                                     670
        IF(J.GT.MRIC) GO TO 1700                                           671
        DO 1695 K=1,M                                                      672
        C(J+MIC,K)=-C(J+MIC,K)                                             673
 1695 C(K,J+MIC)=-C(K,J+MIC)                                               674
        PC1(J+MIC)=-PC1(J+MIC)                                             675
        PC2(J+MIC)=-PC2(J+MIC)                                             676
        GO TO 1680                                                         677
 1700 DO 1705 K=1,M                                                        678
        C(J-MIC,K)=-C(J-MIC,K)                                             679
 1705 C(K,J-MIC)=-C(K,J-MIC)                                               680
        PC1(J-MIC)=-PC1(J-MIC)                                             681
        PC2(J-MIC)=-PC2(J-MIC)                                             682
 1680 LF(J)=1                                                              683
        IF(J.GT.MRIC) GO TO 1720                                           684
        LF(J+MIC)=1                                                        685
        GO TO 1656                                                         686
 1720 LF(J-MIC)=1                                                          687
        GO TO 1656                                                         688
 1650 CONTINUE                                                             689
 1655 CONTINUE                                                             690
CC                                                                        691
CC    IF MRC=0, THEN THE OPTIMAL MATRIX IS PROVIDED BY TWO OPPOSITE       692
CC      COLLECTIONS OF ORIENTATIONS. THE CHOICE OF ONE OF THEM IS         693
CC    PROVIDED BY THE LEXICOGRAPHIC OPTIMIZATION OF THE PETROVSKII        694
CC    CLASSES.                                                            695
```

```
CC                                                                      696
      IF(M.NE.2*MIC) GO TO 1607                                         697
      DO 1609 I=1,M                                                     698
      IF(PC1(I)) 1608,1604,1607                                         699
 1604 IF(PC2(I)) 1608,1609,1607                                         700
 1609 CONTINUE                                                          701
 1608 DO 1603 I=1,M                                                     702
      PC1(I)=-PC1(I)                                                    703
 1603 PC2(I)=-PC2(I)                                                    704
 1607 MRC=M-2*MIC                                                       705
C                                                                       706
C    COMPARISON AND WRITING A VIRTUAL MORSIFICATION IN THE QUEUE        707
C                                                                       708
 9501 FORMAT(1H,'COINC',I5)                                            709
  531 KA1=KA+1                                                          710
      DO 99 I=1,M                                                       711
      DO 81 J=1,M                                                       712
   81 IC(J)=C(I,J)                                                      713
   99 CALL ECON(A(I,2,KA1),A(I,1,KA1),IC,M)                             714
      CALL ECON(P1(2,KA1),P1(1,KA1),PC1,M)                              715
      CALL ECON(P2(2,KA1),P2(1,KA1),PC2,M)                              716
      CALL ECON(IND(2,KA1),IND(1,KA1),INDC,M)                           717
      PARAM(KA1)=MIC+16*NPOZC                                           718
      IF(KA.EQ.0) GO TO 931                                             719
 4477 FORMAT(2I21)                                                      720
      DO 130 K=KN,KA                                                    721
      DO 230 J=1,2                                                      722
      DO 330 I=1,M                                                      723
      IF(A(I,J,KA1).NE.A(I,J,K)) GO TO 130                              724
  330 CONTINUE                                                          725
      IF(P1(J,KA1).NE.P1(J,K)) GO TO 130                                726
      IF(P2(J,KA1).NE.P2(J,K)) GO TO 130                                727
      IF(IND(J,KA1).NE.IND(J,K)) GO TO 130                              728
  230 CONTINUE                                                          729
      IF(PARAM(KA1).NE.PARAM(K)) GO TO 130                              730
      NAIS(K)=NAIS(K)+2**(INVC-1)                                       731
      GO TO 640                                                         732
  130 CONTINUE                                                          733
  931 IF(KA.GE.NMASS-1) GO TO 541                                       734
      KA=KA1                                                            735
      NAIS(KA)=0                                                        736
      IF(INVC.NE.0) NAIS(KA)=2**(INVC-1)                                737
```

```
C                                                                      738
C       KSTEP=1 DENOTES THE INVESTIGATION OF THE ORIGINAL              739
C                 SINGULARITY F(X_1, ..., X_N),                        740
C       KSTEP=2 OF ITS STABILIZATION F-X_{N+1}^2+X_{N+2}^2,            741
C       KSTEP=3 OF F-X_{N+1}^2,   AND                                  742
C       KSTEP=4 OF F+X_{N+2}^2.                                        743
C                                                                      744
        KSTEP=1                                                        745
        IF(P1(1,KA1).NE.NULL1.OR.P1(2,KA1).NE.NULL2) GO TO 102         746
        IF(N.LT.0) GO TO 549                                           747
        KSTEP=2                                                        748
        GO TO 549                                                      749
C                                                                      750
C       HERE COROLLARY TO THEOREM V.1.3 IS USED: BOTH PETROVSKII       751
C                 CYCLES DO NOT VANISH SIMULTANEOUSLY                  752
C                                                                      753
    102 IF(P2(1,KA1).NE.NULL1.OR.P2(2,KA1).NE.NULL2) GO TO 103         754
        IF(N.GT.0) GO TO 549                                           755
        KSTEP=2                                                        756
        GO TO 549                                                      757
    103 KSTEP=3                                                        758
        LAA=LA(3)+LA(4)                                                759
        IF(LAA.EQ.-2) GO TO 640                                        760
        DO 745 I=1,M                                                   761
        IF(I.GT.NPOZC) GO TO 743                                       762
        PC1(I)=-N*PC1(I)                                               763
        PC2(I)=-N*PC2(I)                                               764
        GO TO 745                                                      765
    743 IF(I.GT.MRC) GO TO 745                                         766
        PC1(I)=N*PC1(I)+2*N2*INDC(I)                                   767
        PC2(I)=N*PC2(I)+2*N2*INDC(I)                                   768
    745 CONTINUE                                                       769
        DO 746 I=1,M                                                   770
        IF(PC1(I).NE.0) GO TO 106                                      771
    746 CONTINUE                                                       772
        IF(N.GT.0) GO TO 549                                           773
        KSTEP=4                                                        774
        GO TO 549                                                      775
    106 DO 747 I=1,M                                                   776
        IF(PC2(I).NE.0) GO TO 640                                      777
    747 CONTINUE                                                       778
        IF(N.LT.0) GO TO 549                                           779
```

```
      KSTEP=4                                                      780
  549 DO 217 I=1,MD                                                781
  217 INEUL(I)=INEU4(I,KSTEP)                                      782
      LAC=LA(KSTEP)                                                783
      CALL NLAC(M,NPOZC,INDC,INEUL,MIC,LAC,MD,KSTEP)               784
      IF(LAC.NE.LA(KSTEP)) CALL PECH(M,C,PC1,PC2,INDC,NOP,KA,      785
     *KN,MRC,NPOZC)                                                786
      LA(KSTEP)=LAC                                                787
      INEU4(LAC,KSTEP)=INEUL(LAC)                                  788
      INEU4(LAC+M,KSTEP)=INEUL(LAC+M)                              789
      IF(KSTEP.LE.2) GO TO 103                                     790
      GO TO 640                                                    791
 1234 FORMAT(1H,'THERE ARE NO LACUNAS',I3)                         792
  541 CONTINUE                                                     793
      DO 75 K=1,4                                                  794
      IF(LA(K).LE.0) WRITE(6,1234)    K                            795
   75 CONTINUE                                                     796
      WRITE(6,4477) KA,KN                                          797
      STOP                                                         798
      END                                                          799
      SUBROUTINE DISSIP(ISA2,ISA1,IC,M)                            800
      DIMENSION IC(M)                                              801
      ICA2=ISA2                                                    802
      ICA1=ISA1                                                    803
      DO 1 J=1,M                                                   804
      IC(J)=ICA1-32*(ICA1/32)-16                                   805
      ICA1=ICA1/32+(2**25)*(ICA2-32*(ICA2/32))                     806
    1 ICA2=ICA2/32                                                 807
      RETURN                                                       808
      END                                                          809
      SUBROUTINE ECON(ICA2,ICA1,IC,M)                              810
      DIMENSION IC(M)                                              811
      ICA2=0                                                       812
      ICA1=0                                                       813
      MINN=MINO(M,6)                                               814
      DO 1 J=1,MINN                                                815
    1 ICA1=ICA1+(IC(J)+16)*(32**(J-1))                             816
      IF(M.LE.6) GO TO 3                                           817
      DO 2 J=7,M                                                   818
    2 ICA2=ICA2+(IC(J)+16)*(32**(J-7))                             819
    3 CONTINUE                                                     820
      RETURN                                                       821
```

```
         END                                                      822
         SUBROUTINE PECH(M,C,PC1,PC2,INDC,NOP,KA,KN,MRC,NPOZC)     823
         INTEGER C(M,M),PC1(M),PC2(M)                              824
         DIMENSION INDC(M)                                         825
       8 FORMAT (24I5)                                             826
       7 FORMAT (/24I5)                                            827
       1 FORMAT (/I6,I6,I6,I5,I5)                                  828
       3 FORMAT (1H,'  NOP   KA    KN   MRC NPOZC')                829
         WRITE (6,3)                                               830
         WRITE (6,1) NOP,KA,KN,MRC,NPOZC                           831
         DO 2 I=1,M                                                832
       2 WRITE (6,8) (C(I,J),J=1,M)                                833
         WRITE (6,7) (PC1(I),I=1,M)                                834
         WRITE (6,8) (PC2(I),I=1,M)                                835
         WRITE (6,7) (INDC(I),I=1,M)                               836
         RETURN                                                    837
         END                                                       838
         SUBROUTINE LPC(M,C,INDC,N,N2,NPOZC,MRC,PC1,PC2)           839
         INTEGER C(M,M),PC1(M),PC2(M)                              840
         DIMENSION INDC(M)                                         841
         DO 6 I=1,MRC                                              842
         IF(I.GT.NPOZC) GO TO 4                                    843
         PC1(I)=-N*N2*(1+INDC(I))                                  844
         PC2(I)=N*N2*(1-INDC(I))                                   845
         J=I+1                                                     846
       7 IF(J.GT.NPOZC) GO TO 6                                    847
         PC1(I)=PC1(I)+C(I,J)                                      848
         PC2(I)=PC2(I)-C(I,J)                                      849
         J=J+1                                                     850
         GO TO 7                                                   851
       4 PC1(I)=-N2*(1+N*INDC(I))                                  852
         PC2(I)=N2*(1-N*INDC(I))                                   853
         J=I-1                                                     854
       8 IF(J.LE.NPOZC) GO TO 6                                    855
         PC1(I)=PC1(I)+N*C(J,I)                                    856
         PC2(I)=PC2(I)-N*C(J,I)                                    857
         J=J-1                                                     858
         GO TO 8                                                   859
       6 CONTINUE                                                  860
         RETURN                                                    861
         END                                                       862
         SUBROUTINE BOUNDARY(M,C,B,PC1,PC2,PB1,N,LA,KSTEP)         863
```

```
       INTEGER C(M,M),B(M,M),PC1(M),PC2(M),PB1(M),LA(4)              864
C                                                                    865
C      IF FOR AT LEAST ONE MORSIFICATION OF A GIVEN SINGULARITY      866
C      THE BOUNDARY OF THE PETROVSKII CLASS IS A NONZERO HOMOLOGY    867
C      CLASS (OR, EQUIVALENTLY, THIS CLASS DOES NOT LIE IN THE       868
C      IMAGE OF THE ABSOLUTE HOMOLOGY GROUP), THEN THE SAME IS TRUE  869
C      FOR ANY OTHER MORSIFICATION, AND HENCE THE GIVEN SINGULARITY  870
C      HAS NO LACUNAS                                                871
C                                                                    872
       DO 29 I=1,M                                                   873
       DO 39 J=1,M                                                   874
    39 B(I,J)=C(I,J)                                                 875
       PB1(I)=PC1(I)                                                 876
       IF(N.GT.0) PB1(I)=PC2(I)                                      877
    29 CONTINUE                                                      878
     1 DO 49 J=1,M                                                   879
       DO 12 I=1,M                                                   880
       IF(B(I,J).GE.0) GO TO 12                                      881
       DO 11 JJ=J,M                                                  882
    11 B(I,JJ)=-B(I,JJ)                                              883
    12 CONTINUE                                                      884
     3 IM=1                                                          885
       DO 4 I=2,M                                                    886
       IF(B(IM,J).EQ.0.OR.B(I,J).NE.0.AND.B(I,J).LT.B(IM,J)) IM=I    887
     4 CONTINUE                                                      888
       DO 7 I=1,M                                                    889
       IF(B(I,J).EQ.0.OR.I.EQ.IM) GO TO 7                            890
       DO 8 JJ=J,M                                                   891
     8 B(I,JJ)=B(I,JJ)-B(IM,JJ)                                      892
       GO TO 3                                                       893
     7 CONTINUE                                                      894
   112 FORMAT(1H,'THE BOUNDARY IS NONTRIVIAL',I3)                    895
       IF(B(IM,J).EQ.0.AND.PB1(J).NE.0) GO TO 69                     896
       IF(B(IM,J).EQ.0.AND.PB1(J).EQ.0) GO TO 49                     897
       IF(MOD(PB1(J),B(IM,J)).NE.0) GO TO 69                         898
       K=PB1(J)/B(IM,J)                                              899
       DO 19 JJ=J,M                                                  900
       PB1(JJ)=PB1(JJ)-K*B(IM,JJ)                                    901
    19 B(IM,JJ)=0                                                    902
       GO TO 49                                                      903
    69 WRITE(6,112) KSTEP                                            904
       LA(KSTEP)=-1                                                  905
```

```
      GO TO 18
   49 CONTINUE
   18 KSTEP=KSTEP+1
      IF(KSTEP.EQ.3.OR.KSTEP.EQ.5) GO TO 17
      DO 28 I=1,M
      DO 38 J=1,M
   38 B(I,J)=C(I,J)
      PB1(I)=PC2(I)
      IF(N.GT.0) PB1(I)=PC1(I)
   28 CONTINUE
      GO TO 1
   17 CONTINUE
      RETURN
      END
      SUBROUTINE NLAC(M,NPOZC,INDC,INEUL,MIC,LAC,MD,KSTEP)
      DIMENSION INDC(M),INEUL(MD)
    9 FORMAT (1H,'LACUNA!',I3,I5,I5,I5)
    8 FORMAT (1H,'      INDNEG NLAC  MIC STEP')
      J=0
      IF(NPOZC.EQ.0) GO TO 2
      DO 4 I=1,NPOZC
    4 J=J+INDC(I)
    2 DO 6 I=1,M
      IF(J.NE.INEUL(I)) GO TO 6
      IF(MIC.NE.0.OR.INEUL(I+M).EQ.J) GO TO 1
      INEUL(I+M)=J
    6 CONTINUE
      DO 3 I=1,M
      IF(INEUL(I).NE.MD) LAC=I
    3 CONTINUE
      LAC=LAC+1
      WRITE (6,8)
      WRITE (6,9)  J,LAC,MIC,KSTEP
      INEUL(LAC)=J
      IF(MIC.EQ.0) INEUL(LAC+M)=J
    1 CONTINUE
      RETURN
      END
CCCCCCCCCCCCCCC
      SUBROUTINE DATA(C,INDC,M,N,N2)
      INTEGER C(M,M)
      DIMENSION INDC(M)
```

```
1 FORMAT (///1H,'    THE SINGULARITY Y(1R)')
  WRITE(6,1)
  N=1
  N2=-1
  C(1,2)=-1
  C(1,3)=-1
  C(1,4)=-1
  C(1,5)=-1
  C(1,6)=-1
  C(1,7)=1
  C(1,8)=1
  C(1,9)=1
  C(1,10)=1
  C(1,11)=1
  C(2,7)=-1
  C(2,8)=-1
  C(3,8)=-1
  C(3,9)=-1
  C(4,9)=-1
  C(4,10)=-1
  C(5,10)=-1
  C(5,11)=-1
  C(6,7)=-1
  C(6,11)=-1
  INDC(2)=-1
  INDC(3)=-1
  INDC(4)=-1
  INDC(5)=-1
  INDC(6)=-1
  RETURN
  END
```

FINAL PRINT-OUT FOR THE SINGULARITY $+E_6$

```
 THE SINGULARITY +E(6)
NOP    KA    KN   MRC NPOZC
```

```
   0     0     0    6    3
  0   -1   -1   -1    1    1
  1    0    0    0   -1    0
  1    0    0    0   -1   -1
  1    0    0    0    0   -1
 -1    1    1    0    0    0
 -1    0    1    1    0    0

  0    0    0    0    2    1
  2   -2   -2   -2    0    1

  1   -1   -1   -1    1    1
     INDNEG NLAC  MIC STEP
LACUNA!  1    1    0    4
   NOP    KA    KN   MRC NPOZC

   9    10     4    6    1
  0   -1   -1   -1    1    1
  1    0    0    0   -1    0
  1    0    0    0   -1   -1
  1    0    0    0    0   -1
 -1    1    1    0    0    0
 -1    0    1    1    0    0

 -2    2    2    2   -2   -2
  0    0    0    0    0    0

  1   -1   -1   -1    1    1
     INDNEG NLAC  MIC STEP
LACUNA! -3    1    0    3
   NOP    KA    KN   MRC NPOZC

   8   156   117    6    3
  0    0    0   -1   -1    0
  0    0    0   -1    0    0
  0    0    0   -1    0   -1
  1    1    1    0    0    0
  1    0    0    0    0    0
  0    0    1    0    0    0

  0    0    0    0    0    0
  2    2    2   -2   -2   -2
```

```
  -1   -1   -1    1    1    1
THERE ARE NO LACUNAS  1
THERE ARE NO LACUNAS  2
                 456                    457
Stop - Program terminated
```

Bibliography

[A'Campo 75] N. A'Campo, *Le groupe de monodromie du déploiement des singularités isolées de courbes planes. I*, Math. Ann. 213:1 (1975), 1–32.

[A'Campo 75'] N. A'Campo, *Le groupe de monodromie du déploiement des singularités isolées de courbes planes. II*, In: Actes du Congres Internationale des Mathematiciens (Vancouver, 1974), 1975, vol.1, S.1, 395–404.

[A'Campo 79] N. A'Campo, *Tresses, monodromie et le groupe symplectique.* Comm. Math. Helv. 54 (1979), 318-327.

[Arnold 71] V.I. Arnold, *On the disposition of ovals of real plane algebraic curves, involutions of fourdimensional smooth manifolds and arithmetics of integer quadratic forms*, Funkts. Anal. i Prilozh., 5:3 (1971), 1–9; Engl. translation in Functional Anal. Appl., 5 (1971).

[Arnold 72] V.I. Arnold, *Normal forms of functions close to degenerate critical points, the Weyl groups A_k, D_k, E_k, and Lagrange singularities.* Funkts. Anal. i Prilozh., 6:4 (1972), 3–25; Engl. translation in Functional Anal. Appl., 6 (1972), 254–272.

[Arnold 73] V.I. Arnold, *Remarks on the stationary phase method and Coxeter numbers*, Uspekhi Mat. Nauk, 28:5, 1973, 17–44; Engl. transl. in Russian Math. Surveys, 28:5 (1973), 19–48.

[Arnold 74] V.I. Arnold, *Normal forms of functions in neighborhoods of degenerate critical points,* Uspekhi Mat. Nauk 29:2 (1974), 11–49, Engl. transl. in Russian Math. Surveys 29:2 (1974), 10–50.

[Arnold 74'] V.I. Arnold, *Critical points of functions and classification of caustics*, Uspekhi Mat. Nauk 29:3 (1974), 243–244.

[Arnold 75] V.I. Arnold, *Critical points of smooth functions and their normal forms*, Uspekhi Mat. Nauk 30 (1975), no.5, 3–65, Engl. transl. in Russian Math. Surveys 30 (1975), no.5, 1–75.

[Arnold 78] V.I. Arnold, *Critical points of functions on manifolds with boundary, the simple Lie groups* B_k, C_k, F_4, *and singularities of evolutes*, Uspekhi Mat. Nauk 33 (1978), no.5, 91–105, Engl. transl. in Russian Math. Surveys 33 (1978), 99–116.

[Arnold 78′] V.I. Arnold, *Additional chapters of the theory of ordinary differential equations*, Nauka, Moscow, 1978, 304 p. Engl. transl.: see [Arnold 88″].

[Arnold 83] V.I. Arnold, *Singularities of systems of rays*, Uspekhi Mat. Nauk 38 (1983), no.2, 77–147, Engl. transl. in Russian Math. Surveys 38 (1983), no.2, 87–176.

[Arnold 83′] V.I. Arnold, *Magnetic field analogues of the theorems of Newton and Ivory.* Uspekhi Mat. Nauk 1983, 38:5, 145–146.

[Arnold 85] V.I. Arnold, *On the Newtonian potential of hyperbolic layers,* Selecta Math. Soviet., 4:2, 1985, 103–106.

[Arnold 87] V.I. Arnold, *Kepler's second law and the topology of Abelian integrals*, Kvant, No. 12, 17–21 (Russian).

[Arnold 87′] V.I. Arnold, *A topological proof of transcendence of Abelian integrals in Newton's Principia*, Quant, 1987, No.12, 1–15.

[Arnold 87″] V.I. Arnold, *The 300-th anniversary of mathematical natural philosophy and celestial mechanics*, Priroda, 1987, No.8, 5–15 (in Russian).

[Arnold 88] V.I. Arnold, *Surfaces defined by the hyperbolic equations*, Matem. Zametki 44:1 (1988), 3–18; Engl. transl.: Math. Notes 44, No.112 (1988), 489–497.

[Arnold 88′] V.I. Arnold, *Ramified covering* $\mathbf{C}P^2 \to S^4$, *hyperbolicity and projective topology*, Sibirsk. Mat. Zh., 29:5 (1988), 36–47; Engl. transl.: Siberian Math. J., 29:5 (1988), 717–726.

[Arnold 88″] V.I. Arnold, *Geometrical methods in the theory of ordinary differential equations*, Springer, New York, 1988.

[Arnold 89] V.I. Arnold, *Mathematical methods in clasical mechanics,* "Nauka", Moscow, 1989; English transl. of 1st ed., Springer-Verlag Berlin and New York, 1978.

[Arnold 89′] V.I. Arnold, *Huygens and Barrow, and Newton and Hooke – the first steps of the mathematical analysis and catastrophe theory, from the evolvents to quasicrystals,* "Nauka", Moscow, 1989; English translation: Birkhauser.

[Arnold & Vassiliev 89] V.I. Arnold and V.A. Vassiliev, *Newton's Principia read 300 years later*, Notices Amer. Math. Soc., 36:9 (1989), 1148–1154.

[ABG 70] M.F. Atiyah, R. Bott, L. Gårding, *Lacunas for hyperbolic differential operators with constant coefficients. I*, Acta Math., 124 (1970), 109–189.

[ABG 73] M.F. Atiyah, R. Bott, L. Gårding, *Lacunas for hyperbolic differential operators with constant coefficients. II*, Acta Math., 131 (1973), 145–206.

[AVG 82] V.I. Arnold, A.N. Varchenko, S.M. Gusein-Zade, *Singularities of differentiable maps*, vol.1, "Nauka", Moscow, 1982; Engl. transl.: Birkhäuser, Basel, 1985.

[AVG 84] V.I. Arnold, A.N. Varchenko, S.M. Gusein-Zade, *Singularities of differentiable maps*, vol.2, "Nauka", Moscow, 1984; Engl. transl.: Birkhäuser, Basel, 1988.

[AVGL 88] V.I. Arnold, V.A. Vassiliev, V.V. Goryunov, O.V. Lyashko, *Singularities, 1.* Dynamical systems, 6, VINITI, Moscow, 1988; English transl.: Encyclopaedia Math. Sci., vol. 6, Springer-Verlag, Berlin and New York, 1993.

[AVGL 89] V.I. Arnold, V.A. Vassiliev, V.V. Goryunov, O.V. Lyashko, *Singularities, 2.* Dynamical systems, 39, VINITI, Moscow, 1989; English transl.: Encyclopaedia Math. Sci., vol. 39, Springer-Verlag, Berlin and New York, 1993.

[Berest & Veselov 94] Yu.Yu. Berest, A.P. Veselov, *The Hadamard problem and Coxeter groups: new examples of Huygens' equations,* Funkts. Anal. i Prilozh., 28:1 (1994), Engl. translation in Functional Anal. Appl., 28:1 (1994).

[Bohlin 11] K. Bohlin, Bull. Astr., 28:144, 1911.

[Borovikov 59] V.A. Borovikov, *Fundamental solutions of linear partial equations with constant coefficients*, Transact. Moscow Math. Soc., No.8 (1959), 199–257. English transl.: Amer. Math. Soc. Translations, Ser.2, vol. 25, 11–76, 1963.

[Borovikov 61] V.A. Borovikov, *Some sufficient conditions of the absence of lacunas*, Mat. Sbornik, 55 (97) (1961), 237–254.

[Bourbaki 68] N. Bourbaki, *Groupes et algèbres de Lie, Chapitres 4, 5, 6*, Hermann, Paris, 1968.

[Brieskorn 70] E. Brieskorn, *Die Monodromie der isolierten Singularitäten von Hyperflächen*, Manuscr. Math., 1970, 2, 103–161.

[BBD 83] A. Beilinson, J. Bernstein, P. Deligne, *Faisceaux pervers*, Asterisque 100, Soc. Mat. de France, 1983.

[Cheniot 72] D. Cheniot, *Sur les sections transversales d'un ensemble stratifié*, C. R. Ac. Sci. Paris, ser.A, 275 (1972), 915-916.

[Chislenko 88] Yu.S. Chislenko, *Decompositions of simple singularities of real functions*, Funkts. Anal. i Prilozh., 22:4 (1988), 52–67; Engl. translation in Functional Anal. Appl., 22:4 (1988), 297–310.

[Chmutov 82] S.V. Chmutov, *Monodromy groups of critical points of functions.* Invent. Math. 67 (1982), 123–131.

[Chmutov 83] S.V. Chmutov, *The monodromy groups of critical points of functions. II.* Invent. Math. 73 (1983), 491-510.

[Davydova 45] A.M. Davydova, *A sufficient condition for the absence of a lacuna for a partial differential equation of hyperbolic type*, Ph.D.Thesis, Moscow State Univ., 1945, 43 p.

[Deligne 70] P. Deligne, *Equations différentielles à points singuliers réguliers.* Lect. Notes Math. 163, Springer, Berlin, 1970.

[Dimca & Gibson 83] A. Dimca, C.G. Gibson, *Contact germs from the plane to the plane*, in: [Singularities 83], Part 1, 277–282.

[Ebeling 87] W. Ebeling, *The monodromy groups of isolated singularities of complete intersections.* Lect. Notes Math., vol.1293, Springer, Berlin a.o., 1987.

[Fulton 84] W. Fulton, *Intersection theory*, Springer-Verlag, New York, 1984.

[Gabrielov 68] A.M. Gabrielov, *On projections of semianalytical sets*, Funkts. Anal. i Prilozh., 2:4 (1968), 18–30; Engl. translation in Functional Anal. Appl., 2 (1968), 282–291.

[Gabrielov 73] A.M. Gabrielov, *Intersection matrices for some singularities*, Funkts. Anal. i Prilozh., 7:3 (1973), 18–32; Engl. translation in Functional Anal. Appl., 7 (1973), 182–193.

[Gabrielov 74] A.M. Gabrielov, *Bifurcations, Dynkin diagrams and modality of isolated singularities*, Funkts. Anal. i Prilozh., 8:2 (1974), 7–12; Engl. translation in Functional Anal. Appl., 8 (1974), 94–98.

[Gabrielov 79] A.M. Gabrielov, *Polar curves and intersection matrices of singularities*, Invent. Math., 54, No.1, 15–22.

[Gabrielov 86] A.M. Gabrielov, *Proof of a theorem of I.G.Petrovskii*, in the book [Petrovskii 86], 456–465.

[Gårding 72] L. Gårding, *Local hyperbolicity*, Israel J. Math., 13 (1972), 65–81.

[Gårding 77] L. Gårding, *Sharp fronts of paired oscillatory integrals*, Res. Inst. for Math. Sci., Kyoto Univ., 12 (1977), 53–68.

[Gelfand 86] I.M. Gelfand, *A general theory of hypergeometric functions*, Dokl. Akad. Nauk SSSR, 288:1 (1986), 14–18; Engl. transl.: Sov. Math., Dokl., 33 (1986), 573–577.

[Giusti 77] M. Giusti, *Classification des singularités isolées d'intersections complètes simples*, C.R. Ac. Sci. Paris, Sér. A, 284, 1977.

[Giusti 83] M. Giusti, *Classification des singularités isolées simples d'intersections complètes*, in [Singularities 83], Part 1, 457–494.

[Givental 84] A.B. Givental, *Polynomiality of the electrostatical potentials,* Uspekhi Mat. Nauk, 39:5 (1984), 253–254 (in Russian).

[Givental 88] A.B. Givental, *Twisted Picard–Lefschetz formulae.* Funkts. Anal. i Prilozh., 22:1 (1988), 12–22; Engl. translation in Functional Anal. Appl., 22:1, 10–18 (1988).

[Goresky & MacPherson 80] M. Goresky and R. MacPherson, *Intersection homology theory,* Topology 19 (1980), 135–162.

[Goresky & MacPherson 86] M. Goresky and R. MacPherson, *Stratified Morse theory*, Springer-Verlag, Berlin and New York, 1986.

[Goryunov 91] V.V. Goryunov, *Monodromy of the image of a map* $\mathbf{C}^2 \to \mathbf{C}^3$, Funkts. Anal. i Prilozh., 25:3 (1991), 12–18; Engl. translation in Functional Anal. Appl., 25, No.3 (1991), 174–180.

[Grauert & Remmert 58] H. Grauert and R. Remmert, *Komplexe Räume*, Math. Ann., 136:2 (1958), 245–318.

[Greuel 80] G.M. Greuel, *Dualität in der lokalen Kogomologie isolierter Singularitäten*, Math. Ann., 250, 1980, 157–173.

[Greuel & Hamm 78] G.M. Greuel, H.A. Hamm, *Invarianten quasihomogener follständiger Durchschnitte*, Invent. Math., 1978, 49:1, 67–86.

[Griffiths & Harris 78] Ph. Griffiths, J. Harris, *Principles of algebraic geometry*, John Wiley & Sons, New York a.o., 1978.

[Gudkov 74] D.A. Gudkov, *Topology of real projective algebraic varieties*, Uspekhi Mat. Nauk, 29:4 (1974), 3–79. English transl. in Russian Math. Surveys 29:4 (1974), 1–79.

[Gusein-Zade 74] S.M. Gusein-Zade, *Intersection matrices for some singularities of functions of two variables*, Funkts. Anal. i Prilozh., 8:1 (1974), 11–15; Engl. translation in Functional Anal. Appl., 8 (1974), 10-13.

[Gusein-Zade 74′] S.M. Gusein-Zade, *Dynkin diagrams of singularities of functions of two variables*, Funkts. Anal. i Prilozh., 8:4 (1974), 23–30; Engl. translation in Functional Anal. Appl., 8 (1974).

[Gusein-Zade 77] S.M. Gusein-Zade, *Monodromy groups of isolated singularities of hypersurfaces*, Uspekhi Mat. Nauk 32 (1977), no.2, 23–65, Engl. transl. in Russian Math. Surveys 32, No.2 (1977), 23–69.

[Gusein-Zade 84] S.M. Gusein-Zade, *Index of a singular point of gradient vectorfield*, Funkts. Anal. i Prilozh., 18:4 (1984), 7–12; Engl. translation in Functional Anal. Appl., 18 (1984), 6–10.

[Hadamard 32] J. Hadamard, *Le problème de Cauchy et les équations aux dérivées partielles linéaires hyperboliques,* Paris, 1932.

[Hamm 71] H. Hamm, *Locale topological Eigenschaften komplexer Räume,* Math. Ann., 191, 1971, 235–252.

[Hamm 72] H. Hamm, *Exotische Sphären als Umgebungsränder in speziellen komplexen Räumen*, Math. Ann., 197, 1972, 44–56.

[Hamm 83] H. Hamm, *Lefschetz theorems for singular varieties,* Singularities. Proc. Symp. Pure Math. 40 (part 2) 575–586. AMS, Providence R.I., 1983.

[Hamm & Lê 73] H. Hamm and Lê Dung Trang, Un théorème de Zariski du type de Lefschetz, Ann. Sci. Ecole Norm. Sup. (3) 6 (1973), 317–355.

[Hartshorne 77] R. Hartshorne, *Algebraic geometry*, Springer, Berlin and New York, 1977.

[Hefez & Lazzeri 74] A. Hefez and F. Lazzeri, *The intersection matrix of Brieskorn singularities*, Invent. Math., 25:2 (1974), 143–157.

[**Hironaka 73**] H. Hironaka, *Subanalytic sets*, in: Number Theory, Algebraic Geometry and Commutative Algebra, volume in honor of A.Akizuki, Kinokunya Tokyo 1973, 453–493.

[**Hörmander 71**] L. Hörmander, *Fourier integral operators. I,* Acta Math. 127 (1971), 71–183.

[**Hörmander 83**] L. Hörmander, *The Analysis of Linear Partial Differential Operators I, II,* Springer, Berlin a.o., 1983.

[**Ivory 1809**] J. Ivory, *On the attraction of homogeneous ellipsoids*, Philos. Trans., 99 (1809), 345–372.

[**Janssen 83**] W.A.M. Janssen, *Skew-symmetric vanishing lattices and their monodromy groups,* Math. Ann., 266, 1983, 115–133.

[**Janssen 85**] W.A.M. Janssen, *Skew-symmetric vanishing lattices and their monodromy groups. II,* Math. Ann., 272, 1985, 17–22.

[**Jaworski 86**] P. Jaworski, *Distribution of critical values of miniversal deformations of parabolic singularities*, Invent. Math., 86:1 (1986), 19–33.

[**Karchauskas 77**] K.K. Karchauskas, *Generalized Lefschetz theorem*, Funkts. Anal. i Prilozh., 11:4 (1977), 80–81; Engl. translation in Functional Anal. Appl., 11 (1977), 312–313.

[**Kharlamov 86**] V.M. Kharlamov, *Topology of real algebraic manifolds*, in the book [Petrovskii 86], 465–493.

[**Kleiman 77**] S.L. Kleiman, *The enumerative theory of singularities*, in Real and Complex Singularities. Nordic Summer School, Oslo, 1976. Sijthoff and Noordhoff, Groningen, 1977, 686 p.

[**Kushnirenko 77**] A.G. Kushnirenko, *On the multiplicity of the solution of a system of holomorphic equations*, Optimal Control, no.2, Izdat. Moskov. Univ., Moscow, 1977, 62–65.

[**Lê 73**] Lê D.T., *Calcul du nombre de cycles évanouissants d'une hypersurface complexe,* Ann. Inst. Fourier, Grenoble, 23:4, 1973, 261–270.

[**Lê 74**] Lê D.T., *Calculation of Milnor number of isolated singularity of complete intersection,* Funkts. Anal. i Pril., 8:2, 1974, 45–49; Engl. transl. in Functional Anal. Appl., 8 (1974), 127–131.

[Lê 75] Lê D.T., *Vanishing cycles of analytic spaces.* Kokyuroku series of Res. Inst. Math. Sc., Kyoto Univ., Proc. of symp. on algebraic analysis, 1975.

[Leray 62] J. Leray, *Un prolongement de la transformation de Laplace qui transforme la solution unitaire d'un opérateur hyperbolique en sa solution élémentaire (Probleme de Cauchy, IV)*, Bull. Soc. Math. France, No.90 (1962), 39–156.

[Lojasiewicz 65] S. Lojasiewicz, *Triangulation of semi-analytic sets*, Ann. Scu. Norm. di Pisa (1965), 449–474.

[Lojasiewicz 72] S. Lojasiewicz, *Ensembles sémi-analytiques*, Inst. Hautes Ètudes Sci., Publ. Math., 1972.

[Looijenga 74] E.J.N. Looijenga, *The complement of the bifurcation variety of a simple singularity*, Invent. Math., 23:2 (1974), 105–116.

[Looijenga 77–78] E.J.N. Looijenga, *Semi-universal deformations of a simple elliptic singularity*, Topology 16 (1977), 257–262, and 17 (1978), 23–40.

[Looijenga 78] E.J.N. Looijenga, *The discriminant of a real simple singularity*, Compositio Math., 37:1 (1978), 51–62.

[Looijenga 84] E.J.N. Looijenga, *Isolated singular points of complete intersections*, Cambridge Univ. Press, Cambridge, 1984, 200 p.

[Lyashko 76] O.V. Lyashko, *Decompositions of simple singularities of functions*, Funk. Anal. i Prilozh., 10:2 (1976), 49–56; Engl. translation in Functional Anal. Appl., 10 (1976), 122–128.

[MATHER] J. Mather, *Stability of C^∞-mappings. I-VI.* Ann. Math. 1968, 87:1, 89–104; 1969, 89:2, 254–291; Publ. Sci. IHES, 1969, 35, 127–156; 1970, 37, 223–248; Adv. Math., 1970, 4, 301–335; Lect. Notes Math, vol.192, 1971, 207–253.

[Mather 70] J. Mather, *Notes on topological stability.* Mimeographed notes, Harvard University 1970.

[Milnor 63] J. Milnor, *Morse theory*, Princeton Univ. Press, Princeton, NJ, 1963.

[Milnor 65] J. Milnor, *Lectures on the h-cobordism theorem*, Princeton Univ. Press, Princeton, NJ, 1965.

[Milnor 68] J. Milnor, *Singular points of complex hypersurfaces*, Princeton Univ. Press, Princeton, NJ, and Univ. of Tokyo Press, Tokyo, 1968.

[Newton 1687] I. Newton, *Philosophiae Naturalis Principia Mathematica*, London, 1687.

[Nikulin 79] V.V. Nikulin, *Integral symmetric bilinear forms and some their applications,* Izv. Akad. Nauk SSSR Ser. Mat. 43:1, 1979, 111:177; Engl. transl. in Math. USSR Izv. 14:1, 1980, 103–167.

[Nuij 68] W. Nuij, *A note on hyperbolic polynomials*, Math. Scand. 23 (1968), 69–72.

[Palamodov 67] V.P. Palamodov, *On the multiplicity of a holomorphic map*, Funkts. Anal. i Prilozh., 1:3 (1967), 54–65; Engl. translation in Functional Anal. Appl., 1 (1967), 218–226.

[Palamodov 91] V.P. Palamodov, *Generalized functions and harmonical analysis*, in: Sovremennye Problemy Matematiky, Fundamentalnye Napravlenija, vol. 72; VINITI, 1991, 5–134. English transl.: Encyclopaedia Math. Sci., vol. 72, Springer-Verlag, Berlin and New York, to appear.

[Petrovskii 45] I.G. Petrovsky, *On the diffusion of waves and the lacunas for hyperbolic equations*, Matem. Sbornik, 1945, 17(59), 289–370.

[Petrovskii 86] I.G. Petrovskii, *Selected works. Systems of partial differential equations. Algebraic geometry*, Moscow, "Nauka", 1986.

[Pham 65] F. Pham, *Formules de Picard-Lefschetz génèralisées et ramification des intégrales*, Bull. Soc. Math. France, 93 (1965), 333–367.

[Pham 67] F. Pham, *Introduction à l'étude topologique des singularités de Landau*, Gauthier-Villars, Paris, 1967.

[Pickl 85] A. Pickl, *Die Homologie der Enhängung eines vollständigen Durchschnitts mit isolierter Singularität,* Preprint, Math. Göttingensis Scriftenr., Sonderforschungbereich Geom. Anal. 15, 1985.

[Remmert 56] R. Remmert, *Projectionen analytischer Mengen,* Math. Ann., 130 (1956), 410–441.

[Remmert 57] R. Remmert, *Holomorphe und meromorphe Abbildungen komplexer Räume*, Math. Ann., 133 (1957), 338–370.

[Rokhlin 78] V.A. Rokhlin, *Complex topological characteristics of real algebraic curves,* Uspekhi Mat. Nauk 33 (1978), no.5, 77–89, Engl. transl. in Russian Math. Surveys 33 (1978), no.5, 85–98.

[Schechtman & Varchenko 90] V. Schechtman and A. Varchenko, *Quantum groups and homology of local systems*, preprint, Inst. for Advanced Study, 1990, 1–16.

[Schechtman & Varchenko 91] V. Schechtman and A. Varchenko, *Arrangements of hyperplanes and Lie algebra homology*, Invent. Math. 106 (1991), 139–194.

[Smale 61] S. Smale, *Generalized Poincaré conjecture in dimensions greater than four,* Ann. Math., 74:1, 1961, 391–406.

[Smale 62] S. Smale, *On the structure of manifolds,* Amer. J. Math., 84:3, 1962, 387–399.

[Singularities 83] *Singularities.* Proceedings of Symposia in Pure Mathematics, Vol. 40, Parts I, II. Amer. Math. Soc., Providence, RI, 1983.

[Svensson 70] L. Svensson, *Necessary and sufficient conditions for the hyperbolicity of polynomials with hyperbolic principal part*, Ark. Math., 8 (1970).

[Teissier 77] B. Teissier, *Variétés polaires. I. Invariants polaires des singularités d'hypersurfaces,* Invent. Math., 40, 1977, 267–292.

[Thom 69] R. Thom, *Ensembles and morphismes stratifiés*, Bull. Amer. Math. Soc., 75 (1969), 240–284.

[Trotman 77] D. Trotman, *Counterexamples in stratification theory: two discordant horns*, Proc. Nordic Summer School, Oslo 1976. Sijthoff and Noordhoff, Groningen 1977.

[Trotman 83] D. Trotman, *Comparing regularity conditions of stratifications*, Singularities. Proc. Symp. Pure Math. 40 (part 2) 575–586. AMS, Providence R.I., 1983.

[Tyurina 69] G.N. Tyurina, *Locally semiuniversal flat deformations of isolated singularities of complex spaces,* Izv. Akad. Nauk SSSR Ser. Mat. 33 (1969), 1026–1058; Engl. transl. in Mat. USSR Izv. 3 (1970), 967–999.

[Vainshtein & Shapiro 85] A.D. Vainshtein and B.Z. Shapiro, *Multidimensional analogues of the Newton and Ivory theorems*, Funkts. Anal. i Prilozh., 19:1 (1985), 20–24; Engl. translation in Functional Anal. Appl., 19:1 (1985), 17–20.

[Vainshtein & Shapiro 88] A.D. Vainshtein and B.Z. Shapiro, *Singularities of the boundary of a domain of hyperbolicity*, Itogi Nauki i Tekhn., Ser. Sovrem. Probl. Mat., Nov. Dostizh. 33, 55-78; Engl. translation: J. Sov. Math., 52 (1990), 3326–3337.

[Varchenko 72] A.N. Varchenko, *Theorems on the topological equisingularity of families of algebraic varieties and families of polynomial maps,* Izv. Akad. Nauk SSSR Ser. Mat. 36 (1972), 957–1019; Engl. transl. in Mat. USSR Izv. 6 (1972), 949–1008.

[Varchenko 87] A.N. Varchenko, *Normal forms of nonsmoothness of solutions of hyperbolic equations*, Izv. Akad. Nauk SSSR Ser. Mat. 51:3 (1987), 652–665; Engl. transl. in Mat. USSR Izv. 30, No.3 (1988), 615–628.

[Vassiliev 86] V.A. Vassiliev, *Sharpness and the local Petrovskii condition for hyperbolic equations with constant coefficients,* Izv. Akad. Nauk SSSR Ser. Mat. 50 (1986), 242–283; Engl. transl. in Mat. USSR Izv. 28 (1987), 233–273.

[Vassiliev 88 & 93] V.A. Vassiliev, *Lagrange and Legendre characteristic classes*, Adv. Studies in Contemp. Math., vol.3, Gordon and Breach, New York, 1988; Second edition, 1993.

[Vassiliev 88′] V.A. Vassiliev, *Sharp and diffuse fronts of hyperbolic equations*, in: Partial Differential Equations-II, Itogi nauki VINITI, Fundamentalnye Napr., vol. 31; VINITI, Moscow, 1988, 246–257. English transl.: Encyclopaedia Math. Sci., vol. 31, Springer-Verlag, Berlin and New York.

[Vassiliev 89] V.A. Vassiliev, *Lacunas of hyperbolic partial equations and singularity theory,* in: Theory of Operators in Functional Spaces, Saratov Univ. Press, Kujbyshev dept.; Kujbyshev, 1989; Engl. transl: Singularity Theory and some Problems of Functional Analyzis. S.G.Gindikin, ed. Amer. Math. Soc. Translations, Ser.2, vol. 153, 1992.

[Vassiliev 92] V.A. Vassiliev, *Geometry of the local lacunas of hyperbolic operators with constant coefficients*, Mat. Sbornik, 183:1 (1992), 114–129. English transl.: Russian Acad. Sci. Sbornik Math. 75 (1993), 111–123.

[Vassiliev 92 & 94] V.A. Vassiliev, *Complements of discriminants of smooth maps: topology and applications*, Transl. of Math. Monographs, vol. 98, Amer. Math. Soc., 1992, 214 p. Revised edition, 1994, 274 p.

[VGZ 87] V.A. Vassiliev, I.M. Gelfand, and A.V. Zelevinskii, *General hypergeometric functions on complex Grassmannians,* Funkts. Anal. Prilozh., 21:1, 1987, 23–37; Engl. transl.: Funct. Anal. Appl., 21:1 (1987), 19–31.

[Vassiliev & Serganova 91] V.A. Vassiliev and V.V. Serganova, *On the number of real and complex moduli of singularities of smooth functions and realizations of matroids*, Matematicheskie Zametki 49:1 (1991), 19–27; Engl. transl. in Math. Notes, 49:1, 1991, 15–20.

[Wajnryb 80] B. Wainryb, *On the monodromy group of plane curve singularities,* Math. Ann., 246, 1980, 141–154.

[Wall 75] C.T.C. Wall, *Regular stratifications*, Lecture Notes in Math., vol. 468, Springer-Verlag, Berlin, 1975, 332–344.

[Wall 83] C.T.C. Wall, *Classification of unimodular isolated singularities of complete intersections*, in: [Singularities 83], Part 2, 625–640.

[Zak 87] F.L. Zak, *Structure of Gauss maps*, Funkts. Anal. i Prilozh., 21:1 (1987), 39–50; Engl. translation in Functional Anal. Appl., 21:1 (1987), 39–50.

[Zakalyukin 76] V.M. Zakalyukin, *On Lagrange and Legendre singularities,* Funkts. Anal. i Prilozh., 10:1 (1976), 26–36; Engl. translation in Functional Anal. Appl., 10 (1976), 23–31.

[Zariski 37] O. Zariski, *On the Poincaré group of a projective hypersurface*, Ann. Math. 38 (1937), 131–141.

Subject index

Other *Mathematics and Its Applications* titles of interest:

A.F. Filippov: *Differential Equations with Discontinuous Righthand Sides.* 1988, 320 pp. ISBN 90-277-2699-X

A.T. Fomenko: *Integrability and Nonintegrability in Geometry and Mechanics.* 1988, 360 pp. ISBN 90-277-2818-6

G. Adomian: *Nonlinear Stochastic Systems Theory and Applications to Physics.* 1988, 244 pp. ISBN 90-277-2525-X

A. Tesar and Ludovt Fillo: *Transfer Matrix Method.* 1988, 260 pp. ISBN 90-277-2590-X

A. Kaneko: *Introduction to the Theory of Hyperfunctions.* 1989, 472 pp. ISBN 90-277-2837-2

D.S. Mitrinovic, J.E. Pecaric and V. Volenec: *Recent Advances in Geometric Inequalities.* 1989, 734 pp. ISBN 90-277-2565-9

A.W. Leung: *Systems of Nonlinear PDEs: Applications to Biology and Engineering.* 1989, 424 pp. ISBN 0-7923-0138-2

N.E. Hurt: *Phase Retrieval and Zero Crossings: Mathematical Methods in Image Reconstruction.* 1989, 320 pp. ISBN 0-7923-0210-9

V.I. Fabrikant: *Applications of Potential Theory in Mechanics. A Selection of New Results.* 1989, 484 pp. ISBN 0-7923-0173-0

R. Feistel and W. Ebeling: *Evolution of Complex Systems. Selforganization, Entropy and Development.* 1989, 248 pp. ISBN 90-277-2666-3

S.M. Ermakov, V.V. Nekrutkin and A.S. Sipin: *Random Processes for Classical Equations of Mathematical Physics.* 1989, 304 pp. ISBN 0-7923-0036-X

B.A. Plamenevskii: *Algebras of Pseudodifferential Operators.* 1989, 304 pp. ISBN 0-7923-0231-1

N. Bakhvalov and G. Panasenko: *Homogenisation: Averaging Processes in Periodic Media. Mathematical Problems in the Mechanics of Composite Materials.* 1989, 404 pp. ISBN 0-7923-0049-1

A.Ya. Helemskii: *The Homology of Banach and Topological Algebras.* 1989, 356 pp. ISBN 0-7923-0217-6

M. Toda: *Nonlinear Waves and Solitons.* 1989, 386 pp. ISBN 0-7923-0442-X

M.I. Rabinovich and D.I. Trubetskov: *Oscillations and Waves in Linear and Nonlinear Systems.* 1989, 600 pp. ISBN 0-7923-0445-4

A. Crumeyrolle: *Orthogonal and Symplectic Clifford Algebras. Spinor Structures.* 1990, 364 pp. ISBN 0-7923-0541-8

V. Goldshtein and Yu. Reshetnyak: *Quasiconformal Mappings and Sobolev Spaces.* 1990, 392 pp. ISBN 0-7923-0543-4

Other *Mathematics and Its Applications* titles of interest:

I.H. Dimovski: *Convolutional Calculus.* 1990, 208 pp. ISBN 0-7923-0623-6

Y.M. Svirezhev and V.P. Pasekov: *Fundamentals of Mathematical Evolutionary Genetics.* 1990, 384 pp. ISBN 90-277-2772-4

S. Levendorskii: *Asymptotic Distribution of Eigenvalues of Differential Operators.* 1991, 297 pp. ISBN 0-7923-0539-6

V.G. Makhankov: *Soliton Phenomenology.* 1990, 461 pp. ISBN 90-277-2830-5

I. Cioranescu: *Geometry of Banach Spaces, Duality Mappings and Nonlinear Problems.* 1990, 274 pp. ISBN 0-7923-0910-3

B.I. Sendov: *Hausdorff Approximation.* 1990, 384 pp. ISBN 0-7923-0901-4

A.B. Venkov: *Spectral Theory of Automorphic Functions and Its Applications.* 1991, 280 pp. ISBN 0-7923-0487-X

V.I. Arnold: *Singularities of Caustics and Wave Fronts.* 1990, 274 pp. ISBN 0-7923-1038-1

A.A. Pankov: *Bounded and Almost Periodic Solutions of Nonlinear Operator Differential Equations.* 1990, 232 pp. ISBN 0-7923-0585-X

A.S. Davydov: *Solitons in Molecular Systems. Second Edition.* 1991, 428 pp. ISBN 0-7923-1029-2

B.M. Levitan and I.S. Sargsjan: *Sturm-Liouville and Dirac Operators.* 1991, 362 pp. ISBN 0-7923-0992-8

V.I. Gorbachuk and M.L. Gorbachuk: *Boundary Value Problems for Operator Differential Equations.* 1991, 376 pp. ISBN 0-7923-0381-4

Y.S. Samoilenko: *Spectral Theory of Families of Self-Adjoint Operators.* 1991, 309 pp. ISBN 0-7923-0703-8

B.I. Golubov A.V. Efimov and V.A. Scvortsov: *Walsh Series and Transforms.* 1991, 382 pp. ISBN 0-7923-1100-0

V. Laksmikhantam, V.M. Matrosov and S. Sivasundaram: *Vector Lyapunov Functions and Stability Analysis of Nonlinear Systems.* 1991, 250 pp. ISBN 0-7923-1152-3

F.A. Berezin and M.A. Shubin: *The Schrödinger Equation.* 1991, 556 pp. ISBN 0-7923-1218-X

D.S. Mitrinovic, J.E. Pecaric and A.M. Fink: *Inequalities Involving Functions and their Integrals and Derivatives.* 1991, 588 pp. ISBN 0-7923-1330-5

Julii A. Dubinskii: *Analytic Pseudo-Differential Operators and their Applications.* 1991, 252 pp. ISBN 0-7923-1296-1

V.I. Fabrikant: *Mixed Boundary Value Problems in Potential Theory and their Applications.* 1991, 452 pp. ISBN 0-7923-1157-4

Other *Mathematics and Its Applications* titles of interest:

A.M. Samoilenko: *Elements of the Mathematical Theory of Multi-Frequency Oscillations.* 1991, 314 pp. ISBN 0-7923-1438-7

Yu.L. Dalecky and S.V. Fomin: *Measures and Differential Equations in Infinite-Dimensional Space.* 1991, 338 pp. ISBN 0-7923-1517-0

W. Mlak: *Hilbert Space and Operator Theory.* 1991, 296 pp. ISBN 0-7923-1042-X

N.J. Vilenkin and A.U. Klimyk: *Representations of Lie Groups and Special Functions. Volume 1: Simplest Lie Groups, Special Functions, and Integral Transforms.* 1991, 608 pp. ISBN 0-7923-1466-2

K. Gopalsamy: *Stability and Oscillations in Delay Differential Equations of Population Dynamics.* 1992, 502 pp. ISBN 0-7923-1594-4

N.M. Korobov: *Exponential Sums and their Applications.* 1992, 210 pp. ISBN 0-7923-1647-9

Chuang-Gan Hu and Chung-Chun Yang: *Vector-Valued Functions and their Applications.* 1991, 172 pp. ISBN 0-7923-1605-3

Z. Szmydt and B. Ziemian: *The Mellin Transformation and Fuchsian Type Partial Differential Equations.* 1992, 224 pp. ISBN 0-7923-1683-5

L.I. Ronkin: *Functions of Completely Regular Growth.* 1992, 394 pp. ISBN 0-7923-1677-0

R. Delanghe, F. Sommen and V. Soucek: *Clifford Algebra and Spinor-valued Functions. A Function Theory of the Dirac Operator.* 1992, 486 pp. ISBN 0-7923-0229-X

A. Tempelman: *Ergodic Theorems for Group Actions.* 1992, 400 pp. ISBN 0-7923-1717-3

D. Bainov and P. Simenov: *Integral Inequalities and Applications.* 1992, 426 pp. ISBN 0-7923-1714-9

I. Imai: *Applied Hyperfunction Theory.* 1992, 460 pp. ISBN 0-7923-1507-3

Yu.I. Neimark and P.S. Landa: *Stochastic and Chaotic Oscillations.* 1992, 502 pp. ISBN 0-7923-1530-8

H.M. Srivastava and R.G. Buschman: *Theory and Applications of Convolution Integral Equations.* 1992, 240 pp. ISBN 0-7923-1891-9

A. van der Burgh and J. Simonis (eds.): *Topics in Engineering Mathematics.* 1992, 266 pp. ISBN 0-7923-2005-3

F. Neuman: *Global Properties of Linear Ordinary Differential Equations.* 1992, 320 pp. ISBN 0-7923-1269-4

A. Dvurecenskij: *Gleason's Theorem and its Applications.* 1992, 334 pp. ISBN 0-7923-1990-7

Other *Mathematics and Its Applications* titles of interest:

D.S. Mitrinovic, J.E. Pecaric and A.M. Fink: *Classical and New Inequalities in Analysis.* 1992, 740 pp. ISBN 0-7923-2064-6

H.M. Hapaev: *Averaging in Stability Theory.* 1992, 280 pp. ISBN 0-7923-1581-2

S. Gindinkin and L.R. Volevich: *The Method of Newton's Polyhedron in the Theory of PDE's.* 1992, 276 pp. ISBN 0-7923-2037-9

Yu.A. Mitropolsky, A.M. Samoilenko and D.I. Martinyuk: *Systems of Evolution Equations with Periodic and Quasiperiodic Coefficients.* 1992, 280 pp. ISBN 0-7923-2054-9

I.T. Kiguradze and T.A. Chanturia: *Asymptotic Properties of Solutions of Nonautonomous Ordinary Differential Equations.* 1992, 332 pp. ISBN 0-7923-2059-X

V.L. Kocic and G. Ladas: *Global Behavior of Nonlinear Difference Equations of Higher Order with Applications.* 1993, 228 pp. ISBN 0-7923-2286-X

S. Levendorskii: *Degenerate Elliptic Equations.* 1993, 445 pp. ISBN 0-7923-2305-X

D. Mitrinovic and J.D. Kečkić: *The Cauchy Method of Residues, Volume 2.* Theory and Applications. 1993, 202 pp. ISBN 0-7923-2311-8

R.P. Agarwal and P.J.Y Wong: *Error Inequalities in Polynomial Interpolation and Their Applications.* 1993, 376 pp. ISBN 0-7923-2337-8

A.G. Butkovskiy and L.M. Pustyl'nikov (eds.): *Characteristics of Distributed-Parameter Systems.* 1993, 386 pp. ISBN 0-7923-2499-4

B. Sternin and V. Shatalov: *Differential Equations on Complex Manifolds.* 1994, 504 pp. ISBN 0-7923-2710-1

S.B. Yakubovich and Y.F. Luchko: *The Hypergeometric Approach to Integral Transforms and Convolutions.* 1994, 324 pp. ISBN 0-7923-2856-6

C. Gu, X. Ding and C.-C. Yang: *Partial Differential Equations in China.* 1994, 181 pp. ISBN 0-7923-2857-4

V.G. Kravchenko and G.S. Litvinchuk: *Introduction to the Theory of Singular Integral Operators with Shift.* 1994, 288 pp. ISBN 0-7923-2864-7

A. Cuyt (ed.): *Nonlinear Numerical Methods and Rational Approximation II.* 1994, 446 pp. ISBN 0-7923-2967-8

G. Gaeta: *Nonlinear Symmetries and Nonlinear Equations.* 1994, 258 pp. ISBN 0-7923-3048-X

V.A. Vassiliev: *Ramified Integrals, Singularities and Lacunas.* 1995, 289 pp. ISBN 0-7923-3193-1